Related Titles

Functional Polymer Blends and Nanocomposites
ISBN: 978-0-6482205-6-5 (e-book)
ISBN: 978-0-6482205-7-2 (hardcover)

Polymers in Oil and Gas Industry
ISBN: 978-0-6482205-0-3 (e-book)
ISBN: 978-0-6482205-1-0 (softcover)

Functional Nanomaterials and Nanotechnologies: Applications for Energy & Environment
ISBN: 978-0-6482205-2-7 (e-book)
ISBN: 978-0-6482205-3-4 (softcover)

Biopolymers Based Advanced Materials

Biopolymers Based Advanced Materials

Dr. Vikas Mittal
Editor and Lead Author

CWP
Central West Publishing

This edition has been published by Central West Publishing, Australia

For more information about the books published by Central West Publishing, please visit https://centralwestpublishing.com

Disclaimer
Every effort has been made by the publisher, editor and authors while preparing this book, however, no warranties are made regarding the accuracy and completeness of the content. The publisher, editor and authors disclaim without any limitation all warranties as well as any implied warranties about sales, along with fitness of the content for a particular purpose. Citation of any website and other information sources does not mean any endorsement from the publisher and authors. For ascertaining the suitability of the contents contained herein for a particular lab or commercial use, consultation with the subject expert is needed. In addition, while using the information and methods contained herein, the practitioners and researchers need to be mindful for their own safety, along with the safety of others, including the professional parties and premises for whom they have professional responsibility. To the fullest extent of law, the publisher, editor and authors are not liable in all circumstances (special, incidental, and consequential) for any injury and/or damage to persons and property, along with any potential loss of profit and other commercial damages due to the use of any methods, products, guidelines, procedures contained in the material herein.

NATIONAL LIBRARY OF AUSTRALIA

A catalogue record for this book is available from the National Library of Australia

ISBN (print): 978-0-6482205-5-8
ISBN (e-book): 978-0-6482205-4-1

Contents

Preface

Biopolymers have gained increasing research attention over the years due to advancements in the synthesis strategies leading to more efficient control over structure-property correlations. In addition, a large variety of biopolymer based blends and composites have been developed which present optimal combinations of the constituents' properties. These developments open opportunities for the large-scale use of these biopolymer based advanced materials not only for commodity applications, but also in many engineering and structural uses. It is envisaged that the fundamental insights gained into the tuning of property profiles of biopolymer based systems would result in the replacement of many conventional polymers by biopolymers in near future. The current volume aims to present some of these advanced biopolymers based systems demonstrating the use of both blending and composite generation for enhancing the performance, along with the development of structural modification strategies. All aspects related to synthesis, property analysis as well as applications have been targeted.

Chapter 1 reports melt processed poly(butylene succinate) (PBS)/chitosan blend bio-nanocomposites with a variety of fillers such as silica, alumino-silicate and thermally reduced graphene. The mechanical, rheological, morphological and thermal degradation properties were evaluated. In Chapter 2, biocomposites of biopolyesters (poly(butylene adipate-co-terephthalate) (PBAT) and polycaprolactone (PCL)) have been generated with psyllium seed husk with an aim to generate cost-efficiency. In Chapter 3, to explore the potential of the poly(hydroxybutyrate-co-hydroxyvalerate) (PHBV)/PBAT blend based composites, bio-nanocomposites have been developed with graphene and vermiculite as filler materials, using melt mixing method. Chapter 4 reports about the morphology and physical properties of ternary blends of PCL, PBAT and PBS. Similarly, the morphology and properties of ternary blends based on PCL, poly(lactic acid) (PLA) and PBS have been reported in Chapter 5. Chapter 6 presents further insights into the impact of chain extension and hyperbranching on the structure and properties of PLA. In Chapter 7, successful attempt to enhance the mechanical properties of PLA has been made by combining the polymer blending (PBAT) and filler (graphene) incorporation approaches. Chapter 8 presents a detailed study for generating thin stable chitosan coatings through sol-gel silica synthesis method, followed by a poly(vinyl butyral) (PVB) over-coating. Effect of nano-fillers like graphene nano-platelets on the coating performance was also evaluated using electrochemical impedance spectroscopy and polarization measurements. In Chapter 9, composites of PLA with different types of graphene based materials, such as thermally reduced graphene, graphene oxide, graphite, expanded graphite and graphite oxide have been reviewed briefly. Chapter 10 deals with the review of preparation and properties of PBS/graphene nano-sheets nanocomposites so as to explore their potential use in various applications as well as to replace the existing non-degradable polymers.

The book would not have been successfully accomplished without the continuous support of all the chapter contributors. The book is dedicated to my family for their unswerving motivation and encouragement.

Vikas MITTAL
Australia

1

Melt Processed PBS/Chitosan Bio-nanocomposites with Silica, Alumino-silicate and Thermally Reduced Graphene

1.1 Introduction

In order to achieve useful set of properties, natural polymers like chitin, starch and cellulose are usually blended with synthetic polymers [1,2]. Biodegradable aliphatic polyesters (e.g. poly(lactic acid) (PLA), poly(hydroxybutyrate-co-valerate), poly(butylene succinate) (PBS), etc.) are generally used synthetic polymers in order to achieve biomaterial blends [3]. Further reinforcement of these biomaterials with functional nano-fillers leads to generation of bio-nanocomposites with enhanced performance, which also makes them capable of replacing conventional non-biodegradable materials in different commercial applications [4-7].

Chitosan (CS) is the alkaline deacetylated form of chitin and is biodegradable and non-toxic [8]. It is second most abundant natural polymer after cellulose [4]. Solution blending of chitosan with synthetic polymers like methyl methacrylate, methacrylic acid, polyethylene glycol and polyvinyl alcohol has been reported [9,10]. However, little work has been reported on the melt processed blends, especially blends of chitosan and biodegradable aliphatic polyesters [3]. Composites of chitosan with various fillers have also been generated using solution mixing method [6,11-17]. In reported studies, aqueous solution of CS in acetic acid and silica/tetraethyl orthosilicate (TEOS) were mixed by stirring followed by NaOH neutralization to generate CS-silica nanocomposites [13,14]. Similarly, nanocomposite films were produced by solution casting method by first dispersing graphene in aqueous acetic acid with the help of ultrasonication followed by dissolution of chitosan [17]. Similar to the blends, little research effort has been devoted to the melt processed composites of blends of chitosan with biodegradable aliphatic polyesters. Correlo *et al.* [3] reported the properties of melt processable

Fakhruddin Patwary and Vikas Mittal, The Petroleum Institute (part of Khalifa University of Science and Technology), Abu Dhabi, UAE*
**Current address: Bletchington, Wellington County, Australia*

blends of chitosan with biodegradable polyesters like poly-ε-caprolactone, PBS, PLA, poly(butylene terephthalate adipate) and poly(butylene succinate adipate), however, no composites of these blends were reported. Mixing PLA with chitosan improved the water barrier properties and reduced the water sensitivity of chitosan film [4]. However, the tensile strength and elastic modulus of chitosan decreased. PBS:CS-hydroxyapatite composite with 5% plasticizer (glycerol) was extruded in a co-rotating twin-screw extruder [18]. The addition of hydroxyapatite decreased the crystallinity, tensile strength and elongation in chitosan/PBS blends.

The goal of the current study was to generate melt processed nanocomposites of PBS:CS blend with fillers like silica, alumino-silicate and thermally reduced graphene and evaluate their mechanical, rheological, morphological and thermal degradation properties. The ratio of PBS and chitosan in the blend was optimized based on the processing ease and mixing quality. The filler nanoparticles were of 0-dimensional (0-D) and 2-dimensional (2-D) geometrical forms and were chosen because of their benefits as reinforcing fillers. The amount of fillers was also varied in order to study the effect on the nanocomposite properties. The morphology of the composites was analyzed using scanning and transmission electron microscopy as well as optical microscopy. Crystal structure changes and other interactions between the components were studied using X-ray diffraction, differential scanning calorimetry and infra-red spectroscopy. Mass spectroscopy analysis coupled with thermogravimetric analysis was performed to study the thermal degradation behavior. The mechanical properties were also modeled using factorial design statistical methods.

1.2 Experimental

1.2.1 Materials

Chitosan flakes, procured from Bio21, Thailand, were provided by Prof. Wang at The Petroleum Institute. Chitosan had a molecular weight of 180000 g/mol and 90% degree of deacetylation (DD). PBS (Bionelle 1050 from Showa Highpolymer Co., Japan) was kindly donated by Prof. Shih from Chaoyang University, Taiwan. Synthetic silicon dioxide powder (ZEOFREE® 5161 S) and synthetic alumino-silicate (ZEOLEX® 23) were supplied by J. M. Huber Private Limited, India. Thermally reduced graphene was produced by thermal exfoliation of precursor graphite oxide using modified Hummer's method, as reported earlier [19,20].

1.2.2 Generation of Bio-nanocomposites

Melt mixing using mini twin conical screw extruder (MiniLab HAAKE Rheomex CTW5, Germany) was carried out to generate blends and nanocomposites. The weight ratio of PBS and chitosan in the blend was optimized by preparing blends of different weight ratios and using different processing conditions of temperature and screw speed. Bio-nanocomposites were prepared by melt mixing of the biopolymers with fillers at a temperature of 170 °C and rotational speed of 100 rpm for 15 minutes. Composites with filler content of 2, 5, 7.5 and 10% were prepared. Tetrahedron MTP-10 hot press was used to compression mold the blend and composite sheets. A temperature of 170 °C and a pressure of 800 bars were used (holding pressure 290 bars). The samples were cooled to 50 °C at a rate of 5 °C/min. Sheets of 12*12 cm with a thickness of 1.5 mm were produced and disc, dumbbell and bar-shaped test specimens for rheological and mechanical characterization were stamped out from these sheets.

1.2.3 Nanocomposites Characterization

Differential scanning calorimetric measurements on the polymer blend and bio-nanocomposites were performed using a Netzsch DSC under nitrogen atmosphere. The scans were obtained from 30-200-30 °C using heating and cooling rates of 15 °C/min and 5 °C/min respectively. Thermal properties of the samples were analyzed using Netzsch thermogravimetric analyzer (TGA). Nitrogen was used as a carrier gas and the scans were obtained from 50 to 700 °C at a heating rate of 20 °C/min. LINSEIS STA PT1600 simultaneous thermogravimetric analysis - differential thermal analysis (TGA-DTA) system was also used to measure the weight change during heating in air atmosphere. This system was coupled with a Phipher mass spectrometer (MS) which allowed to determine the elimination of H_2O (m/z=18), CO_2 (m/z=44) and other fractions during heating. All measurements were realized with a heating rate of 3 °C/min, using flow rate of 20 ml/min.

Wide angle X-ray diffraction analysis of the composites was performed on Panalytical Powder Diffractometer (X'Pert PRO) using

CuKα radiation (λ = 1.5406 A°) in reflection mode. Zero-background holder was used to minimize the noise. The samples were step-scanned from 10-70° 2Θ at room temperature using a step size of 0.02° 2Θ and a step time of 10 s. IR spectra were collected on Nicolet iS10 spectrometer equipped with SmartiTR diamond ATR accessory (angle of incidence of 45°), DTGS KBr detector and KBr beam splitter. It had a diamond ATR crystal (index of refraction 2.4 at 1000 cm^{-1}) and a depth penetration of 2 μm at 1000 cm^{-1} for sample with refractive index of 1.5. Spectra were recorded by the OMNIC software in the 4000-525 cm^{-1} region with a resolution of 4 cm^{-1} from 16 scans.

Universal testing machine Testometric M-350 10CT was used to analyze the mechanical properties. The dumbbell-shaped samples of 53 mm length, 4 mm width and 2 mm thickness conforming to standard ASTM-638-V were used. A loading rate of 10 mm/min was used and the tests were carried out at room temperature. Win Test Analysis software was used for the calculation of tensile modulus and yield stress properties. Secant modulus calculated at a stress value of 8 MPa was reported as Young's modulus and an average of six values was calculated. DMA (dynamic mechanical analyzer) RSA3 of TA Instruments was used to study the change in modulus with temperature (25-100 °C). Dynamic temperature ramp test was used at a frequency of 5 rad/s on bar-shaped samples, using a ramp rate of 3 °C/min and 1% strain for all nanocomposites. AR 2000 rheometer from TA Instruments was used to characterize rheological properties such as storage modulus (G′), loss modulus (G″), viscosity (η′) and elasticity (η″). Disc shaped samples of 25 mm diameter and 2 mm thickness were characterized at 170 °C using a gap opening of 1 mm. Frequency sweep scans (dynamic testing) of all polyester nanocomposites were recorded at 1% strain from ω = 0.01 to 600 rad/s.

For the bright field transmission electron microscopy analysis of the composite samples, Philips CM 20 (Philips/FEI, Eindhoven) electron microscope at 120 kV and 200 kV accelerating voltages was used. Thin sections of 70-90 nm thickness were microtomed from the sample block and were supported on 100 mesh grids sputter coated with a 3 nm thick carbon layer. The macroscopic features of nanocomposites were also observed under an optical microscope Olympus BX 51. The cross-section of the samples was characterized at a magnification of 20x and both bright field and dark field reflection modes were used. Prior to the analysis, the surface of the block face of the samples was smoothened using a knife. The microstructure of the blend was also characterized using scanning electron microscopy (FEI Quanta, FEG250, USA) at accelerating voltages of 10-20 kV. The sample surfaces were sputter-coated with 3 nm thick gold layer.

MINITAB 16 program was utilized to model and predict the storage modulus of PBS:CS nanocomposites. The DoE (design of experiments) model was used to generate factorial designs.

1.3 Results and Discussion

Melt blending approach was used to generate blends of chitosan and PBS and corresponding nanocomposites in order to retain the commercial attractiveness of melt mixing process in composite manufacturing. In order to ascertain an optimum weight ratio of PBS and chitosan in the blend based on ease of processing, efficiency of mixing as well as optimal mechanical performance, a number of blends with different weight ratios were generated using different processing conditions. Table 1.1 demonstrates the mechanical properties of these blends. Processing at 160 °C and 80 rpm for 15 min caused very high buildup of torque in the extruder which resulted in inefficient mixing. On the other hand, the output was much better at higher processing temperature and rpm. 60:40 blend was chosen as

Table 1.1 Mechanical properties of PBS:CS blends containing different weight ratios, generated using different processing conditions

PBS:CS blend weight ratio	T (°C)	Screw speed (rpm)	Young's modulus[a] (N/mm^2)	Stress @yield[b] (N/mm^2)	Stress @peak[c] (N/mm^2)	Strain @yield[d] (%)
100:00	170	100	500	17	37	1.40
50:50	160	80	1848	17	25	0.66
50:50	160	100	2064	16	33	0.36
50:50	170	100	2025	16	33	0.50
60:40	170	100	1691	16	32	0.52
70:30	170	100	1282	12	28	0.43

[a]Relative probable error 3%.
[b]Relative probable error 3%.
[c]Relative probable error 5%.
[d]Relative probable error 5%.

the final composition based on best processing and mixing behavior, though it had slightly lower mechanical performance than the 50:50 blend. Blending PBS with chitosan (60:40) resulted in enhancement of modulus by >300% as compared to pure PBS. The yield stress was unchanged on chitosan addition. There was a reduction in the yield strain in the blend due to the chitosan flakes hindering the deformation/elongation of PBS chains, as shown in Table 1.1.

PBS:CS blend was earlier reported to have slightly depressed (approx. 2.5 °C) peak melting temperature than pure PBS [3], which was also observed in the current study. Interactions between the chitosan and polyester chains leading to thinner lamellar thickness crystals were reported to cause this depression [21]. The degree of crystallinity of PBS was estimated from the theoretical value of ΔH_m of 110 $J.g^{-1}$ for 100% crystalline PBS, as proposed earlier by Van Krevelen [3,22]. The pure PBS had a degree of crystallinity of 44% in the as-processed samples, which was reduced to 30% in the 60:40 PBS:CS blend. The peak melting temperature of PBS:CS blend was measured to be 111 °C. Introducing 2 wt% fillers marginally enhanced the peak melting temperature by 2 °C. It was also reflected in 5-8% increment in the relative degree of crystallinity, with graphene having the highest impact. Further addition of fillers did not change the peak melting point. At the highest filler fraction, an average reduction of 1-2 °C in peak melting point was observed, which was also supported by a reduction in overall degree of crystallinity due to hindrance caused by large amount of filler particles present in the matrix. Figure 1.1 shows the DSC thermograms of the PBS:CS nanocomposites with 10% filler content. The hump around 125 °C in the 60:40 PBS:CS was due to the glass transition temperature of chitosan. The silica composite with 10% filler content was also observed to have bi-modal melting peaks probably due to different crystal sizes and types. The peak crystallization temperature in the PBS:CS blend was measured to be 79 °C. It was observed to increase to around 86 °C on the addition of silica indicating its nucleation effect (Figure 1.1). The increase in the crystallization temperature was not affected by the filler content. Graphene also aided in nucleation, but its extent was not as significant as silica particles. Addition of silicate did not affect the peak crystallization temperature indicating that the filler particles did not aid in nucleation. Thus, different fillers impacted the blend behavior differently, but the effect of processing conditions on the observed calorimetric properties should also not be ignored.

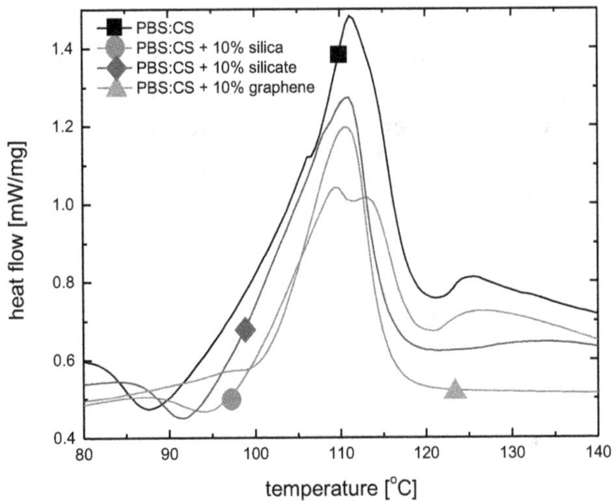

(a)

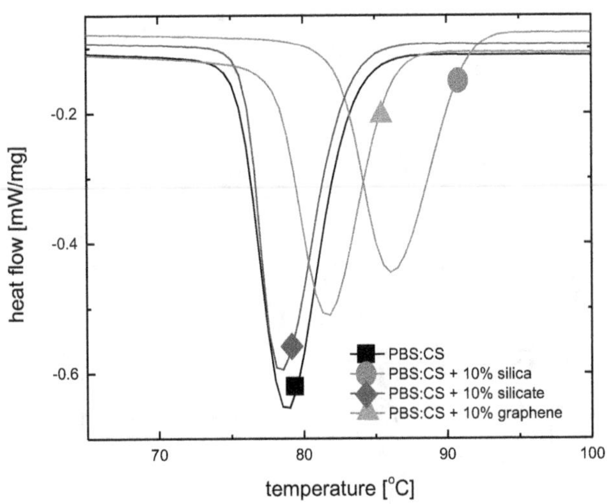

(b)

Figure 1.1 DSC (a) melting and (b) crystallization thermograms of PBS:CS nanocomposites with 10% filler content. PBS:CS represent the blend of PBS and CS in 60:40 weight proportion. The heating runs represent second heating.

Figure 1.2a shows the X-ray diffractograms of pure polymers and their 60:40 blend. PBS displayed a (002) peak at 2Θ = 19.8° and (110) peak at 2Θ = 22.6° (monoclinic crystal lattice). There were also shoulder peaks at the higher angles. On incorporation of chitosan flakes in PBS, no change in the position of diffraction peaks in PBS was observed indicating there were no molecular-level interactions between the two polymers. There was a relative decrease in the intensity in case of blend as compared to pure PBS due to reduction in

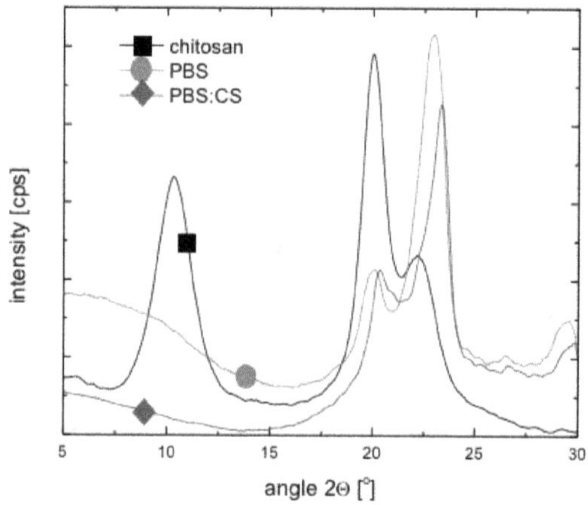

(a)

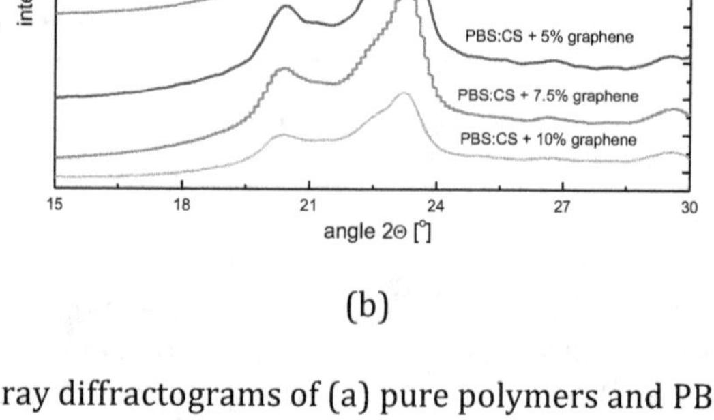

(b)

Figure 1.2 X-ray diffractograms of (a) pure polymers and PBS:CS blend; (b) nanocomposites of graphene.

crystallinity. Figure 1.2b shows the diffractograms of the graphene nanocomposites in comparison with PBS:CS blend. Other fillers also exhibited similar behavior. The nanocomposites showed no significant shift in the peak positions. Also, though silica and graphene were observed in DSC to have a nucleation effect, but it did not impact the crystal type. On increasing the filler content, the peaks became diffuse and broad. The intensities were observed to qualitatively reduce confirming that the crystallinity reduced in the samples.

In the IR spectrum of chitosan in Figure 1.3a, the peaks at 1647 cm^{-1}, 1585 cm^{-1} and 1149 cm^{-1} are from C=O stretching, N-H bending and C-O-C asymmetric stretching, which are characteristic of chitosan's polysaccharide structure [23,24]. The peak at 1377 cm^{-1} is the characteristic band of CH_3 symmetrical deformation mode [24,25] and the broad band around 3300 cm^{-1} is presumably from stretching of O-H and N-H bonds [24]. In PBS, the band at the 1044 cm^{-1} region is assigned to O-C-C stretching vibrations, whereas the band at the 1715 cm^{-1} region is from C=O stretching vibrations of the ester group [26]. The band around 2940 cm^{-1} is from C-H absorption of CH_2 groups [27]. The 60:40 PBS:CS blend and pure PBS exhibited similar spectra with peaks at the same frequency indicating physical mixing of two polymers. Specifically, the broad band corresponding to N-H bond disappeared in the PBS:CS blend spectrum confirming that the surface of the blend contained mostly PBS. The characteristic peak of silica at 1065 cm^{-1} contributed to the appearance of a hump around 1090 cm^{-1} in the PBS:CS-silica nanocomposites spectra (Figure 1.3b). This hump gradually became more prominent as the silica content was increased. Incorporation of silicate influenced the PBS:CS blend in similar way as silica (Figure 1.3c). The characteristic peak in silicate was observed at 1030 cm^{-1} and it resulted in a depression around 1090 cm^{-1} in the spectrum of nanocomposite with 2% silicate. The depression gradually turned into a hump with increasing silicate content confirming the presence of silicate. Spectrum of thermally reduced graphene showed presence of O-H bonds (broad spectra around 3300 cm^{-1} and peak at 1390 cm^{-1}). Peaks at 1630 cm^{-1} and 1714 cm^{-1} are from C=C and C=O stretching [28,29]. By the inclusion of graphene, a new shoulder peak appeared in the nanocomposites at 1570 cm^{-1} (Figure 1.3d) which requires further analysis to confirm the possibility of any chemical interactions between the polar groups (hydroxyl, carboxyl and epoxy) on the filler surface with polymer.

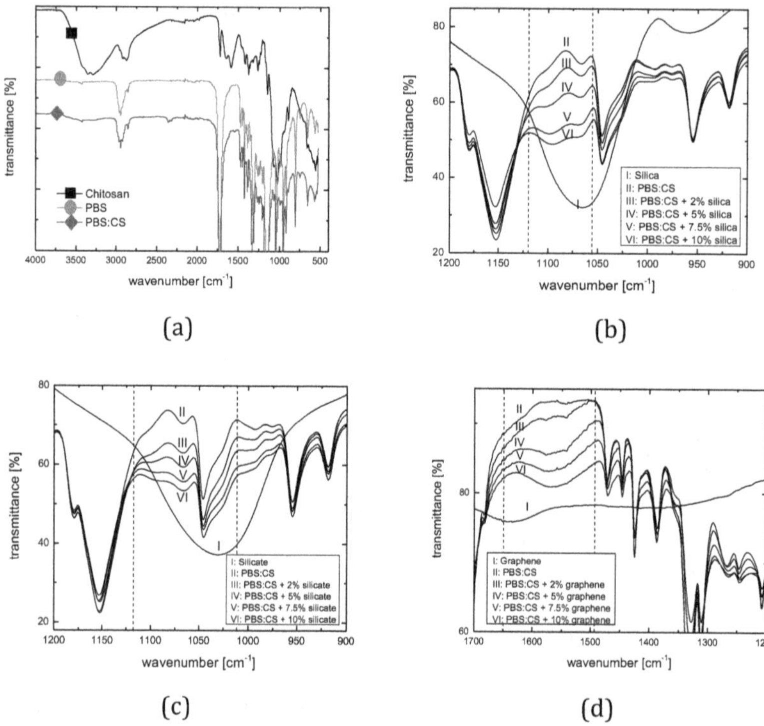

Figure 1.3 IR spectra of (a) pure polymers and PBS:CS blend; (b) silica nanocomposites; (c) silicate nanocomposites; (d) graphene nanocomposites. Scale of x-axis has been varied to effectively present the features of respective composite system.

Figure 1.4 demonstrates the morphological analysis of the blend and composites. Figures 1.4a-c show the scanning electron micrographs of the PBS:CS blend. As the chitosan phase did not melt during the melt mixing process, it exhibited morphology like reinforcing filler incorporated in the continuous matrix. Thus, large agglomerates of the chitosan fibers forming sheath like structures were observed. The average thickness of these structures is 20-30 μm, though thinner and thicker sheaths were also observed. Similar sheaths have also been reported by Correlo *et al.* [3]. The shear during compounding was probably insufficient to disperse the chitosan phase into individual fibers, however, these were observed to be uniformly distributed in the PBS matrix. Also, the sheaths did not debond from the PBS matrix and no pull-out was observed. The ten-

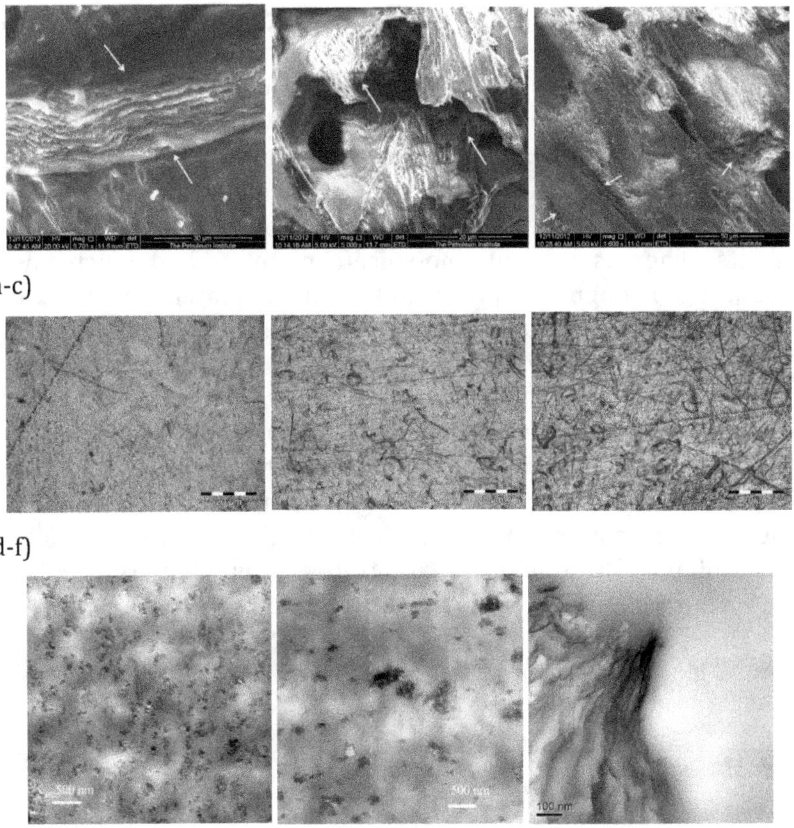

(a-c)

(d-f)

(g-i)

Figure 1.4 (a)-(c) SEM micrographs of PBS:CS blend representing sheathing of chitosan particles and PBS fibrils (shown by arrows); (d)-(f) optical images of nanocomposites with 5% silica, silicate and graphene respectively; (g)-(i) TEM images of nanocomposites with 5% silica, silicate and graphene respectively.

sile surfaces of the samples also exhibited fibrous structures formed by the PBS matrix due to elongation of its domains into fibrils (Figure 1.4b). The optical characterization of nanocomposites (Figures 1.4d-f) also revealed that the external surfaces were not smooth and some random pits and channels were present on the surface. No chitosan sheaths were observed on the surface, which confirmed the SEM and IR findings that the chitosan was present inside the PBS phase and the outside skin of the composites contained primarily PBS. The transmission electron microscopy characterization of the nanocomposites (Figures 1.4g-i) exhibited uniform distribution of fillers in the blend matrix. Though no special filler surface modification processes were applied, the polar surface of the polymers and fillers can be expected to interact positively thus leading to good dispersion. It also indicated that the torque in the compounder was sufficient to attain good mixing of the fillers. The graphene platelets were observed to have highest aspect ratio due to long and thin platelets, whereas the alumino-silicate platelets had much lower aspect ratio. Though the silicate and graphene platelets were extensively exfoliated, some aggregates as well as stacks of varying thicknesses were also observed.

As observed in Table 1.1, incorporation of chitosan in PBS enhanced its modulus by more than 300%. Also, the tensile strength of PBS was not impacted negatively indicating that even though phase domains of chitosan were in the range of μm, no extensive pull-out or debonding occurred. As shown in Table 1.2, addition of 2% graphene enhanced the modulus to 3409 MPa as compared to 1691 MPa for pure blend. Silica particles exhibited an increase of 20% in

Table 1.2 Mechanical properties of PBS:CS blend and nanocomposites with silica, silicate and graphene

Composition	Young's modulus[a] (MPa)	Tensile strength[b] (MPa)	Elongation at break[c] (mm)
PBS:CS (60:40)	1691	16	0.14
PBS:CS + 2% silica	2018	19	0.14
PBS:CS + 2% silicate	1713	15	0.10
PBS:CS + 2% graphene	3409	17	0.08
PBS:CS + 5% silica	2687	16	0.10
PBS:CS + 5% silicate	2070	15	0.13
PBS:CS + 5% graphene	4300	15	0.05
PBS:CS + 7.5% silica	2928	14	0.06
PBS:CS + 7.5% silicate	2319	15	0.11
PBS:CS + 7.5% graphene	5293	15	0.05
PBS:CS + 10% silica	3012	13	0.12
PBS:CS + 10% silicate	2793	14	0.12
PBS:CS + 10% graphene	5513	12	0.03

[a]Relative probable error 3%.
[b]Relative probable error 3%.
[c]Relative probable error 10%.

the modulus, whereas silicate at 2% content did not have any effect. At 5% filler content, the modulus in graphene composites was enhanced by >150%, whereas the silica and silicate particles resulted in increments of 60% and 20% respectively. The enhancement crossed >225% for 10% graphene content, whereas it was 80% and 65% for silica and silicate nanoparticles. Thus, all the three fillers enhanced the modulus of the PBS:CS blend in varying degrees confirming efficient stress transfer and the enhancement did not level off even at 10% filler content. Figure 1.5a demonstrates the relative tensile modulus of the nanocomposites as a function of filler fraction. Figure 1.5b also shows the stress-strain curves for PBS:CS blend and graphene nanocomposites. The ductile behavior characteristic of pure PBS was modified and the samples showed brittle behavior. The strain at break was observed to decrease as the graphene content increased in the composites (Table 1.2). The tensile strength in the composites marginally increased at lower filler fraction, and subsequently retained the same value as pure polymer at higher filler fractions. Only in the nanocomposites with 10% filler content, there was a reduction of 12-25% in the tensile strength of the nanocomposites. These findings, thus, underline successful enhancement of PBS by addition of both chitosan as well as fillers.

The storage modulus from dynamic mechanical analysis of nanocomposites with 5% filler content is demonstrated in Figure 1.6. Similar to tensile modulus, the storage modulus of nanocomposites was higher than the PBS:CS blend throughout the temperature range. Graphene addition resulted in maximum enhancement of storage modulus of the composites. The composites containing platelets with high aspect ratio are reported to exhibit much higher storage moduli and viscosity as compared to particles with lower aspect ratio at the same mass fraction due to formation of solid-like network structures [30]. Though silicate platelets had higher aspect ratio than silica particles, however, their impact on storage moduli of the composites was the least indicating that the silicate platelets may not be exfoliated and thick stacks may exist which lead to reduction in overall surface to volume ratio. This is also supported by the lowest extent of increment in tensile modulus in the case of silicate nanocomposites. The storage modulus was also observed to increase with the filler content. For nanocomposite with 10% graphene content, the storage modulus at room temperature was 1.8 times the modulus of pure blend. All the bio-nanocomposites failed between 90 °C and 95 °C.

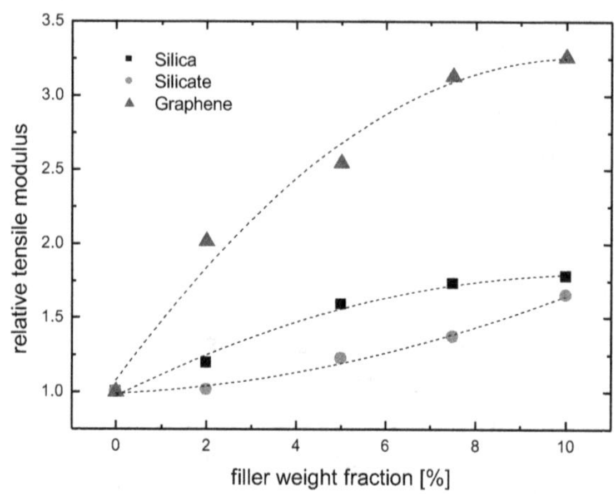

(a)

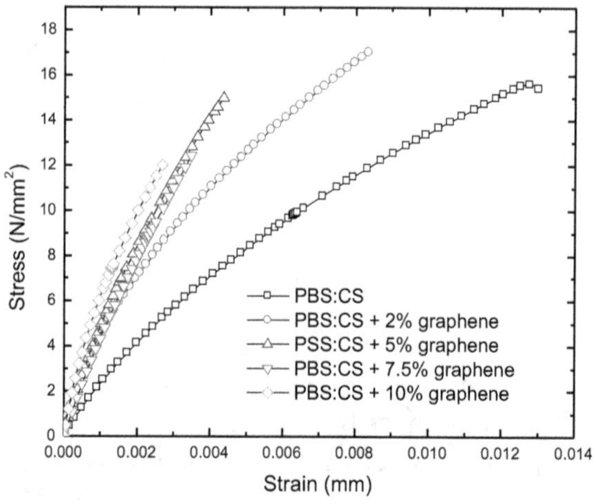

(b)

Figure 1.5 (a) Relative tensile modulus of the nanocomposites as a function of filler weight fraction; (b) stress-strain curves of nanocomposites with graphene.

Figure 1.7 also shows the characterization of the network structure using shear rheology. The polymer blend exhibited a viscous response as G" > G' for all frequencies, though the difference was not

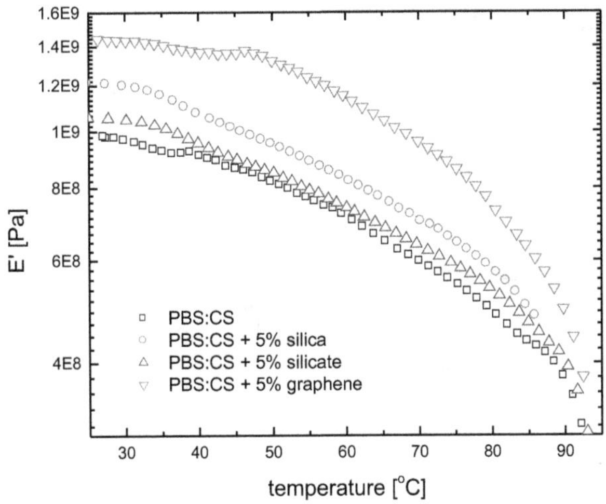

Figure 1.6 Storage modulus of nanocomposites with 5% filler content, obtained from DMA analysis.

significant and the curves deviated from typical response of a viscous liquid due to the incorporation of chitosan in PBS. The silica and silicate nanocomposites with 5% filler content had elastic dominance as G' > G" at low frequencies. For graphene nanocomposites, the liquid like low frequency scaling disappeared and G' was almost constant at lower frequencies (G' >> G" for all frequencies). This rheological state of constant G' for a large low frequency region is also termed as gel concentration [30]. The gel concentration for graphene composites was around 2%. For silica and silicate nanocomposites, such a behavior was observed around 5% filler content, thus confirming the network structure was formed in graphene based composites at a much lower filler amount. Figure 1.7c also shows the relative storage modulus of the nanocomposites at a frequency of 1 rad/s as a function of filler concentration. The impact of graphene in reinforcing the blend matrix is clearly observed. The impact of increasing graphene content beyond 5 wt% was less significant indicating its effectiveness at lower weight fraction of filler. The silica and silicate fillers were observed to enhance the modulus even at 10% filler weight fraction. The corresponding impact on viscosity of the graphene nanocomposites is also demonstrated in Figure 1.7d. Nearly four orders of magnitude increment was observed for the composites with 10% graphene content, which may pose processing difficulties. However, effectiveness of graphene at much lower concentrations would lead to optimization of nanocomposite processing and properties.

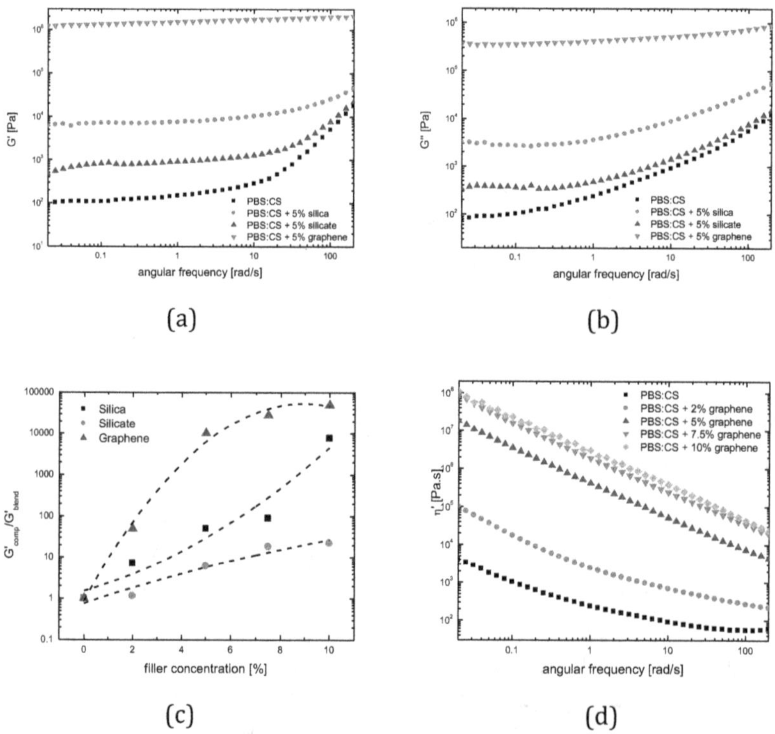

Figure 1.7 Shear rheology analysis to generate (a) storage modulus and (b) loss modulus of the nanocomposites (with 5% filler content), as a function of angular frequency; (c) relative G' of nanocomposites as a function of filler concentration (1 rad/s angular frequency); (d) viscosity of the nanocomposites with graphene as a function of angular frequency.

In order to ascertain the thermal stability, degradation characterization of the nanocomposites under nitrogen and air atmospheres was performed. As shown in Figure 1.8, the degradation of the blend under nitrogen atmosphere followed a two-step weight loss profile (the first degradation corresponding to chitosan, whereas the second one due to PBS), which was also reported earlier [31]. The nanocomposites exhibited similar degradation pattern as the polymer blend and the first degradation step started between 250-300

°C for all samples. At around 350 °C, second weight loss event was observed which continued till 430 °C. Peak degradation temperature of the second degradation step was higher for all the nanocomposites as compared to PBS:CS blend, indicating the stability of chitosan was not affected whereas PBS stability was enhanced in the composites. Moreover, for silica nanocomposites, there was a gradual increase in peak degradation temperature with increasing filler amount (408 °C for composites with 10% silica content as compared to 303 °C for polymer blend). For silicate and graphene containing composites, this temperature was observed to be 398 °C and 404 °C respectively, which did not show significant variation as a function of filler fraction.

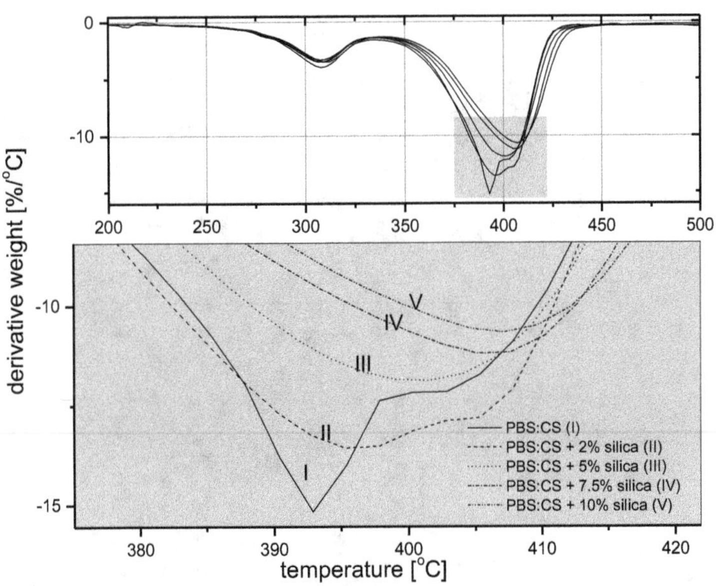

Figure 1.8 Differential TGA thermograms of silica containing nanocomposites. The analysis was performed under nitrogen.

In air atmosphere, however, the graphene nanocomposites were observed to be less stable than other fillers as well as pure blend (Figure 1.9a) and the effect was independent of filler concentration. In the mass spectroscopy analysis, water (m/z=18) was observed to evolve during three different stages (Figure 1.9b). Peaks at 130 °C and 300 °C could be assigned to mass loss due to the free and bound

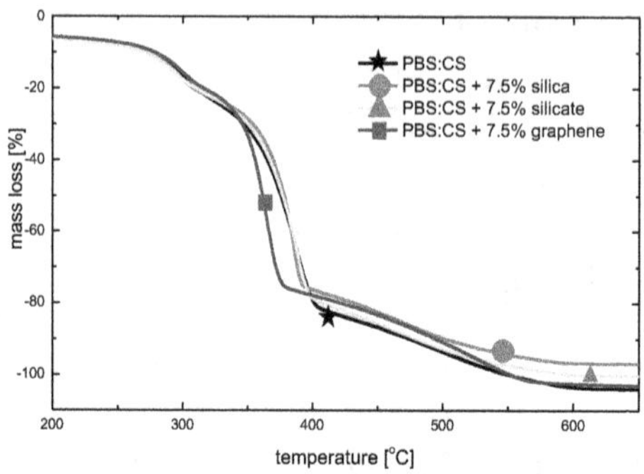

(a)

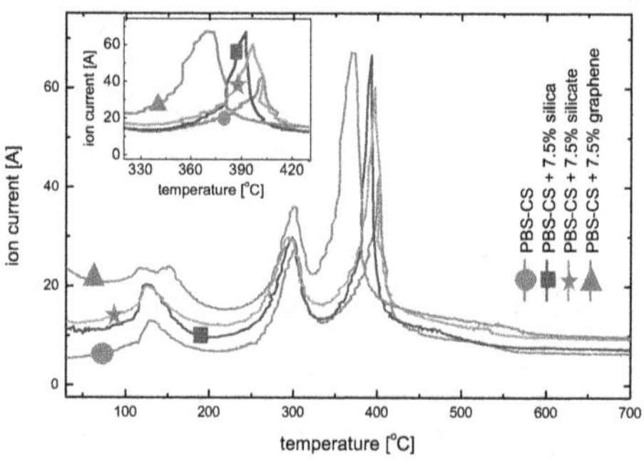

(b)

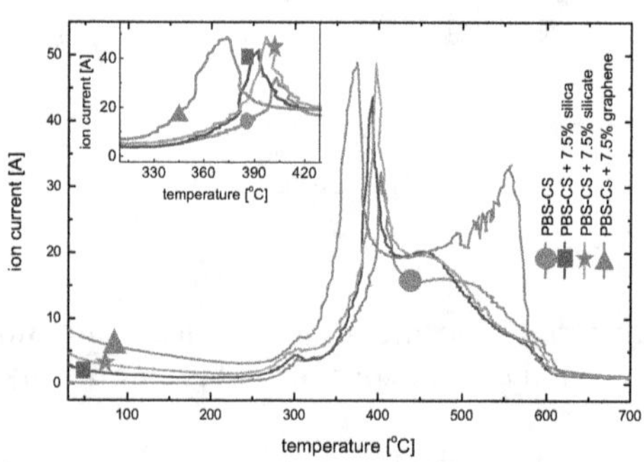

(c)

Figure 1.9 TGA-MS (in air) analysis of the polymer blend and nanocomposites with 7.5% filler content: (a) cumulative mass loss, (b) m/z = 11 and (c) m/z = 10.

moisture release from the composites. Mass loss between 325 °C and 450 °C due to the moisture release also coincided with release of CO_2 (m/z = 44) indicating thermal degradation (Figure 1.9c). The release of moisture and CO_2 from the graphene composites was observed to occur at lower temperature. Cipiriano *et al.* [31] reported an increase in the rate of mass loss, when the amount of multi-walled nanotubes was enhanced in the composites. This behavior was attributed to increase in the thermal conductivity of the samples. In the present case, this effect may not be responsible for the faster degradation observed for graphene based nanocomposites as this effect was not observed under nitrogen atmosphere. Thus, the interaction of functional groups on the graphene surface with air would have resulted in less thermally stable products. The defects/impurities present in the graphene platelets could also contribute to the earlier thermal degradation in the presence of air. The degradation of graphene platelets was also observed by a broad signal with peak at 580 °C.

Design of Experiments factorial design methodologies were used to model the storage modulus from DMA and shear rheology. For modeling storage modulus from DMA, the factors considered were type of filler, filler loading (2-10%) and temperature (25-90 °C). The storage modulus from shear rheology was modeled based on factors of type of filler, filler loading (2-10%) and angular frequency (0.1-100 rad/s). In Pareto chart of standardized effects for storage modulus from DMA, all the three factors were found to be statistically significant along with the interaction between filler and loading. Temperature was observed to be the most significant factor. For the storage modulus from shear rheology, the Pareto chart indicated that the significant factors were type of filler and filler loading. It was expected as the modulus was observed in Figure 1.7 to become independent of angular frequency for all nanocomposites at filler lading of 2-5%. Figure 1.10 shows the model predictions at different temperature and frequency conditions for graphene nanocomposites along with comparison with the experimental data. The model predictions were observed to be in fairly good agreement with the experimental data thus confirming the validity of the generated models in predicting the properties.

1.4 Conclusions

Melt processed bio-nanocomposites of the PBS:CS blend with silica,

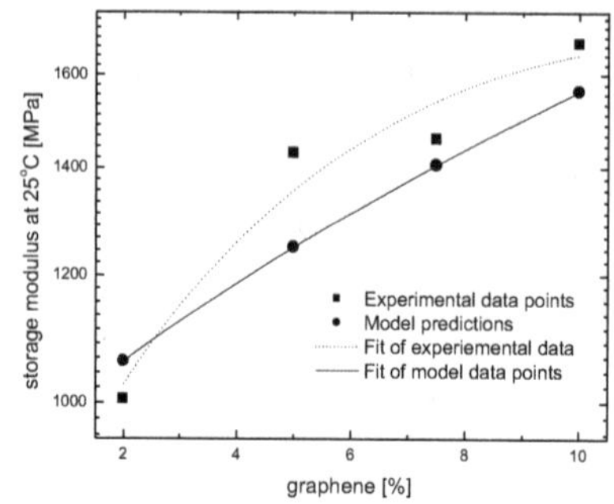

(a)

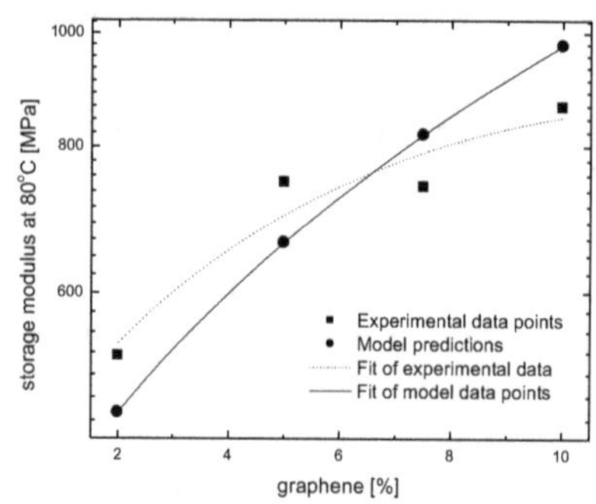

(b)

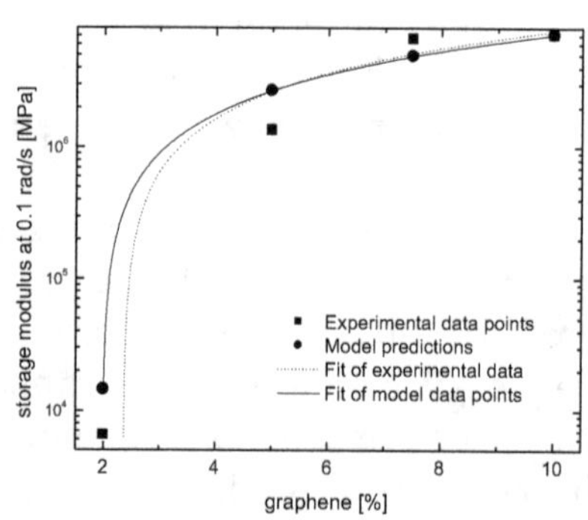

(c)

Figure 1.10 Modeling of storage modulus of the composites obtained from (a, b) DMA and (c) shear rheology, using factorial design method. The dotted lines are the fit of experimental data, whereas the solid lines represent the predictions.

alumino-silicate and graphene could be successfully generated. Incorporation of chitosan to PBS resulted in significant improvement in its modulus without decreasing tensile strength. The fillers marginally enhanced the peak melting point and crystallinity at lower weight fractions, however, at higher fractions, the similar extent of reduction in both parameters was observed. Graphene and silica also had a nucleation effect on the polymer. No change in the diffraction peak positions was observed in composites indicating no changes in crystal type, but the diffraction peaks became broad and less intense at higher filler fraction due to decrease in overall crystallinity. New IR peaks were detected in the composites containing graphene indicating the possibility of a polymer-filler chemical interaction.

Chitosan was present in the nanocomposites as well dispersed sheaths which did not debond or pull out during the tensile testing. The graphene platelets were observed to have much higher aspect ratio than silicate platelets. 10% fraction of graphene enhanced the tensile modulus of the nanocomposites by >225%, whereas it was 80% and 65% for corresponding silica and silicate containing composites. The tensile strength was also only marginally decreased at the highest filler content. Graphene nanocomposites had much higher storage modulus due to the formation of network structures of high aspect ratio graphene platelets at much lower filler weight fraction. These composites were observed to exhibit constant G′ for a large low frequency region at much lower concentration of 2%, whereas this gel concentration for silica and silicate nanocomposites was around 5%.

The thermal degradation behavior of the nanocomposites under nitrogen was similar and improved from the pure blend. However, in air atmosphere, the graphene nanocomposites exhibited faster degradation due to interaction of air with the functional groups present on the filler surface. The degradation resulted in the generation of water and CO_2 molecules predominantly.

Statistical models could be developed to predict the properties of the nanocomposites which were also validated by comparison with the experimental data.

Acknowledgements

The authors are indebted to Dr. N. B. Matsko at Graz Centre of Electron Microscopy for the TEM analysis.

References

1. Blair, H. S., Guthrie, J., Law, T. K., and Turkington, P. (1987) Chitosan and modified chitosan membranes. I. Preparation and characterization. *Journal of Applied Polymer Science*, **33**, 641-656.
2. Gong, P., Zhang, L., Zhuang, L., and Lu, J. (1998) Synthesis and characterization of polyurethane-chitosan interpenetrating polymer networks. *Journal of Applied Polymer Science*, **68**, 1321-1329.
3. Correlo, V. M., Boesel, L. F., Bhattacharya, M., Mano, J. F., Neves, N. M., and Reis, R. L. (2005) Properties of melt processed chitosan and aliphatic polyester blends. *Materials Science and Engineering A*, **403**, 57-68.
4. Yu, L., Dean, K., and Li, L. (2006) Polymer blends and composites from renewable resources. *Progress in Polymer Science*, **31**, 576-602.
5. Ray, S. S., Okamoto, K., and Okamoto, M. (2003) Structure-property relationship in biodegradable poly (butylene succinate)/layered silicate nanocomposites. *Macromolecules*, **36**, 2355-2367.
6. Wang, X., Bai, H., Yao, Z., Liu, A., and Shi. G. (2010) Electrically conductive and mechanically strong biomimetic chitosan/reduced graphene oxide composite films. *Journal of Materials Chemistry*, **20**, 9032-9036.
7. Al Zaabi, K., and Mittal, V. (2013) Biodegradable polyester nanocomposites: Phase miscibility and properties. *Journal of Applied Polymer Science*, **130**, 516-525.
8. Chandy, T., and Sharma, C. P. (1990) Chitosan as a biomaterial. *Biomaterials, Artificial Cells and Artificial Organs*, **18**(1), 1-24.
9. Singh, D. K., and Roy, A. R. (1997) Graft copolymerization of 2-hydroxyethylmethacrylate onto chitosan films and their blood compatibility. *Journal of Applied Polymer Science*, **66**, 869-877.
10. Chandy, T., and Sharma, C. P. (1992) Prostaglandin E1-immobilized poly (vinyl alcohol)-blended chitosan membranes: Blood compatibility and permeability properties. *Journal of Applied Polymer Science*, **44**, 2145-2156.
11. Liu, Y.-L., Hsu, C.-Y., Su, Y.-H., and Lai, J.-Y. (2005) Chitosan-silica complex membranes from sulfonic acid functionalized silica nanoparticles for pervaporation dehydration of ethanol-water solutions. *Biomacromolecules*, **6**, 368-373.
12. Park, S.-B., You, J.-O., Park, H.-Y., Haam, S.-J., and Kim, W.-S. (2001) A novel pH-sensitive membrane from chitosan-TEOS IPN; preparation and its drug permeation characteristics. *Biomaterials*, **22**, 323-330.
13. Vijaya, Y., and Krishnaiah, A. (2009) Sorptive response profile of chitosan coated silica in the defluoridation of aqueous solution. *E-Journal of Chemistry*, **6**, 713-724.

14. Molvinger, K., Quignard, F., Brunel, D., Boissiere, M., and Devoisselle, J.-M. (2004) Porous chitosan-silica hybrid microspheres as a potential catalyst. *Chemistry of Materials*, **16**, 3367-3372.
15. Rafiee, M. A., Rafiee, J., Wang, Z., Song, H., Yu, Z.-Z., and Koratkar, N. (2009) Enhanced mechanical properties of nanocomposites at low graphene content. *ACS Nano*, **3**, 3884-3890.
16. Fan, H., Wang, L., Zhao, K., Li, N., Shi, Z., Ge, Z., and Jin, Z. (2010) Fabrication, mechanical properties, and biocompatibility of graphene-reinforced chitosan composites. *Biomacromolecules*, **11**(9), 2345-2351.
17. Fang, M., Long, J., Zhao, W., Wang, L., and Chen, G. (2010) pH-responsive chitosan- mediated graphene dispersions. *Langmuir*, **26**, 16771-16774.
18. Correlo, V. M., Boesel, L. F., Bhattacharya, M., Mano, J. F., Neves, N. M., and Reis, R. L. (2005) Hydroxyapatite reinforced chitosan and polyester blends for biomedical applications. *Macromolecular Materials and Engineering*, **290**, 1157-1165.
19. Chaudhry, A. U., and Mittal, V. (2013) High-density polyethylene nanocomposites using masterbatches of chlorinated polyethylene/graphene oxide. *Polymer Engineering and Science*, **53**, 78-88.
20. McAllister, M. J., Li, J. L., Adamson, D. H., Schniepp, H. C., Abdala, A. A., Liu, J., Herrera-Alonso, M., Milius, D. L., Car, R., Prud'homme, R. K., and Aksay, I. A. (2007) Single sheet functionalized graphene by oxidation and thermal expansion of graphite. *Chemistry of Materials*, **19**, 4396-4404.
21. Cheung, M. K., Wan, K. P. Y., and Yu, P. H. (2002) Miscibility and morphology of chiral semicrystalline poly-(R)-(3-hydroxybutyrate)/chitosan and poly-(R)-(3-hydroxybutyrate-co-3-hydroxyvalerate)/chitosan blends studied with DSC, H-1 T-1 and T-1 rho CRAMPS. *Journal of Applied Polymer Science*, **86**, 1253-1258.
22. van Krevelen D. W. (1990) *Properties of Polymers*, Elsevier, The Netherlands.
23. Lai, J.-Y., Li, Y.-T., and Wang, T.-P. (2010) In-vitro response of retinal pigment epithelial cells exposed to chitosan materials prepared with different cross-linkers. *International Journal of Molecular Sciences*, **11**, 5256-5272.
24. Mobarak, N. N., and Abdullah, M. P. (2010) Synthesis and characterization of several lauryl chitosan derivatives. *Malaysian Journal of Analytical Sciences*, **14**, 82-99.
25. Bhattarai, N., Ramay, H. R., Chou, S.-H., and Zhang, M. (2006) Chitosan and lactic acid-grafted chitosan nanoparticles as carriers for prolonged drug delivery. *International Journal of Nanomedicine*, **12**, 181-187.

26. Kim, H.-S., Kim, H.-J., Lee, J.-W., and Choi, I.-G. (2006) Biodegradability of bio-flour filled biodegradable poly(butylene succinate) bio-composites in natural and compost soil. *Polymer Degradation and Stability*, **91**, 1117-1127.
27. Li, Y.-D., Zeng, J.-B., Wang, X.-L., Yang, K.-K., and Wang, Y.-Z. (2008) Structure and properties of soy protein/poly (butylene succinate) blends with improved compatibility. *Biomacromolecules*, **9**, 3157-3164.
28. Guo, H.-L., Wang, X.-F., Qian, Q.-Y., Wang, F.-B., and Xia, X.-H. (2009) A green approach to the synthesis of graphene nanosheets. *ACS Nano*, **3**, 2653-2659.
29. Mao, Y., Bao, Y., Wang, W., Li, Z., Li, F., and Niu, L. (2011) Layer-by-layer assembled multilayer of graphene/Prussian blue toward simultaneous electrochemical and SPR detection of H_2O_2. *Talanta*, **85**, 2106-2112.
30. Cipiriano, B. H., Kashiwagi, T., Raghavan, S. R., Yang, Y., Grulke, E. A., Yamamoto, K., Shields, J. R., and Douglas, J. F. (2007) Effects of aspect ratio of MWNT on the flammability properties of polymer nanocomposites. *Polymer*, **48**, 6086-6096.
31. Pandey, J. K., Chu, W. S., Kim, C. S., Lee, C. S., and Ahn, S. H. (2009) Bio-nano reinforcement of environmentally degradable polymer matrix by cellulose whiskers from grass. *Composites, Part B: Engineering*, **40**, 676-680.

2

Biopolyester Composites with Psyllium Husk

2.1 Introduction

Realization about the limitation of fossil based resources and environmental issues has diverged the focus towards biodegradable materials based on renewable sources. Biopolymers are materials of rising interest because of their several economic and environmental benefits. Contrary to the conventional petrochemicals based plastics, the renewable resources based biopolymers provide solutions to many problems associated with environmental degradability and fast depletion of fossil resources [1]. Due to many advancements in the synthesis and structural modification processes in the recent years, biopolymers are competing with conventional plastics in many application areas, specifically traditional commodity and medical fields [2].

Poly(lactic acid) (PLA) and poly (butylene adipate-co-terephthalate)(PBAT) are biodegradable polyesters and belong to family of thermoplastics having combination of comparable properties such as excellent processability, thermal stability, mechanical and physical properties [3]. Specifically, PBAT is an aliphatic/aromatic copolyester based on the monomers 1,4-butanediol, adipic acid and terephthalic acid and is useful in applications requiring flexibility, resistance to chemicals and good set of physical properties. Polycaprolactone (PCL) is another biodegradable and semi-crystalline polyester, with high commercial potential. Though its high elongation and low modulus are barriers to its applications, it is thermally very stable.

Due to inherent challenges and property limitations, efforts are also needed to enhance the properties of the biopolymers so as to enable them to compete with non-biodegradable commodity plastics for pragmatic and engineering applications [4]. On many occasions, high cost of biopolymers is also a significant deterrent towards their

Vikas Mittal and Ali U. Chaudhry**, The Petroleum Institute (part of Khalifa University of Science and Technology), Abu Dhabi, UAE*
**Current address: Bletchington, Wellington County, Australia; **Current address: Texas A&M University, Qatar*

application and strategies to reduce the cost are expected to ensure large scale application. The hybridization of polymers and fillers to generate blends, alloys or composites is a quick and practical technique for development of new materials having improved properties. In general, composites with biopolymer matrices are formed to fulfill various requirements for different applications where cost and weight reduction, dimensional stability, thermal stability and processability are required [5-7]. For instance, in an attempt to improve the glass transition temperature of PLA for extrusion purposes, composites with talc as nucleating agent were fabricated [8]. Increased crystallization speed was observed due to the reduction in the activation energy. Similarly, addition of graphene oxide and polyurethane improved the antibacterial activity of PLA, thus, making it useful for clinical applications [9]. Flame retardancy is another issue affecting the application of PLA in the electronics industry. In order to address this important property, PLA was compounded with aluminum hypophosphite (AHP) and expanded graphite (EG) [10]. The retardancy mechanism was attributed to the formation of a protective char layer where AHP and EG acted as three-dimensional network structure and carbonization agents [10]. In a similar manner, high brittleness and low toughness of PLA were improved by using blends of PLA and PBAT. As addition of PBAT reduces modulus and tensile strength of the blend, hence calcium carbonate was also additionally added [11]. As mentioned above, PBAT is practically suitable for packaging, agriculture and disposable bags, but cost is a significant obstruction towards its commercialization. Hence, many attempts were made to overcome this limitation, without sacrificing the required properties, such as addition of poly(butylene succinate) (PBS) and organo-modified montmorillonite (OMMT) [12], (PLA)/(PBAT) blend [13], PVC/PBAT/clay composite [14] etc. Similarly, for PCL, previous studies have also shown that PCL/thermoplastic starch (TPS) blends are readily biodegradable and that blending TPS with PCL increases the biodegradation rate of PCL. Apart from its blends with TPS, biodegradable blends with PLA have also been reported to mutually complement the advantages of each individual polymer and to come across their limitations.

Further, "true" biocomposites or "green" composites belong to a special class of hybrid materials having all the phases of the composite biodegradable in nature. The advantages of using biofiller over conventional fillers are their renewability, low cost, complete or partial recyclability, biodegradability, compatibility with biopolymer and adequate physical and mechanical properties. Examples of such composites are PLA, PBS with bamboo fiber [15], flax fiber/PLA [16], PLA/hydroxyapatite fibers [17], PLA/wood [18], cellulose/PBAT [19], oil palm empty fruit bunch fiber/PBAT [20], etc. Recently, Mittal *et al.* [21,22] reported the "true" biocomposites based on PLA and PBAT with date seed powder, specifically to reduce the cost of the polymer, without sacrificing the useful properties. It has also been noted earlier that the addition of biofillers for the composite generation is needed in order to retain the biodegradable nature of the polymer matrix as the conventionally used inorganic fillers like silicates may hinder the biodegradability of the polymer, thus, inducing unwanted reduction in the polymer performance [23]. The use of the date seed powder enhanced the rate of biodegradation in both PLA and PBAT as a function of its content (Figure 2.1). In addition, the filler also enhanced the physical properties of the polymers significantly. Thus, there is a need to explore more varieties of biofillers so as to achieve functional biocomposites with enhanced/retained biodegradation, reduced cost, as well as enhanced properties so as to effectively compete with the conventional polymer/polymer composite systems.

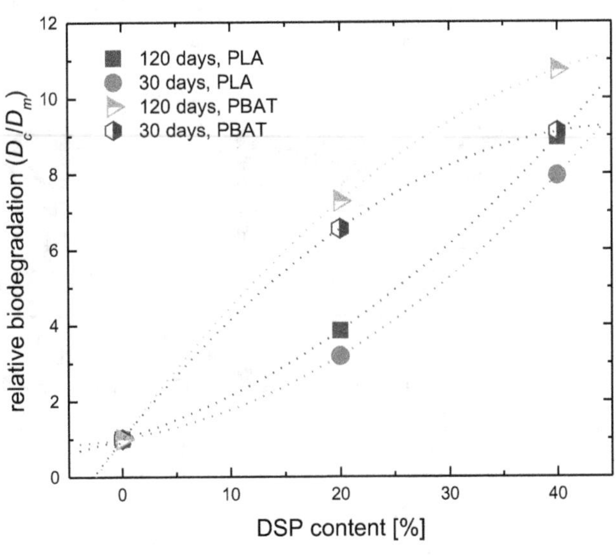

Figure 2.1 Biodegradation behavior of the PLA and PBAT based composites as a function of date seed powder content. Reproduced from Reference 22 with permission from Wiley.

Mechanically removed natural polysaccharides such as psyllium husk seeds of *Plantago ovata Forsk* are known for use in many applications as binding, thickening, emulsifying, gelling and controlled release agent [24,25]. These are also known for usefulness in numerous medicinal applications. The complex structure of psyllium seed husk was described by Fischer *et al.*, Edwards *et al.*, and Marlett & Fischer, among other authors (Figure 2.2) [24,26-29]. The main part is a complex heteroxylan having monosaccharides, arabinose and xylose, collectively called arabinoxylan. It is highly branched non-starch polysaccharide composed of b-(1,4)-linked xylopyranose residues, arabinofuranose and xylopyranose residues, rhamnose and galacturonic acid [28]. Composites of psyllium husk with polymers have also reported in literature, however, only a few studies are available. Singh *et al.* [30] also reported the modification of psyllium husk with polyacrylamide forming pH sensitive hydrogels through polymerization of polyacrylamide on to husk using ammonium persulphate (APS). It was reported that APS generated free radicals on the surface of husk, hence, promoting three-dimensional structure with polyacrylamide [30]. Similarly, husk was also grafted with poly(N-vinyl-2-pyrollidone) using microwave-assisted free-radical polymerization using APS. Overall, there is a need to explore the use of psyllium husk further as a filler in the biopolymers so as to gain more insights into its potential as reinforcing phase, which can also result in the accomplishment of "true" biocomposites.

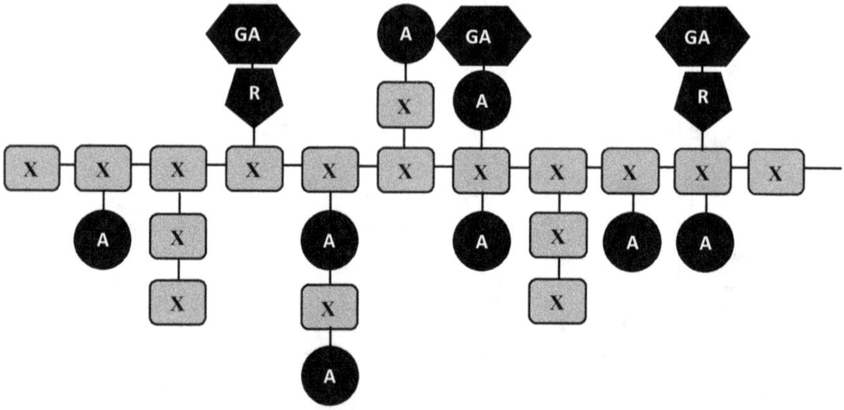

Figure 2.2 Schematic representation of possible segment of psyllium seed husk [24,26], where X represents xylopyranose, A is arabinofuran, R is rhamnose and GA is galacturonic acid.

In the current study, biocomposites of biopolyesters (PBAT and PCL) have been generated with psyllium seed husk with an aim to generate cost-efficiency. The amount of the filler was varied from 10-40 wt%. The composites were characterized for thermal, mechanical and rheological properties, along with morphological analysis to gain extensive insights into the effect of the biofiller on the polymer performance.

2.2 Experimental

2.2.1 Materials

PBAT biopolyester with trade name Ecoflex® F Blend C1200 was supplied by BASF SE, Germany. As mentioned earlier, it is a biodegradable aliphatic-aromatic copolyester based on the monomers 1,4-butanediol, adipic acid and terephthalic acid in the polymer chain, which biodegrades to the basic monomers and eventually to carbon dioxide, water and biomass when metabolized in the soil or compost under standard conditions. PCL with a trade name of CAPA 6506 S was procured from Perstorp Polyols, USA. It had weight average molecular weight (M_w) of 50,000 g/mol and peak melting temperature (T_m) of 60 °C (supplier information). Psyllium husk particles were procured locally. No further treatment was carried out on the materials and these were used as received. The husk particles had an average size of 100 μm and spherical, though non-uniform shape.

2.2.2 Preparation of Biocomposites

Biocomposites were prepared by melt mixing of the biopolymers with husk using mini twin conical screw extruder (MiniLab HAAKE Rheomex CTW5, Germany). For generating PCL composites, compounding temperature of 100 °C was used, whereas PBAT composites were compounded at 160 °C. Rotational speed of 80 rpm were used and the mixing was performed for 7 min. The screw length and screw diameter were 109.5 and 5/14 mm conical respectively. Composites with filler content of 10, 20, 30 and 40 wt% were generated. Pure polymers were also processed under same thermal conditions to maintain consistency. Disc and dumbbell-shaped test specimens for rheological and mechanical characterization were prepared by mini injection molding machine (HAAKE MiniJet II, Germany) at the same processing temperature as the melt mixing. The injection pressure was

820 bar for 6 s whereas holding pressure was 400 bar for 3 s. The temperature of the mold was kept at 40 °C.

2.2.3 Characterization Methods

Calorimetric properties of the pure polymer and biocomposites were recorded on a Netzsch Differential Scanning Calorimeter (DSC 200 F3 Maia) under nitrogen atmosphere. The scans were obtained from 50 to 170 to 50 °C using heating and cooling rates of 15 and 5 °C/min respectively. Thermal properties of the samples were analyzed using Netzsch Thermogravimetric Analyzer (STA 449 F1 Jupiter). Nitrogen was used as a carrier gas and the scans were obtained from 50 to 700 °C at a heating rate of 20 °C/min. The heat enthalpies used to calculate the extent of crystallinity were recorded in a narrow error range (±0.1%), which were also confirmed by repeated runs.

AR 2000 rheometer from TA Instruments was used to characterize rheological properties such as storage modulus (G'), loss modulus (G"), and complex viscosity (η*). Disc shaped samples of 25-mm diameter and 2-mm thickness were characterized at 130 °C and 70 °C for PBAT and PCL systems respectively, using a gap opening of 1 mm. Frequency sweep scans (dynamic testing) of all polyester composites were recorded at 1% strain from ω 0.1 to 100 rad/s.

Mechanical characterization of pure polymers and biocomposites was performed on universal testing machine (Testometric M350, UK). The dumbbell shaped samples with 53 mm length, 4 mm width, and 2 mm thickness were used. A loading rate of 5 mm/min was employed and the tests were carried out at room temperature. Win Test Analysis software was used for the calculation of tensile modulus. An average of five values was calculated. DMA (dynamic mechanical analyzer) RSA3 from TA Instruments was used to study the change in the modulus with temperature (25-100 °C). Dynamic temperature ramp test was used at a frequency of 5 rad/s, ramp rate 3 °C/min, and 1% strain for all samples.

IR spectra were collected on Nicolet iS10 spectrometer equipped with SmartiTR diamond ATR accessory (angle of incidence of 45°), DTGS KBr detector and KBr beam splitter. It had a diamond ATR crystal (index of refraction = 2.4 at 1000 cm^{-1}) and a depth penetration of 2 μm at 1000 cm^{-1} for sample with refractive index of 1.5. Spectra were recorded by the OMNIC software in the 4000-525 cm^{-1} region with a resolution of 4 cm^{-1} from 16 scans. For SEM analysis, sample blocks were fixed on the sample holders followed by sputter coating with 3 nm platinum. The analysis was performed in Hitachi field emission in-lens high-resolution (HR)-SEM S-900 microscope and investigated at 10-20 kV. The block faces of the specimens were also investigated at ambient conditions using ZEISS Axioplan light microscope (Zeiss, Germany) equipped with Zeiss Axio Cam ICc 1 CCD camera in reflected polarized light.

2.3 Results and Discussion

The melt compounding approach, used for the generation of the composites, could be effectively employed to incorporate psyllium husk in the biopolymers. PBAT and PCL were chosen in this study due to their extensive set of useful properties as well as ease of processing. By the addition of the filler, it was envisaged to enhance the application of these polymers by reducing the cost, while maintaining (or enhancing) other properties. The chosen compounding temperature was based on the peak melting point of the pure polymers. Specifically, the compounding was carried out at temperature roughly 40 °C higher than the peak melting point value so as to ensure good mixing. However, care was taken not to choose too high compounding temperature to avoid the degradation of the composite constituents. Similarly, the compounding time was also kept long enough to ensure good mixing, however, overexposure of the composite components to high temperature was avoided. The conditions chosen for the processing of the pure polymers could be used for the composites irrespective of the filler content, thus, confirming that the processing of the polymer was not affected by the addition of the filler.

Figure 2.3 shows the calorimetric, thermal degradation and IR characterization of the pure husk particles. The DSC thermogram of the filler (Figure 2.3a) exhibited a broad endothermic transition with peak maximum at 100 °C, however, the transition was completely absent in the second heating cycle. As the filler did not liquefy at higher temperature and as no crystallization peak was observed in DSC, the broad transition at 100 °C can be attributed to the loss of moisture. The TGA thermogram (in nitrogen) of husk in Figure 2.3b also indicated a weight loss of ~15% in the temperature range of 50-200 °C probably due to loss of moisture, indicating that the powder had large amount of moisture content. This phenomenon was retained even though the powder was dried overnight at 60 °C under vacuum, which confirmed that the moisture may not be present only on the surface of particles, but more tightly bound in the structure. Similar findings have also been reported for the date seed powder filler in earlier studies [21,22]. Figure 2.3c also demonstrates the FTIR spectrum of husk particles. Intense signal related to the –OH groups also confirmed the presence of water molecules in the sample. The TGA thermogram also indicated a sharp degradation signal around 310 °C, with initiation of degradation around 250 °C. It confirmed the thermal stability of husk for high temperature compounding with the polymers. Further studies using mass spectroscopy coupled with TGA would be beneficial to elucidate the fractions generated during degradation to gain insights into the degradation mechanism of husk particles.

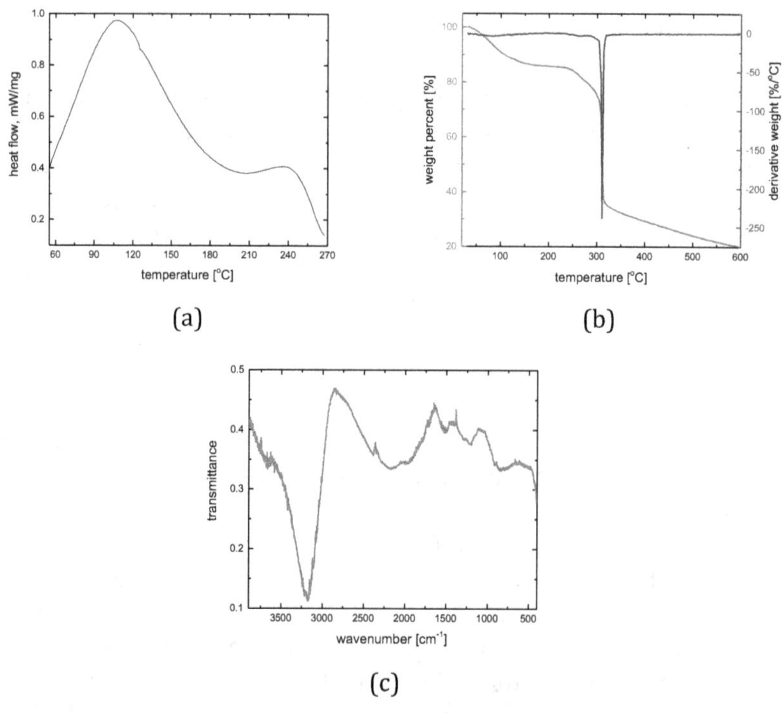

Figure 2.3 (a) DSC and (b) TGA thermograms of husk; (c) IR spectrum of husk particles.

Figure 2.4 shows the FTIR spectra of PBAT and PCL composites with 10% and 40% filler fraction. The other composites also exhibited similar patterns. For comparison, the IR spectra of pure husk and pure polymers are also plotted. The IR spectra of the composites resembled that of pure polymers. The signal corresponding to the –OH groups in the husk spectrum was also absent in the polymer composites, indicating that the moisture contained in the husk sample would have evaporated during high temperature melt compounding process. Due to the physical mixing of the husk with the biopolymer, no new peak was observed in the composite spectra. This is specifically important to ensure that the filler did not interact negatively with the polymer. As the biofillers comprise of large number of components, there is a possibility of these fillers disintegrating inside the

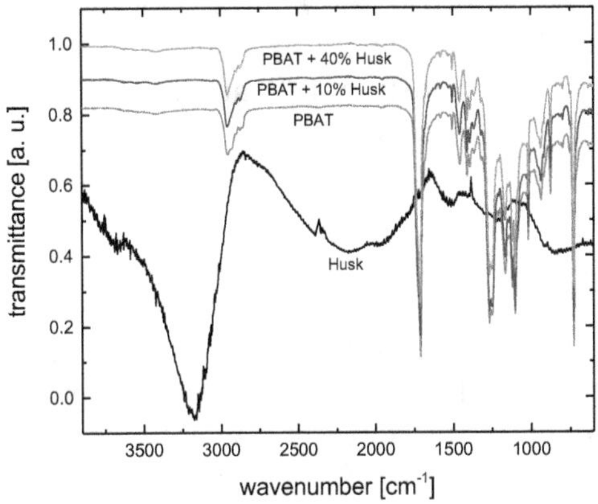

(a)

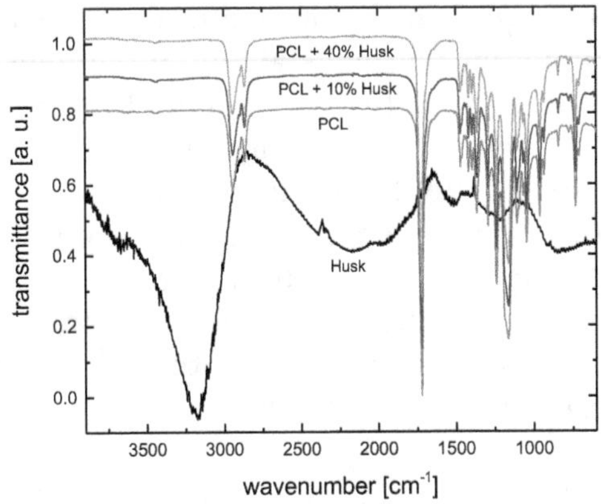

(b)

Figure 2.4 FTIR spectroscopy analysis of (a) PBAT and its composites with 10% and 40% filler content; (b) PCL and its composites with 10% and 40% filler content. For comparison, the FTIR spectrum of husk is also shown.

polymer and some fractions having unwanted chemical interactions with the polymers, thereby decreasing the performance. The FTIR findings are also supported by the TGA of the husk presented in Figure 2.3b which indicated that the husk had a well-defined degradation temperature around 310 °C. As the compounding temperature (100 °C for PCL and 160 °C for PBAT composites) was much lower than the peak degradation temperature of the husk, it ensured the dimensional stability of husk particles in the polymer matrices.

As mentioned earlier, retaining the thermal stability of the polymer after incorporation of the biofiller is very important to ensure effective service applications. To further confirm the thermal stability of the polymer in the composites, TGA analysis was carried out and compared with the pure polymer response, as demonstrated in Figure 2.5. Both polymers exhibited peak degradation temperatures above 400 °C. The TGA thermograms of the composites exhibited dual degradation peaks corresponding to the polymer and filler. It indicated that the components were not molecularly mixed, which is expected due to the large size of the husk particles. However, the degradation peak corresponding to husk did not exhibit the actual fraction of the filler added to the polymer, indicating some extent of uniform filler mixing at micro- or possibly nano-scale would have resulted due to the shear in the melt mixer. Overall, even though the TGA thermograms of the composites exhibited early degradation due to the presence of the filler, however, the composites were still thermally stable for subsequent high temperature processing or applications, due to the inherent thermal stability of the filler particles. In addition, the incorporation of the filler did not lead to any change in the degradation mechanism of the polymers. The degradation observed in the thermogram of the pure husk sample due to the presence of the moisture was not observed in the composite, thus, confirming the loss of filler's moisture during processing. Also, the dynamic TGA analysis at the compounding temperature as a function of time did not indicate any weight loss for both polymer systems, thus, confirming the absence of any degradation phenomenon at the compounding temperature.

As noticed earlier in Figure 2.3, husk particles exhibited a broad endothermic transition around 100 °C, which was not observed in the second heating. This transition was also absent in the first and second heating curves of the polymer composites (Figure 2.6), confirming the transition to be due to the presence of moisture. If the observed peak was due to the crystalline melting of the filler, it would have also

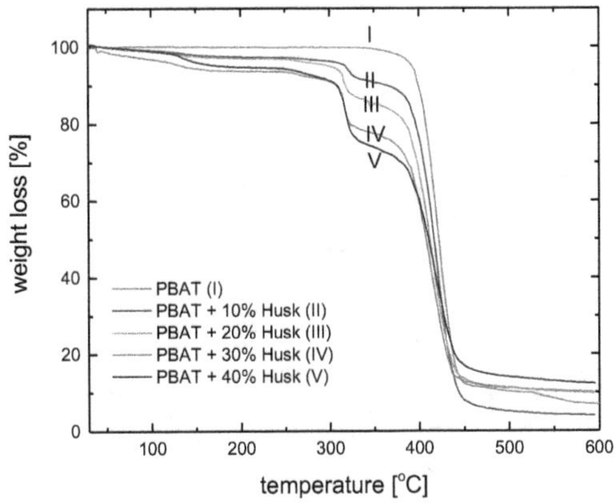

(a)

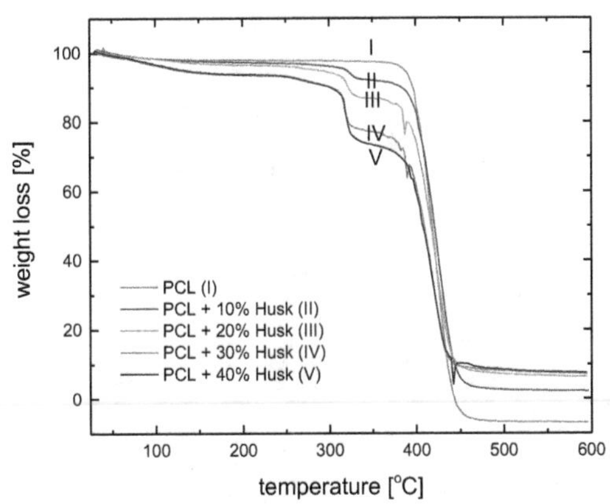

(b)

Figure 2.5 TGA cumulative thermograms of (a) PBAT and (b) PCL composites.

been observed in the DSC thermograms of the composites due to the lack of molecular mixing with the polymer. For PBAT composites, there was no change in the peak melting temperature as a function of filler fraction (Table 2.1). For instance, the peak melting temperature of polymer increased from 123°C to 125 °C in the composite with 40% filler content. There was an increase in crystallinity as the filler content was increased indicating the nucleation effect of the filler on the polymer. This effect was evident from the significant increase in

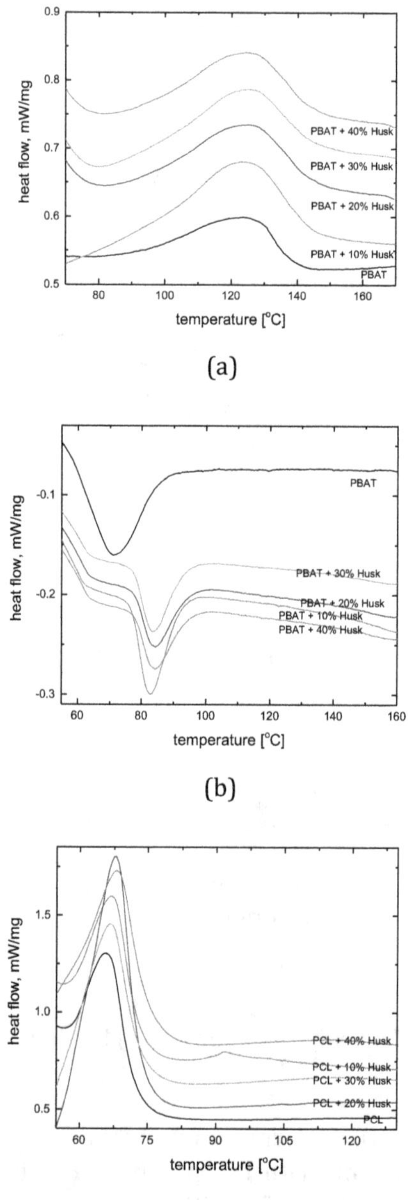

Figure 2.6 (a) crystalline melting thermograms of PBAT composites; (b) crystallization behavior of PBAT composites; and (c) crystalline melting thermograms of PCL composites. 2nd heating curves are shown for both composites systems.

the peak crystallization temperature. For instance, the peak crystallization temperature of the pure polymer was measured to be 71 °C, which was significantly enhanced to 84 °C in the composites with 20% or more filler content. Similar to the PBAT system, the peak melting point in the case of PCL composites was not significantly

Table 2.1 Calorimetric properties of the PBAT composites as a function of filler fraction

	Peak melting temperature [°C]	Melt enthalpy [J/g]	Peak crystallization temperature [°C]
PBAT	123	12.4	71
PBAT + 10% Husk	123	14.3	82
PBAT + 20% Husk	125	15.1	84
PBAT + 30% Husk	126	20.0	84
PBAT + 40% Husk	125	20.2	84

affected by the addition of husk. For instance, the temperature of the pure polymer was observed to be 66 °C, which was enhanced marginally to 68 °C for the composite with 40% filler content. The filler also did not have any nucleation effect on the PCL polymer.

After ascertaining the thermal stability of the composites on the incorporation of the biofiller, the mechanical performance of the composites was analyzed. Figure 2.7a demonstrates the relative tensile modulus of the composites as a function of filler content. The modulus of the pure polymers was measured to be 78 and 393 MPa for PBAT and PCL respectively. The composites with 10% filler content exhibited a decrease in the tensile modulus as compared to the pure polymer, probably due to some plasticization of the matrix in the presence of biofiller. The effect was more pronounced in the case of PBAT composite. However, on increasing the filler content, the modulus was observed to increase gradually, indicating overcoming of the plasticization by the reinforcement effect. The modulus was observed to increase for both the polymers even with 40% filler indicating that the polymers were able to hold large extent of filler particles due to their rubbery nature (the PBAT and PCL polymers are rubbery at room temperature). PCL composite with 40% filler fraction exhibited an increase of 60% in the modulus as compared to the pure polymer. On the other hand, an increment of more than 65% was observed for the PBAT composite containing the same filler amount.

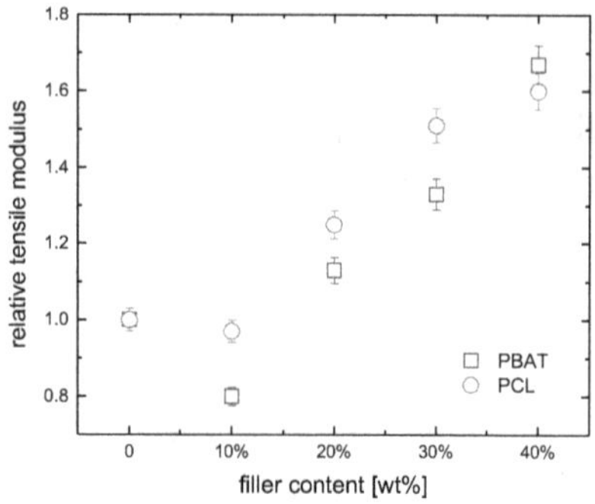

(a)

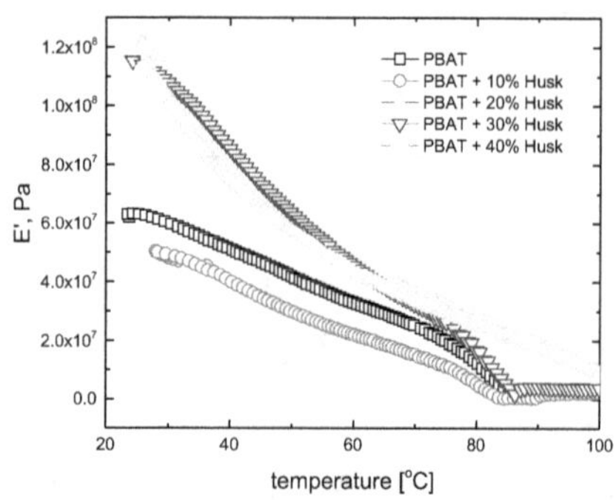

(b)

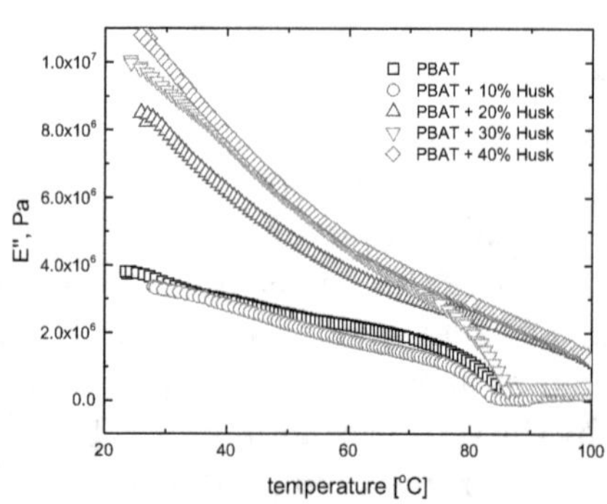

(c)

Figure 2.7 (a) Relative tensile modulus of the PBAT and PCL composites as a function of filler concentration; dynamic mechanical analysis depicting the (b) storage modulus and (c) loss modulus of PBAT composites as a function of temperature.

These findings corroborated the earlier studies on the generation of the "true" biocomposites, where the biofiller positively impacted the mechanical properties of the biopolymers [2122]. These findings also confirmed that the biofiller was not present in the polymers as a passive filler, but had active contribution towards performance enhancement of the polymers. The addition of the filler did not impact the elongation of the polymers significantly, thus, allowing the polymers to retain flexibility. Thus, even though TGA analysis indicated the lack of molecular mixing among the composite components, the mechanical response indicated that a good degree of filler dispersion could be achieved due to the shear applied during the compounding process which resulted in enhanced mechanical properties due to the efficient stress transfer from the polymer to the filler particles. Further insights into the mechanical response as a function of temperature was also gained through dynamic mechanical analysis, as shown in Figure 2.7b and 2.7c for PBAT composites. Similar to the results of tensile testing, the composite with 10 wt% was observed to have lower storage modulus than the pure polymer throughout the measured temperature range, indicating the plasticization of the polymer matrix in the presence of lower amount of biofiller particles. The composites with 20 or higher weight percent were observed to have significantly enhanced modulus as compared to the pure polymer and the trend was maintained over the whole temperature range. For instance, the storage modulus of the pure polymer at 30 °C was 5.9E7 Pa, which was enhanced to 1.9E8 for the composite with 40% filler content. Though the extent of enhancement in the modulus was decreased with increasing temperature, however, the composites still performed better than the pure PBAT even at 90 °C. This confirmed that the addition of the biofiller to PBAT enhanced its applicability for high temperature uses. This also clearly indicated that the biofiller was not only helpful in reducing the overall cost of the composite, but was also instrumental in significantly enhancing the composite properties. The loss modulus of the PBAT composites also exhibited the same trend as the storage modulus. Overall, the loss modulus values of the pure polymer and the composites were lower than the corresponding storage modulus values over the whole temperature range, indicating dominance of the elastic behavior.

Network structure of the polymer composites was evaluated with shear rheology and the storage modulus, loss modulus, and complex viscosity of the samples as a function of angular frequency are demonstrated in Figures 2.8 and 2.9. Strain sweep was conducted and

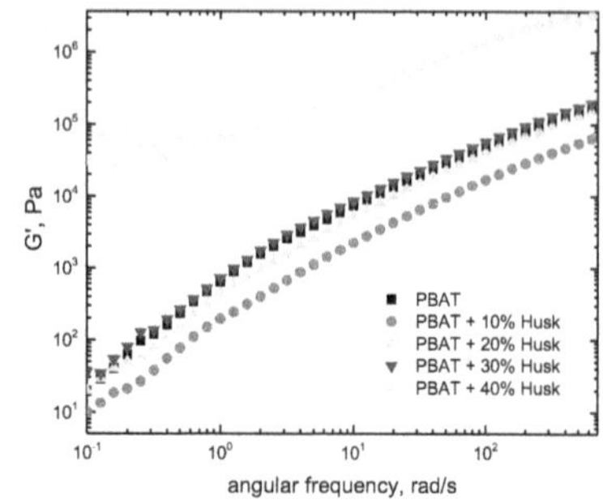

(a)

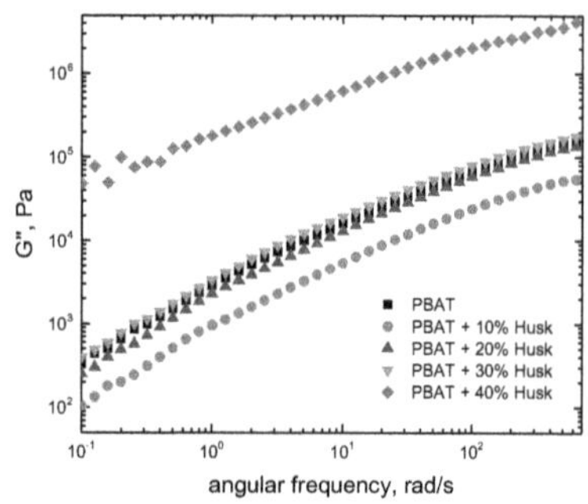

(b)

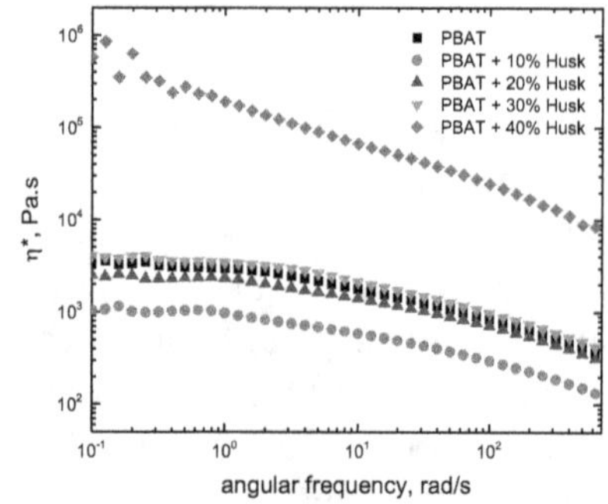

(c)

Figure 2.8 (a) Storage modulus and (b) loss modulus of the PBAT composites in comparison with pure PBAT as a function of angular frequency; (c) complex viscosity of the PBAT composites as a function of angular viscosity.

samples were observed to be safe up to 10% strain. Frequency sweep of the samples was performed with controlled shear strain at 1% using frequency range of 0.1-100 rad/s. As observed in Figure 2.8a, PBAT composite with 10% filler content had lower storage modulus than the pure polymer, as also observed earlier. The composites with 20% and 30% filler content exhibited similar storage modulus as the pure polymer. However, the composite with 40% filler content exhibited significantly higher the storage modulus over the whole angular frequency range, probably due to the presence of filler aggregates. For instance, at angular frequency value of 10 rad/s, the G' of pure PBAT was measured to be 7.4E3 Pa, which enhanced to 2.6E5 for the composite with 40% filler content. All the samples exhibited gradual decline in the extent of modulus enhancement as a function of angular frequency due to shear thinning effect. The loss modulus curves (Figure 2.8b), on comparison with the storage moduli of the samples, revealed that G" was higher than G' for all the samples at any frequency. It indicated that the polymer chains had dominant viscous behavior with long relaxation times. Also to note that the observed behavior of viscous dominance at the melt shearing temperature was different from the elastic dominant behavior earlier observed in DMA for the lower temperature range. Figure 2.8c also shows the complex viscosity of the PBAT composites as a function of angular frequency. Contrary to the modulus, the viscosity decreased on increasing angular frequency due to enhancement in flexibility with frequency. Similar to the modulus, the composite with 10% filler content had the lowest viscosity value. Though the composite with 40% filler exhibited much higher viscosity, however, it could still be processed using the same processing conditions as the pure polymer. Similar to PBAT composites, the storage modulus of the PCL composite with 10% filler was lower than the pure polymer (Figure 2.9a). On increasing the filler fraction, the storage modulus was observed to increase, though marginally. Also, the composite with 40% filler fraction was not stable during the rheological analysis probably due to large filler aggregates. In these composites also, G" > G' for most of the angular frequency range, thus, indicating the viscous dominance. The viscosity of the composites also increased on addition of the biofiller (Figure 2.9c), however, the composites were still processable using the same processing conditions, thus, requiring no additional effort for their manufacturing. Thus, these findings confirmed the positive impact of the psyllium husk based biofiller on the polymer properties and performance.

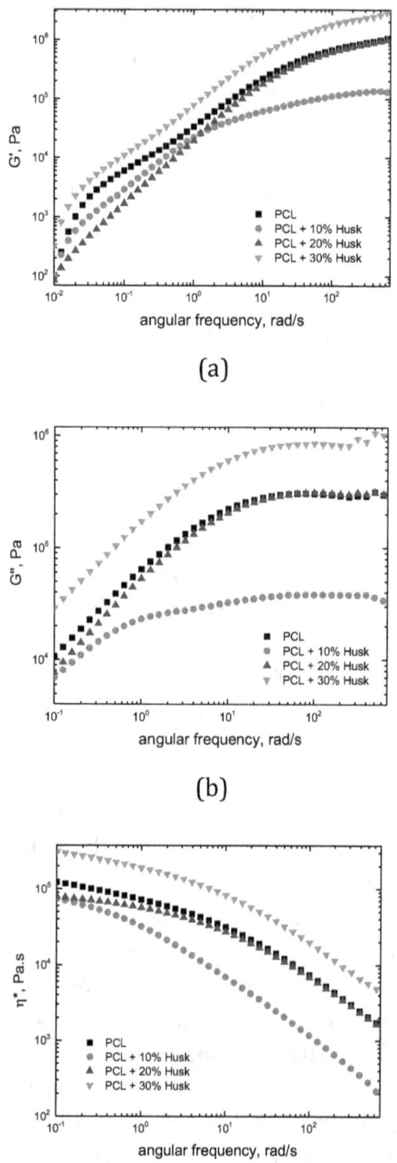

Figure 2.9 (a) Storage modulus and (b) loss modulus of the PCL composites in comparison with pure PCL as a function of angular frequency; (c) complex viscosity of the PCL composites as a function of angular viscosity.

The miscibility of the component phases was further studied using Cole-Cole viscosity plot which develops relationships between real (η') and imaginary (η") parts of complex viscosity [31-33]. A smooth, semi-circular shape of the graph would suggest miscible system with homogenous phase. The deviation from this behavior indicates phase segregation due to immiscibility of the components. Also to note that these analyses have been traditionally developed for gaining insights into the phase miscibility of polymer blends. However, the composites generated in the current study also fit somewhat to this definition due to the organic nature of the biofiller. Figure 2.10a demonstrates the Cole-Cole viscosity plot for the PBAT composites. The composites with filler fraction from 20-30% followed well defined semi-circular path suggesting phase miscibility in the composites. The composite with 10% filler content though exhibited the semi-circular behavior, but the deviation was significant at higher viscosity values. Also, the composite with 40% filler exhibited extensive scatter and could not be plotted, probably due to the presence of a large number of filler aggregates and non-uniform filler dispersion, thus, indicating phase immiscibility. The findings from the Cole-Cole plot support the earlier observed improvements in the mechanical performance of the composites, except the composite with 40% filler content, where the mechanical properties still enhanced even with phase immiscibility. However, it is necessary to further support the findings from Cole-Cole plots, as these findings could sometimes be misleading [34]. Figure 2.10b shows the analysis of the rheological data based on van Gurp plots representing the relationship between complex modulus (G*) and phase angle delta (δ) [35,36]. The van Gurp plots confirmed the findings from Cole-Cole analysis. In the case of composites with 20-30% filler content, the time-temperature superposition principle was observed to hold (indicated by the merging of the curves into a common curve). Some deviation was observed for the composite with 10% filler content, as also observed in Figure 2.10a. Though it is unexpected for the composite with 10% filler content to have phase immiscibility, the observed behavior could have resulted from the plasticization of the polymer and can also be linked to the earlier observed lower mechanical and rheological response for this composite. The rheological behavior of the composites was also analyzed using criteria reported for compatibility by Han and Chuang [37,38], as shown in Figure 2.10c. As G' is plotted vs. G", such analysis generates composition independent correlation for compatible systems, whereas the correlation is composition dependent for

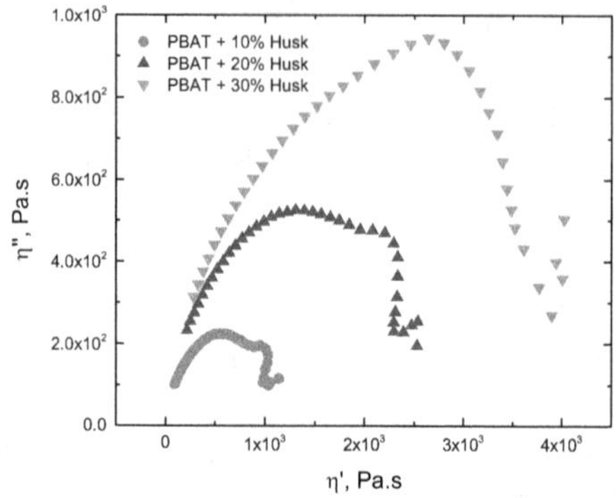

(a)

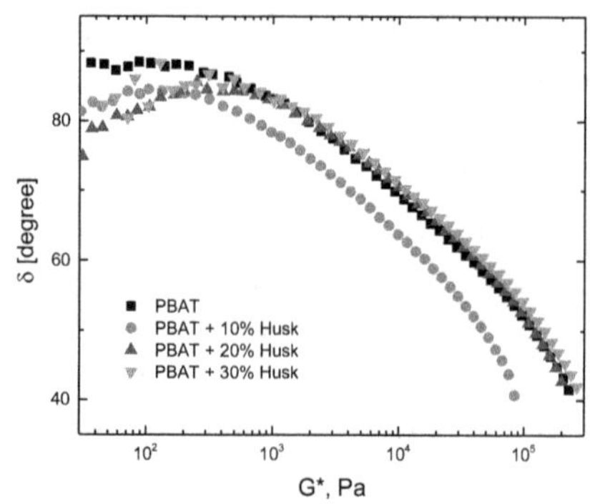

(b)

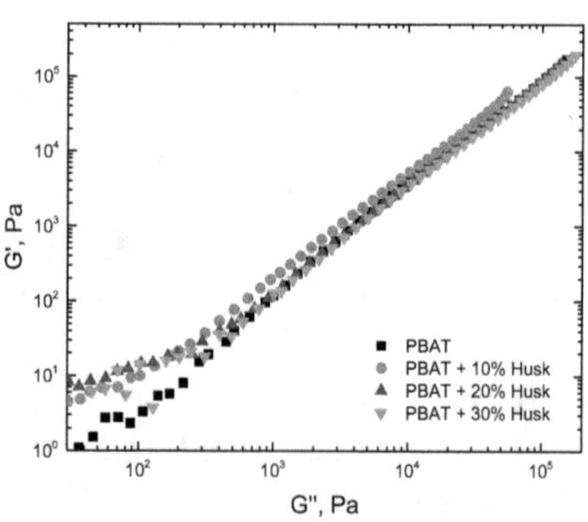

(c)

Figure 2.10 (a) Cole-Cole, (b) van Gurp and (c) Han-Chuang plots of PBAT composites.

incompatible cases. Similar to other analyses, concentration independent correlation was observed for composites with 20 and 30% filler fraction. The correlation was slightly concentration dependent in the case of the composite with 10% filler content. Figure 2.11 also shows the similar Cole-Cole, van Gurp and Han-Chuang miscibility analysis plots for the PCL composites. Higher degree of phase immiscibility was observed in the PCL composites in comparison with the corresponding PBAT composites. For instance, the PCL composite with 10% filler fraction exhibited larger deviation in all three analyses (absence of semi-circular curve, non-adherence to time-temperature superposition principle and concentration dependent behavior). The composites with 20% and 30% filler fractions though indicated good extent of phase miscibility, however, deviations could still be observed. Similar to PBAT system, composite with 40% filler fraction exhibited extensive immiscibility in the phases. Thus, the phase miscibility analysis indicated relatively poorer filler dispersion in the case of PCL composites. However, it is necessary to further confirm these finding through the morphological analysis of the composites.

Figure 2.12 shows the SEM images of the external surface of the PBAT composites with 10% and 30% filler fractions. Filler particles are visible on the surface of the composites, and also appear to be uniformly mixed with the polymer matrix. These images also confirmed that the polymer matrix was able to retain large number of filler particles. Figure 2.12c also demonstrates the polymer-filler interface. The interface appeared to be uniform with no obvious debonding or non-adhesion between the components. On the other hand, the PCL composites demonstrated different morphology, as shown in the SEM images in Figure 2.13. The composites appeared to have some voids and the filler distribution was less optimal. In addition, the polymer formed long strands, probably during processing. The polymer-filler interface demonstrated in Figure 2.13c also indicated lack of adhesion between the components. These findings also confirmed the insights gained from the phase miscibility analysis about the more homogenous distribution of the filler in the PBAT composites.

Figure 2.14 also demonstrates the light microscopy images of the PBAT and PCL composites. As seen in Figure 2.14a for PBAT composite with 20% filler content, the filler particles were uniformly distributed in the matrix. In addition, the particles seemed to be elongated by shearing in the presence of viscous melt in the compounder. The polymer-filler interfaces were also well developed with no filler

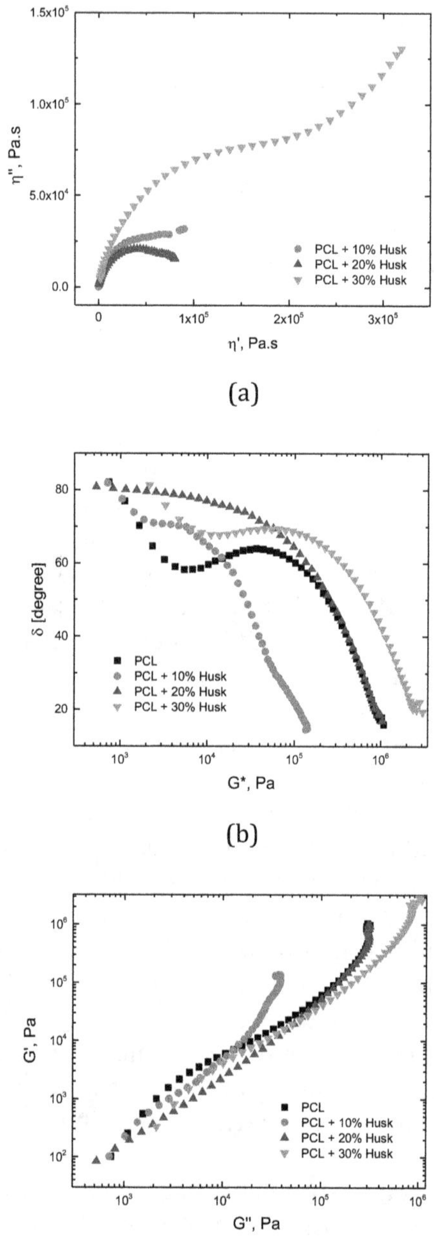

Figure 2.11 (a) Cole-Cole, (b) van Gurp and (c) Han-Chuang plots of PCL composites

pull-out observed during sectioning of the block face. The morphology of the PCL composites was observed to be significantly different from the PBAT composites, as also observed during the SEM analysis. As observed in Figure 2.14d, the polymer composite with 20% filler content had large spherical voids, which were not visible for PBAT composite. In addition, the pure PCL did not exhibit any voids after processing, thus, indicating that the voids resulted during compounding of the filler with the polymer. As the PBAT composites also did not exhibit such voids, it indicated that the formation of the voids was driven by the compounding temperature. As psyllium husk was confirmed to hold a large amount of moisture in the pristine state, it would lose the moisture at the compounding conditions, thus, one possible cause of such voids is due to the escape of high temperature steam from the matrix. As PBAT was compounded at much higher temperature than PCL, thus, the husk may have lost the moisture instantaneously when compounded with PBAT at higher temperature. Consequently, it may have provided enough time for the composite to become homogenous during 7 min of compounding time. On the other hand, as PCL was compounded at 100 °C, it would have resulted in slow release of moisture from the husk particles. It is possible that the moisture was still being released by the time compounding duration was over, which thus resulted in the formation of the escape routes for the steam. However, further analysis would be required to confirm this hypothesis. Such micro-channels in the matrix can be of immense benefit in certain applications related to migration of gases or other materials through the polymer. The PCL composites were also observed to have much larger filler particles/aggregates as compared to PBAT composite, the finding which is in agreement with the phase miscibility analysis. In addition, comparing the PBAT and PCL composites, the filler morphology was also observed to be different, probably due to the different heat treatments the filler was subjected to during compounding with the polymers. Earlier studies on the date seed powder containing biocomposites have also reported the formation of such voids during compounding [21,22].

2.4 Conclusions

In this study, successful incorporation of psyllium husk with biopolyesters PBAT and PCL has been demonstrated. The psyllium husk powder had significant extent of bound moisture which was observed to be evaporated during melt compounding with the polymer

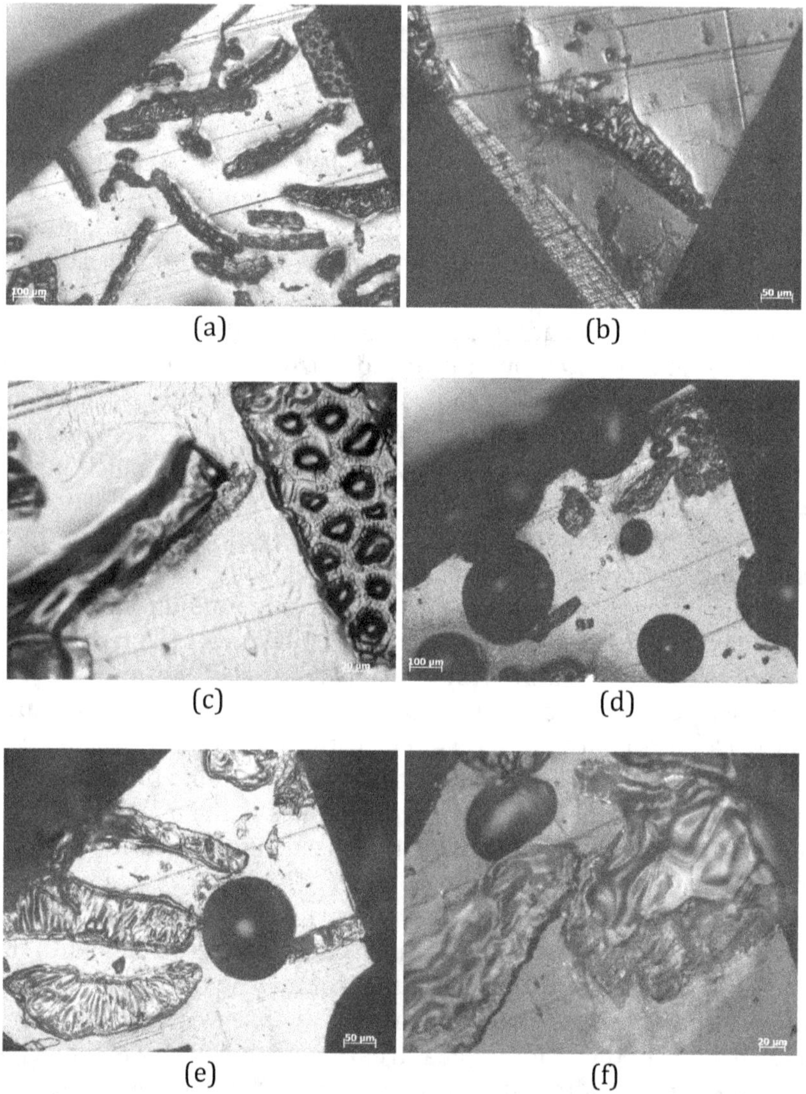

Figure 2.14 Light microscopy images of the block face in reflected light. (a-c) PBAT composite with 20 wt% filler content, and (d-f) PCL composite with 20 wt% filler content.

at high temperature. The husk powder had a well-defined sharp thermal degradation with peak around 310 °C indicating its suitability for high temperature processing. The incorporation of husk did not lead to any chemical interactions with the biopolyesters. The TGA of the composites exhibited two thermal degradation patterns owing to the individual composite components, indicating absence of molecular mixing. However, certain degree of phase mixing was achieved during shearing in the compounder as exhibited by increment in the tensile modulus as a function of filler fraction in the composites. Tensile modulus was improved by more than 60% for both the polymers at 40% filler content, though the composites with 10% filler content were observed to have lower modulus than the pure polymers probably owing to the plasticization effect. The husk particles did not impact the peak melting temperature of the polymers. The filler acted as a nucleating agent for PBAT as an increase of 13 °C in the peak crystallization temperature was observed for the composites with 20% or more filler content. The storage modulus of the composites was also observed to increase with filler fraction over the entire measured temperature range. The storage modulus and viscosity of the composites with 40% filler content was also observed to significant enhanced as a function of angular frequency probably due to aggregates, however, the composites could still be processed using the same processing conditions as the pure polymers. The loss modulus of the materials at the melt shearing temperature was always higher than the storage modulus, thus, indicating the dominance of viscous nature. The phase miscibility analysis of the composites also indicated high degree of phase mixing in PBAT composites with 20-30% filler content, with 10% composite exhibiting some extent of phase immiscibility. The composite with 40% filler was, however, observed to exhibit significant phase immiscibility. In comparison, the miscibility in PCL composites was less optimal. The SEM analysis also confirmed the findings from the phase miscibility analysis. The polymer-filler interface was observed to be more optimal in PBAT composites. The light microscopy analysis also indicated the elongated nature of the husk particles probably due to shear during compounding. The PCL composites were observed to contain more filler aggregation. In addition, the PCL composites had large voids probably due to the escape of the moisture from the filler during compounding. Such voids were absent in PBAT composites, as these were compounded at higher temperature. Thus, in conclusion, the addition of psyllium husk to the biopolymers had a positive impact on the physical properties of the biopolymers. Also to mention that the psyllium husk used in the study contained micron sized particles, which were used without any further modification or size reduction. On suitable size reduction and surface modification of these particles, it is expected that the performance of the composites would improve even further.

As shown in Scheme 2.1, these composites represent a functional framework for the development of eco-friendly materials. It would also be interesting to study the biodegradation behavior of these composites to elucidate the effect of the biofiller on the biodegradation behavior.

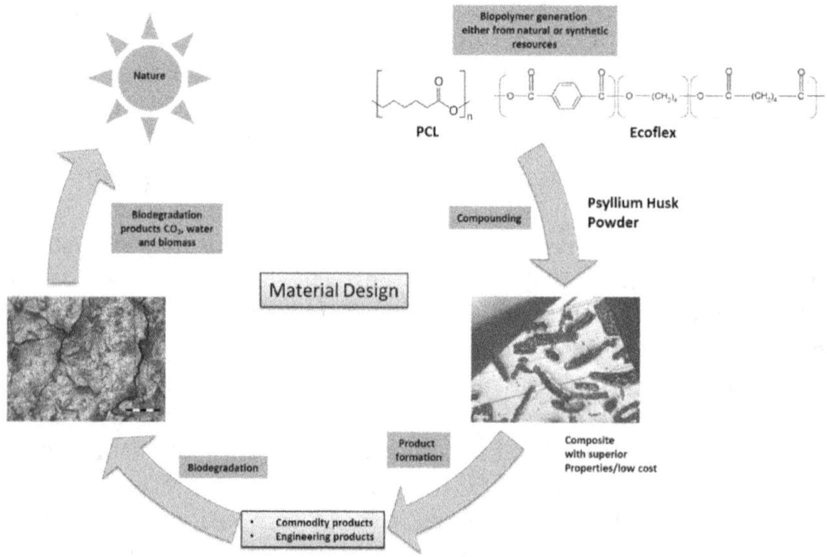

Scheme 2.1 Material design cycle from the generation of the biopolymers and biopolymer composites to the final biodegradation.

Acknowledgements

The authors are indebted to Dr. N. B. Matsko for the light microscopy analysis.

References

1. Vasile, C., and Zaikov, G. E. (2009) *Environmentally Degradable Materials Based on Multicomponent Polymeric Systems*, Brill, The Netherlands.
2. van de Velde, K., and Kiekens, P. (2002) Biopolymers: overview of several properties and consequences on their applications. *Polymer Testing*, **21**(4), 433-442.

3. Garlotta, D. (2001) A literature review of poly(lactic acid). *Journal of Polymers and the Environment*, **9**(2), 63-84.
4. Martin, O., and Averous, L. (2001) Poly(lactic acid): plasticization and properties of biodegradable multiphase systems. *Polymer*, **42**(14), 6209-6219.
5. Mittal, V., and Al Zaabi, K. (2013) Biodegradable polyester nanocomposites: Phase miscibility and properties. *Journal of Applied Polymer Science*, **130**(1), 516-525.
6. Mittal, V., Chaudhry, A. U., and Luckachan, G. E. (2014) Biopolymer-thermally reduced graphene nanocomposites: Structural characterization and properties. *Materials Chemistry and Physics*, **147**, 319-332.
7. Mittal, V., Akhtar, T., and Matsko, N. (2015) Mechanical, thermal, rheological and morphological properties of binary and ternary blends of PLA, TPS and PCL. *Macromolecular Materials and Engineering*, **300**, 423-435.
8. Battegazzore, D., Bocchini, S., and Frache, A. (2011) Crystallization kinetics of poly (lactic acid)-talc composites. *Express Polymer Letters*, **5**(10), 849-858.
9. An, X., Ma, H., Liu, B., and Wang, J. (2013) Graphene oxide reinforced polylactic acid/polyurethane antibacterial composites. *Journal of Nanomaterials*, **2013**, Article ID 373414.
10. Tang, G., Zhang, R., Wang, X., Wang, B., Song, L., Hu, Y., and Gong, X. (2013) Enhancement of flame retardant performance of bio-based polylactic acid composites with the incorporation of aluminum hypophosphite and expanded graphite. *Journal of Macromolecular Science, Part A*, **50**(2), 255-269.
11. Teamsinsungvon, A., Ruksakulpiwat, Y., and Jarukumjorn, K. (2010) Mechanical and Morphological Properties of Poly(lactic acid)/Poly(butylene adipate-co-terephthalate)/Calcium Carbonate Composite. *18th International Conference on Composite Materials*, Korea. Online: http://www.iccm-central.org/Proceedings/ICCM18proceedings/data/3.%20Poster%20Presentation/Aug23(Tuesday)/P2-18~35%20Green%20Composites/P2-27-IF1477.pdf [assessed 3rd May 2017].
12. Chieng, B., Ibrahim, N., and Wan Yunus, W. (2010) Effect of organo-modified montmorillonite on poly (butylene succinate)/poly (butylene adipate-co-terephthalate) nanocomposites. *Express Polymer Letters*, **4**(7), 404-414.
13. Lin, S., Guo, W., Chen, C., Ma, J., and Wang, B. (2012) Mechanical properties and morphology of biodegradable poly(lactic acid)/poly(butylene adipate-co-terephthalate) blends compatibilized by transesterification. *Materials & Design*, **36**, 604-608.

14. Rahim, M. N. M., Ibrahim, N. A., Sharif, J., and Wan Yunus, W. M. Z. (2010) Mechanical and thermal properties of poly(vinyl chloride)/poly(butylene adipate-co-terephthalate) clay nanocomposites. *Journal of Reinforced Plastics and Composites*, **29**(21), 3219-3225.
15. Lee, S.-H., and Wang, S. (2006) Biodegradable polymers/bamboo fiber biocomposite with bio-based coupling agent. *Composites, Part A: Applied Science and Manufacturing*, **37**(1), 80-91.
16. Oksman, K., Skrifvars, M., and Selin, J. F. (2003) Natural fibres as reinforcement in polylactic acid (PLA) composites. *Composites Science and Technology*, **63**(9), 1317-1324.
17. Kasuga, T., Ota, Y., Nogami, M., and Abe, Y. (2000) Preparation and mechanical properties of polylactic acid composites containing hydroxyapatite fibers. *Biomaterials*, **22**(1), 19-23.
18. Gregorova, A., Sedlarik, V., Pastorek, M., Jachandra, H., and Stelzer, F. (2011) Effect of compatibilizing agent on the properties of highly crystalline composites based on poly(lactic acid) and wood flour and/or mica. *Journal of Polymers and the Environment*, **19**(2), 372-381.
19. Mukherjee, T., Czaka, M., Kao, N., Gupta, R. K., Choi, H. J., and Bhattacharya, S. (2014) Dispersion study of nanofibrillated cellulose based poly(butylene adipate-co-terephthalate) composites. *Carbohydrate Polymers*, **102**, 537-542.
20. Siyamak, S., Ibrahim, N. A., Abdolmohammadi, S., Yunus, W. M. Z. W., & Rahman, M. Z. A. (2012) Effect of fiber esterification on fundamental properties of oil palm empty fruit bunch fiber/poly (butylene adipate-co-terephthalate) biocomposites. *International Journal of Molecular Sciences*, **13**(2), 1327-1346.
21. Mittal, V., Chaudhry, A. U., and Matsko, N. B. (2014) "True" biocomposites with biopolyesters and date seed powder: Mechanical, thermal, and degradation properties. *Journal of Applied Polymer Science*, **131**(19), doi: 10.1002/app.40816.
22. Mittal, V., Luckachan, G. E., Chernev, B., and Matsko, N. B. (2015) Biopolyester–date seed powder composites: Morphology and component migration. *Polymer Engineering and Science*, **55**(4), 877-888.
23. Wu, T.-M., and Wu, C.-Y. (2006) Biodegradable poly(lactic acid)/chitosan-modified montmorillonite nanocomposites: Preparation and characterization. *Polymer Degradation and Stability*, **91**, 2198-2204.
24. Singh, B., and Kumar, S. (2008) Synthesis and characterization of psyllium-NVP based drug delivery system through radiation crosslinking polymerization. *Nuclear Instruments and Methods in Physics Research Section B: Beam Interactions with Materials and Atoms*, **266**(15), 3417-3430.

25. Bhatia, M., and Ahuja, M. (2014) Synthesis and evaluation of psyllium-g-poly(N-vinyl-2-pyrrolidone) as mucoadhesive gel. *Der Pharmacia Lettre*, **6**(2), 127-131.
26. Marlett, J. A., and Fischer, M. H. (2002) A poorly fermented gel from psyllium seed husk increases excreta moisture and bile acid excretion in rats. *The Journal of Nutrition*, **132**(9), 2638-2643.
27. Fischer, M. H., Yu, N., Gray, G. R., Ralph, J., Anderson, L., and Marlett, J. A. (2004) The gel-forming polysaccharide of psyllium husk (Plantago ovata Forsk). *Carbohydrate Research*, **339**(11), 2009-2017.
28. van Craeyveld, V., Delcour, J. A., and Courtin, C. M. (2009) Extractability and chemical and enzymic degradation of psyllium (Plantago ovata Forsk) seed husk arabinoxylans. *Food Chemistry*, **112**(4), 812-819.
29. Edwards, S., Chaplin, M. F., Blackwood, A. D., and Dettmar, P. W. (2003) Primary structure of arabinoxylans of ispaghula husk and wheat bran. *Proceedings of the Nutrition Society*, **62**(1), 217-222.
30. Singh, B., Chauhan, G. S., Sharma, D. K., Kant, A., Gupta, I., and Chauhan, N. (2007) The release dynamics of salicylic acid and tetracycline hydrochloride from the psyllium and polyacrylamide based hydrogels (II). *Carbohydrate Polymers*, **67**(4), 559-565.
31. Cho, K., Lee, B. H., Hwang, K. M., Lee, H., and Choe, S. (1998) Rheological and mechanical properties in polyethylene blends. *Polymer Engineering and Science*, **38**(12), 1969-1975.
32. Kim, H. K., Rana, D., Kwag, H., and Choe, S. (2001) Melt rheology of ethylene 1-octene copolymer blends synthesized by Ziegler-Natta and metallocene catalysts. *Korea Polymer Journal*, **8**, 34-43.
33. Kwag, H., Rana, D., Choe, K., Rhee, J., Woo, T., Lee, B. H., and Choe, S. (2000) Binary blends of metallocene polyethylene with conventional polyolefins: Rheological and morphological properties. *Polymer Engineering and Science*, **40**(7), 1672-1681.
34. Utracki, L. A. (1989) *Rheology of Polymer Alloys and Blends*, Hanser Publishers, Germany.
35. Joshi, M., Butola, B. S., Simon, G., and Kukaleva, N. (2006) Rheological and viscoelastic behavior of HDPE/octamethyl-POSS nanocomposites. *Macromolecules*, **39**(5), 1839-1849.
36. van Gurp, M., and Palmen, J. (1998) Time-temperature superposition for polymeric blends. *Rheology Bulletin*, **67**(1), 5-8.
37. Chuang, H. K., and Han, C. D. (1984) Rheological behavior of polymer blends. *Journal of Applied Polymer Science*, **29**(6), 2205-2229.
38. Han, C. D., and Chuang, H. K. (1985) Criteria for rheological compatibility of polymer blends. *Journal of Applied Polymer Science*, **30**(11), 4431-4454.

3

Eco-friendly Bio-nanocomposites of PHBV/PBAT Blend with Graphene and Vermiculite

3.1 Introduction

Bio-based polymers have gained interest due to the need for alternate materials owing to depleting fossil resources and environmental legislations. Various biopolymers are biocompatible and biodegradable which, thus, have no indelible effect on the environment and due to their biological degradation process, greenhouse gas emissions are reduced [1-6]. In addition, significant enhancements in the property profiles of the biopolymers gained in the last few decades have opened many opportunities for their use in a large number of useful application areas, thus, competing with the traditional commodity and specialty polymers. However, many biopolymers still have traditionally higher costs and structural limitations compared to commercial thermoplastic polymers, thus, requiring improvements to make them fully competitive [7-11]. One such route is to generate bio-nanocomposites with enhanced properties as compared to the pure polymer, however, achieving simultaneous enhancement of strength and flexibility for in-field structural applications as well as maintaining the biodegradability is still a challenge. Blending of the biopolymer is also a common approach for combining the useful properties as well as for alleviating their limitations.

In the family of biopolyesters, poly(hydroxybutyrate-co-hydroxyvalerate) (PHBV) presents comparable structural properties as commercial polyolefins along with excellent biodegradability, but suffers from drawbacks like brittleness. Different studies have reported enhancement in its flexibility and mechanical properties by compounding with various fillers [12-23]. For instance, the enhancement in the

Vikas Mittal[a,], Bradley Burren[a,**] and Nadejda Matsko[b]*
[a]The Petroleum Institute (part of Khalifa University of Science and Technology), Abu Dhabi, UAE; [b]Graz Centre of Electron Microscopy, Graz, Austria
**Current address: Bletchington, Wellington County, Australia; **Current address: University of Waterloo, Canada*

tensile modulus was observed in the composites of PHBV with bamboo pulp fibers, without sacrificing ductility [12]. The effect of various biodegradable plasticizers was also studied on the mechanical properties of PHBV matrix [13]. The authors reported a decrease in the tensile modulus due to plasticization, however, significant improvement in the toughness was observed. In another study, a reduction in the tensile modulus by 40% for the composites of PHBV, granular starch and calcium carbonate ($CaCO_3$) was reported, whereas flexural modulus was observed to increase by 25% as compared to pure PHBV [14]. The effect of various organically modified clays was also reported on the mechanical properties of PHBV matrix [16]. The authors observed a decrease in the flexibility on the addition of organo-modified montmorillonite, whereas it was reported to increase on the addition of halloysite. The tensile strength of the composites increased for both the filler systems. In other studies based on the biopolymer blends, PHBV was blended with poly(lactic acid) (PLA) in different proportions [17]. The blends exhibited absence of any significant increment in modulus, however, a small increment in strain was observed. Mechanical properties of PHBV were also analyzed by blending it with thermoplastic polymers like poly(l-lactides) (PLLA) [20], polycaprolactone (PCL) [21], poly(ethylene succinate) (PES) [22] and poly(ethylene glycol) (PEG) [23], however, no studies on biodegradability in the presence of fillers were reported.

Poly(butylene adipate-co-terephthalate) (PBAT) is another useful aliphatic/aromatic copolyester based on the monomers 1,4-butanediol, adipic acid and terephthalic acid. PBAT polymer has high flexibility than other biodegradable polyesters, such as poly(lactic acid) and poly(butylene succinate), therefore, has significant application for food packaging and agricultural flexible films. It was also confirmed from the biodegradation tests on the polymer that the polymer did not indicate any environmental risk (eco-toxicity) when introduced into composting processes [24]. Only a few studies reporting the composites of PBAT with nanofillers like layered silicates have been reported [25-28]. Chivrac *et al.* [26] reported increase in the stiffness of the polymer as a function of clay (Cloisite 20A) content due to the strong interfacial interactions between the polymer and filler phases. The nucleation of the polymer was also enhanced on addition of clay, but simultaneously crystal growth was hindered. Mittal and Al Zaabi [28] further explored the nanocomposites of PBAT with functional fillers like silica and alumino-silicate in order to access their suitability for various applications. Apart from the mechanical performance, their thermal and rheological performance was positively affected. For instance, the addition of 5% of fillers enhanced the modulus of the polymer by 20% and also enhanced or retained the elongation and yield stress. The morphological characterization of the composites confirmed good dispersion of filler particles in the polymer matrix, however, the magnitude of large sized aggregates increased with increasing filler fraction. Mittal *et al.* [29] also studied the nanocomposites of PBAT with varying amounts of thermally reduced graphene, generated by melt mixing. PBAT nanocomposites exhibited extensive exfoliation with large number of single graphene platelets. The average stack thickness was reduced significantly and polymer intercalation in the interlayers was observed. In another study, biocomposites of date seed powder with PBAT were reported [30]. After embedding in compost soil, the composites had significantly enhanced extent of biodegradation as compared to pure polymers, thus, confirming the "true" biocomposite nature. The tensile modulus of PBAT was significantly enhanced by incorporation of date seed powder and showed enhancement even at 40% date seed powder content indicating good filler-polymer interaction.

In studies reporting the combination of PHBV and PBAT for improving the individual polymer performance, Nagarajan *et al.* [18] reported small improvement in structural properties of PHBV/PBAT blend composites with lignocellulosic fibers. The authors also reported the use of switchgrass in the PHBV/PBAT blends [19]. An increase in tensile modulus was observed, however, the composites had a significant reduction in the impact strength. In a recent study, composites of PHBV/PBAT blend with graphene and vermiculite as fillers were generated and studied for their mechanical and biodegradation (using lipase and phosphate buffered saline) behavior [31]. The composites exhibited significant enhancement in the stiffness and toughness. The biodegradation behavior was also enhanced with the addition of PBAT and vermiculite (Figure 3.1). In context, composites with 10 wt% vermiculite had 108 and 1500% higher degradation in lipase and phosphate buffered saline environment after 8 days, respectively, over neat PHBV, which confirmed the importance of blend nanocomposites for achieving functional systems.

Some reports on soil burial tests as a mean of analyzing biodegradibility of composites have also been reported [32-37]. Wu [32] reported polycaprolactone (PCL)/chitosan and acrylic acid grafted polycaprolactone (PCL-g-AA)/chitosan composites. It was found that for both PCL/chitosan and PCL-g-AA/chitosan, the weight loss was high for samples with higher content of chitosan. Kim *et al.* [35] analyzed the biodegradability of bio-flour filled poly(butylene succinate) (PBS) composites both in natural and aerobic compost soil. As expected, biodegradability of these composites increased with increasing content of biodegradable bio-flour, when aerobic composting soil was used. Several test methods and standards for testing biodegradability of polymers like CO_2 evolution, aerobic composting and respirometric tests were also discussed by Pagga [38] and it was suggested that biodegradation test for polymeric material should be aerobic composting test (compost as inoculum, 60 °C, CO_2 measurement). Patwary *et al.* [39] melt processed bio-nanocomposites of PBS-chitosan generated with varying amounts of silica, alumina-silicate and thermally reduced graphene and analyzed for their biodegradation behavior. Nanocomposites had, in general, lower extent of weight loss than the pure blend, but the extensive surface roughness and cracks were observed for all systems indicating the initiation of biodegradation.

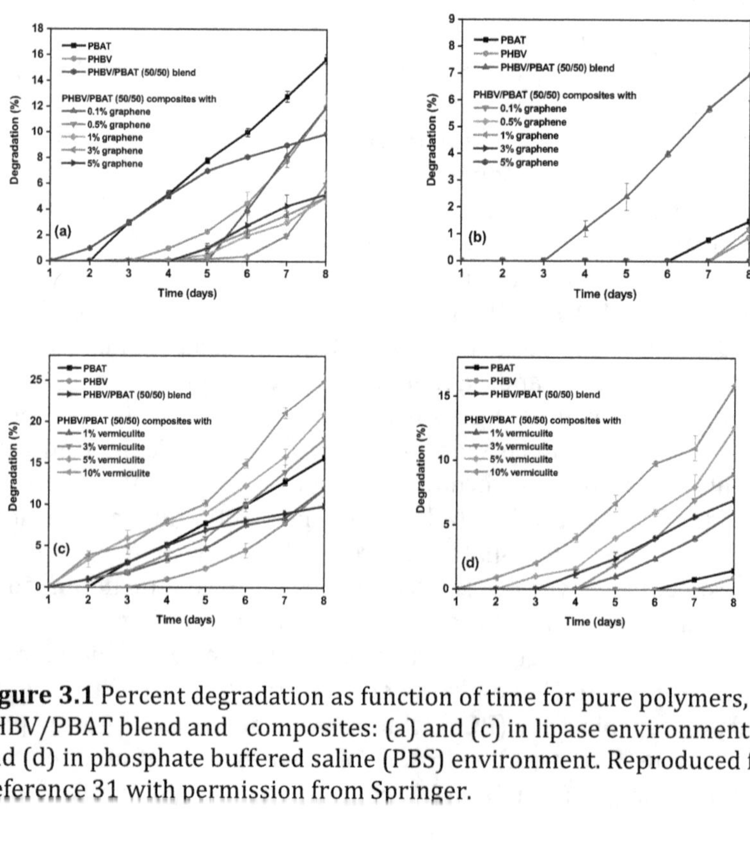

Figure 3.1 Percent degradation as function of time for pure polymers, PHBV/PBAT blend and composites: (a) and (c) in lipase environment; (b) and (d) in phosphate buffered saline (PBS) environment. Reproduced from Reference 31 with permission from Springer.

To further explore the potential of the PHBV/PBAT blend based composites, similar bio-nanocomposites have been generated using PHBV/PBAT blend with graphene and vermiculite as filler materials, as reported in earlier study, using melt mixing method [31]. The earlier study incorporated a 50:50 blend of PHBV and PBAT materials. The experimental findings had signaled the positive impact of PBAT on the composite properties. Thus, to further explore the effect of increasing PBAT concentration on the composite properties, the current study reports the use of 60:40 blend composition of the PBAT and PHBV polymers. Optimum processing parameters were employed to ensure efficient mixing of the composite constituents, but to avoid their thermal degradation when subjected to high temperature and shear during compounding. Thorough analysis of the thermal, rheological, and mechanical properties has been presented. In addition, the morphology of the composites from the perspective of filler dispersion and phase distribution has also been studied. The biodegradation of the composites has also been studied using soil embedding method.

3.2 Experimental

3.2.1 Materials and Generation of Thermally Reduced Graphene

Biopolyester PBAT with trade name Ecoflex® F Blend C1200 was supplied by BASF SE, Germany. As mentioned earlier, it is a biodegradable aliphatic-aromatic copolyester which biodegrades to the basic monomers and eventually to carbon dioxide, water and biomass when metabolized in the soil or compost under standard conditions. PHBV, a compostable resin obtained from corn, was procured from Sigma Aldrich and was used as received. Russian vermiculite with a chemical composition of $(Mg,Al,Fe)_3(Al,Si)_4O_{10}(OH)_2Mgx(H_2O)_n$ was obtained from Thermax, Greinsfurth, Austria. As processed in earlier study [31], the vermiculite flakes were wet ground to the desired particle diameter (partially spherical stacks with average particle size of 5 μm). The grinding operation was performed keeping in mind that the lateral dimensions of the platelets are not milled, however, the thickness of the stacks is reduced so as to generate platelets with high aspect ratio. Thermally reduced graphene was prepared through thermal exfoliation of precursor graphite oxide [40] using modified Hummer's method [41], as described earlier [42]. A short description for the preparation of graphite oxide and graphene is as follows: 5 g of graphite powder was added with 125 ml concentrated H_2SO_4. Subsequently, 2.5 g of $NaNO_3$ was added to this mixture. The mixture was kept in an ice-bath (5 °C) under stirring. After 30 min, 15 g of $KMnO_4$ was added to the mixture and the temperature was allowed to rise gradually to 35 °C. The mixture was stirred for 2 h under these conditions. This was followed by the addition of deionized water till the temperature increased to 100 °C. After further stirring for 15 min while maintaining the same temperature, the mixture was quenched and diluted by pouring it into 1.5 l deionized water. 30% H_2O_2 was slowly added to the dilute solution until the evolution of bubbles of hydrogen stopped. The solution was then filtered using Buchner funnel to remove the non-graphite oxide (GO) waste. The residues were dispersed in 2 l deionized water and were added with dilute HCl (6%) (2 l) to remove the SO_4^{2-} ions. The dispersion was filtered and the filtrate was analyzed for SO_4^{2-} and Cl^- ions using $BaSO_4$ and $AgNO_3$, respectively (generation of white precipitates). The cleaning and filtration was continued until no SO_4^{2-} and Cl^- ions could be observed in the filtrate. The washed GO was dried under vacuum at 60 °C for 24 h. Thermally reduced graphene was generated via thermal exfoliation of dried GO. The process was carried out by placing 1 g GO in a long quartz tube with 25 mm internal diameter and sealed at one end. The other end of the quartz tube was closed using a rubber stopper. The sample was flushed with nitrogen, followed by insertion of the tube in a tube furnace preheated to 1050 °C. The tube was held in the furnace for 30 s. The density of the obtained graphene was measured by tapped density tester to be 0.0161 g/ml. The obtained graphene material was fluffy in nature and required careful handling.

3.2.2 Blend and Nanocomposite Generation

The same procedure was followed for the generation of the PHBV/PBAT blend and the nanocomposites with graphene and vermiculite, as reported earlier [31]. Polymer blend was prepared using a weight ratio of 60:40 of PBAT:PHBV. To generate blend and nanocomposites, melt mixing was carried out using a mini twin screw extruder MiniLab HAAKE Rheomex CTW5, Germany. The screw length and screw diameter were 109.5 mm and 5/14 mm conical, respectively. The amount of vermiculite in the composites was varied as 1, 3, 5, and 10 wt%, whereas graphene content was varied as 0.1, 0.5, 1, 3, and 5 wt%. Approximately 15 g of each sample was prepared and dry mixed prior to extrusion, in order to ensure proper ratios of the components. An initial mass of 6 g was added to the compounder and blended at 175 °C and 60 rpm. After 5 min, 3 g was flushed from the machine so that another 3 g could be added. This process was repeated until 15 g material was blended. This method ensured that the compounder would not be overloaded and also provided small batches to ensure more intense mixing. The pure polymers were also processed similarly, with a lower processing temperature of 160 °C for PBAT. Disc and dumbbell-shaped test specimens were injection molded using a mini injection molding machine (HAAKE MiniJet, Germany) using same processing temperatures as used during compounding. The injection pressure was 700 bar for 6 s, whereas holding pressure was 400 bar for 3 s. The temperature of the mold was kept at 50 °C.

3.2.3 Material Characterization

For the microscopy analysis, the samples were mounted in special sample holders which at the same time fit in the microtome and were suitable for the examination of the block face by atomic force microscopy. Block faces of the samples were obtained using a Leica Ultracut E microtome (Leica, Austria) equipped with a diamond knife (Diatome, Switzerland) at -120 °C. AFM analysis of the block face of samples was obtained in tapping mode at ambient conditions using a Digital Instruments NanoScope III and silicon nitride cantilevers with natural frequencies in the 300 kHz range (force constant 20 N/m, tip radius 10 nm (NT-MDT, Russia)). The block faces of specimens after cryo ultramicrotomy were also investigated at ambient conditions using ZEISS Axioplan light microscope (ZEISS, Germany) equipped with ZEISS Axio Cam ICc 1 CCD camera in reflected polarized light. For the bright field transmission electron microscopy (TEM) analysis of the composite samples, Philips CM 20 (Philips/FEI, Eindhoven) electron microscope at 120 kV and 200 kV accelerating voltages was used. Thin sections of 70-90 nm thickness were microtomed from the sample block and were supported on 100 mesh grids sputter coated with a 3 nm thick carbon layer, followed by the bright field TEM investigation.

Differential scanning calorimetric (DSC) measurements on the polymer blend and bio-nanocomposites were performed using 204 F1 Phoenix Netzsch DSC under nitrogen atmosphere. The sample scans were obtained in the temperature range 30-200-30 °C using heating and cooling rates of 5 °C/min. Thermal degradation of the samples was analyzed using Netzsch STA 409 PC Luxx thermogravimetric analyzer (TGA). Nitrogen was used as a carrier gas and the scans were obtained from 50 to 700 °C temperature at a heating rate of 20 °C/min.

The mechanical testing of the blend and nanocomposites was performed on a universal testing machine (Testometric, UK). The dumbbell shaped samples with 53 mm length, 4 mm width and 2 mm thickness were used. A loading rate of 5 mm/min was employed, and the tests were carried out at room temperature. Win Test Analysis software was used for the calculation of tensile modulus and elongation at break of the materials. An average of six values was calculated. RSA3 dynamic mechanical analyzer from TA Instruments was used to study the change in the modulus with temperature (25-100 °C). Dynamic temperature ramp test was used at a frequency of 5 rad/s, ramp rate 3 °C/min, and 0.1% strain for thin polymer and composite samples.

XRD analyses were conducted on a Panalytical Powder Diffractometer (X'Pert PRO) using Cu-Kα radiation (λ 1.5406 A°, 45 kV, 40 mA). Diffraction patterns were collected from injection molded samples in the range of 5-40° 2θ scale, with a step size of 0.0170° s^{-1}. AR 2000 rheometer from TA Instruments was used to characterize rheological properties such as storage modulus (G'), loss modulus (G'), and complex viscosity (η^*). Disc shaped samples of 25 mm diameter and 2 mm thickness were characterized at 190 °C using a gap opening of 1 mm. Frequency sweep scans (dynamic testing) of all composites were recorded at 1% strain from ω 0.1 to 100 rad/s.

Biodegradability of the blend and nanocomposites was analyzed with soil burial tests, as reported in an earlier study [39]. The tests were carried out under natural environmental conditions and no additional effort was made to modify the conditions. The samples were buried in commercially available compost soil for gardening (Bestgreen Universal potting soil). Two pots of 52*14*10 cm dimension were used for the soil burial test and placed in an open place usually used for gardening. Soil was filled to a height of 10 cm in the pots. Samples with dimension of 17*9.5*1.5 mm were buried 5 cm deep, 4 cm apart in longitudinal direction and 5 cm apart in transverse direction. The soil moisture content was 60 to 70 % on a normal day. On a sunny day, the soil temperature was usually between 42 to 45 °C at mid-day and at night between 30 to 35 °C. The samples were dug out every 15 d, cleaned from soil particles and dried in a vacuum oven at 70 °C, followed by measurement of their weight.

3.3 Results and Discussion

Use of blend composites has evolved into an effective strategy for the simultaneous improvement of stiffness and toughness of polymers. For example, if a rubbery polymer enhances flexibility, the incorporation a filler leads to enhancement in modulus. However, optimizing the blend composition and filler concentration, which leads to optimal enhancement in properties, is not straightforward. In the biopolymer based system, it is even more challenging as such property enhancements may be achieved at the expense of biodegradability, which is the primarily material property of the biopolymers. As mentioned earlier, in the current study, the objective was to enhance the flexibility of PHBV by the addition of PBAT. As a result, the addition of higher modulus PHBV to PBAT also improves the mechanical properties of PBAT. To further consolidate the properties of the PHBV/PBAT blend, different amounts of two 2D fillers were added. Melt mixing method was used to generate nanocomposites, thus, indicating compatibility of the processing of these materials with industrial manufacturing processes.

Different microscopy analyses were employed to gain insights into the phase morphology of the blend and nanocomposites. Figure 3.2 shows the optical images of the PHBV/PBAT composites with 5% filler content. The polymers exhibited phase immiscibility as evident from the presence of distinct domains. The polymers formed co-continuous blend morphology. The domains though were large in size, but were uniformly distributed. As also observed in the earlier study, both composites exhibited distribution of filler solely in PBAT phase [31]. It is unlikely that a specific thermodynamic interaction existed between PBAT and the fillers due to similar nature of PHBV and PBAT chains, thus, the observed behavior of filler dispersion seemed to have resulted from the kinetic factors prevailing during melt mixing. As both the polymers were added to the extruder at the same time, thus, due to the melting of PHBV at higher temperature, PBAT phase would have received earlier opportunity to mix with the filler. In fact, an additional trial was carried out where the filler was loaded in the compounder only after complete melting of the polymers. The resulting composite revealed a different morphology with a fraction of filler also present in the PHBV phase. However, as observed earlier [31], the addition of PHBV reduced the rate of degradation of the blend and incorporation of vermiculite helped to enhance the biodegradation of the blend due to its polar nature, thus, the presence of the filler in the

PBAT phase may be advantageous in achieving enhanced rate of polymer biodegradation.

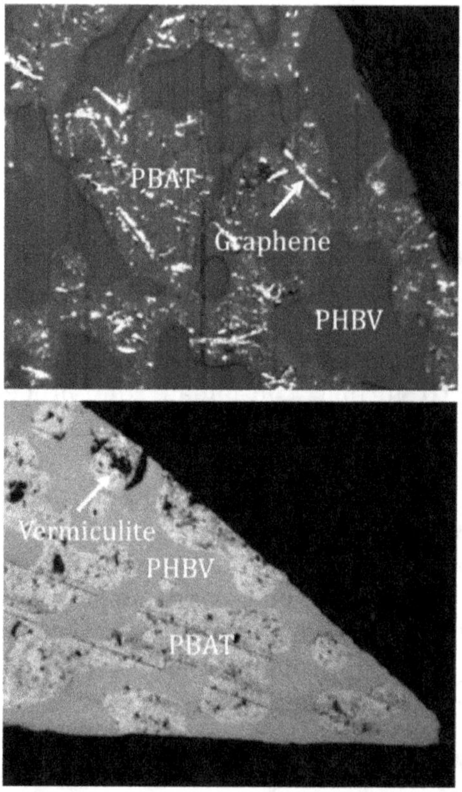

Figure 3.2 Optical images of (top) PHBV/PBAT composite with 5% graphene and (bottom) PHBV/PBAT composite with 5% vermiculite.

Figures 3.3 and 3.4 also demonstrate the AFM height and phase images of the 5% nanocomposite samples. The AFM analysis supported the findings from optical microscopy, however, additional features of the composites were also revealed. In the images, the bright phase corresponded to PHBV, whereas the dark phase contained PBAT. Interestingly, in both the composites, though the filler was present in the PBAT phase, however, there was an additional layer of the bright phase surrounding the filler particles inside the PBAT phase. The layer appeared to contain either solely PHBV or a mixture of PHBV/PBAT. Such a phenomenon can still be traced back to the melt compounding process. At the time when lower temperature melting polymer (PBAT) melted fully in the compounder, a small fraction of higher temperature melting PHBV would be present which also got mixed with the PBAT phase and the filler, thus, resulting in entrapment inside the PBAT phase along with the filler particles. Interestingly, the PHBV inside the PBAT phase was always observed to be present together with the filler, not alone. Further analysis is though needed to confirm if the layer surrounding the filler particles contained only PHBV or PHBV mixed with PBAT. The AFM analysis of the samples was also repeated after 60 days of composite generation so as to gain insights about any polymer relaxation or internal phase mixing as a function of time. No observable difference was observed in the morphology and filler distribution between fresh and aged composite samples, indicating phase stability, even though the phases are only kinetically mixed. In addition, the phase morphology also did not alter even after few processing cycles. However, it should also be noted that over the course of time, some natural changes in the microstructure of the composite can still be expected.

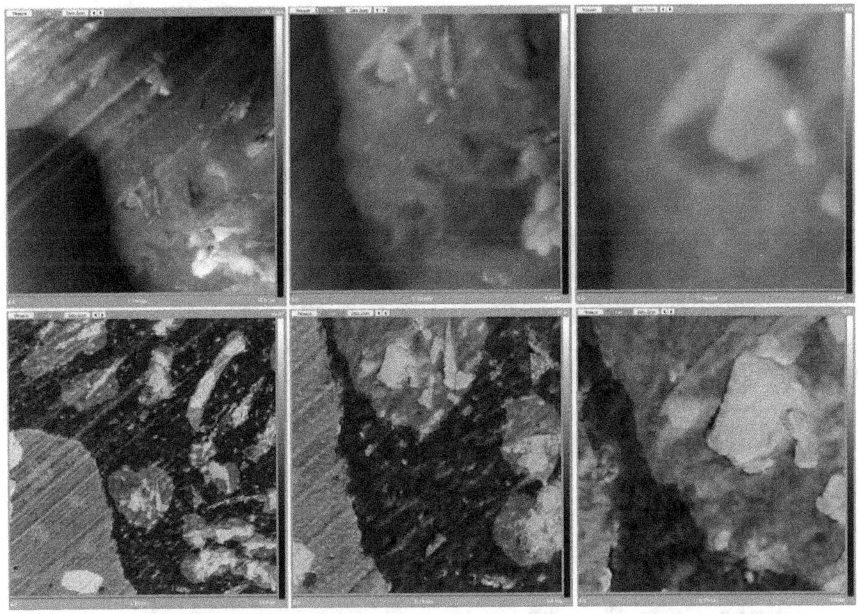

Figure 3.3 AFM height (top row) and phase (bottom row) images of the PHBV/PBAT/5% graphene composite.

TEM analysis of the composites was also carried out to gain further insights in the morphology, especially the layer surrounding the

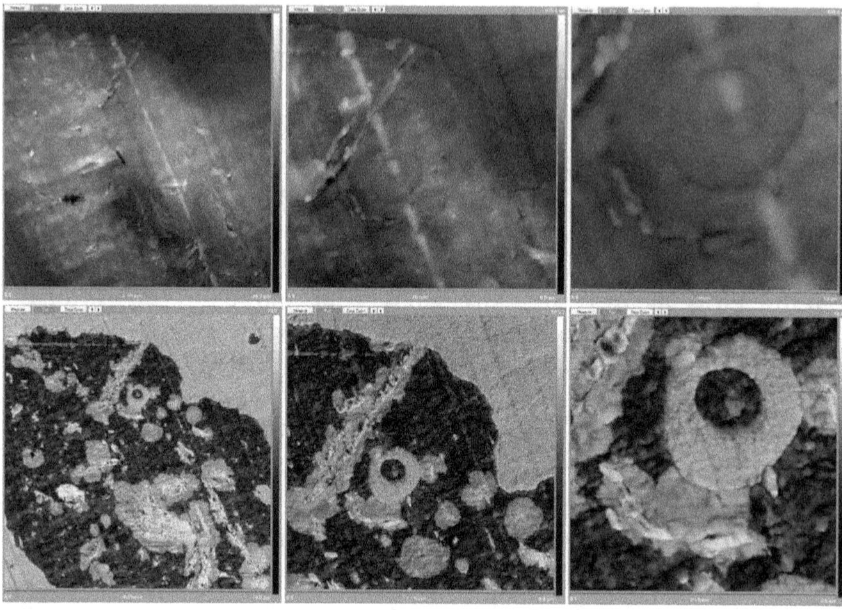

Figure 3.4 AFM height (top row) and phase (bottom row) images of the PHBV/PBAT/5% vermiculite composite.

filler particles in the PBAT phase. Figures 3.5 and 3.6 show the TEM images of the corresponding 5% filler composites in bright field imaging mode. As also seen in AFM and optical analysis, a clear separation between the polymer phases was observed with filler only present in the PBAT phase. It is interesting that the filler particles were also absent from the interface between PHBV and PBAT, thus, indicating that the fillers did not induce commonly observed particle stabilization of blends. Both the fillers were observed to be uniformly dispersed in the polymer phase. In contrast, some graphene aggregation was observed in the earlier study [31], which was not observed here. Probably the use of higher amount of flexible PBAT allowed better intermixing with the filler. Similar observation was also made for the other composites irrespective of the filler content. The X-ray analysis also confirmed the absence of d_{001} 2Θ peak at 5.9° corresponding to vermiculite in the composites. Similarly, 2Θ diffraction peak at 26.4° for graphene was also absent in the corresponding composites, indicating effective delamination. In addition, in Figure 3.5, the interfacial layer surrounding the filler particles observed in AFM was also visible as a lighter phase around the filler particles (for instance, Figure 3.5a). Figure 3.5c also shows the analysis of plasmon energy at different locations in the micrograph (inset) so as to confirm the nature of the interfacial layer. The chosen locations were PHBV phase, PBAT phase, graphene filler localized in the PBAT phase and the interfacial bright layer surrounding the graphene particles in the PBAT phase. As evident in Figure 3.5c, there was a difference in plasmon energy between the dark PBAT and bright PHBV phases. The plasmon

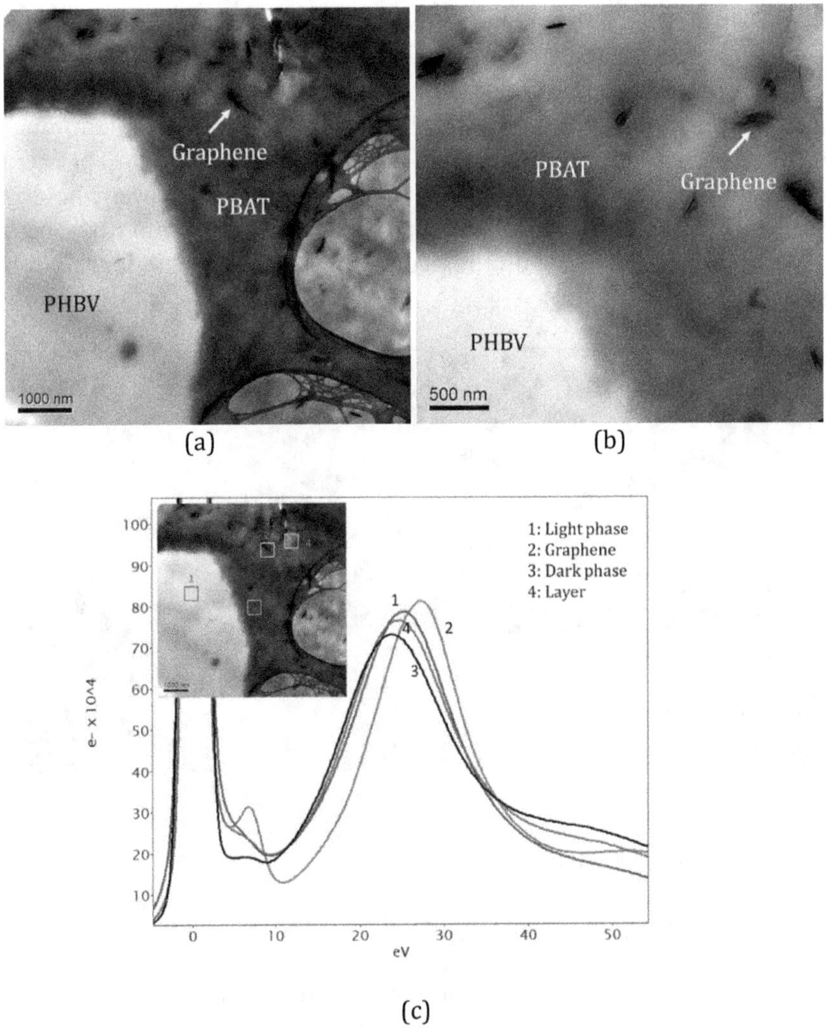

Figure 3.5 (a,b) TEM images of the PHBV/PBAT/5% graphene composite; (c) plasmon energy of the PHBV (light phase), PBAT (dark phase), graphene, and the interfacial layer at the graphene-polymer interface. The exact analysis locations are also marked in the inset of image c.

energy of the interfacial layer was observed just between the two pure polymers, indicating the layer to be a mix of PHBV and PBAT. Though the difference in the energy is small, however, it confirmed the nature of the layer to different from the individual polymers forming it. Similar observation about the plasmon energy change was also made for the vermiculite nanocomposite, which indicated that the observed behavior was not specifically influenced by the nature of the filler. Overall, it can be concluded from the AFM and TEM analysis that some degree of phase intermixing between the PHBV and PBAT polymers took place simultaneously with the fillers during the compounding process, which was subsequently followed by entrapment of this intermixed phase and filler within the PBAT phase. This also indicated that a control on the blend and nanocomposite morphology could be achieved with specific distribution of the components, which can have beneficial effect on the properties, especially biodegradability of the composites.

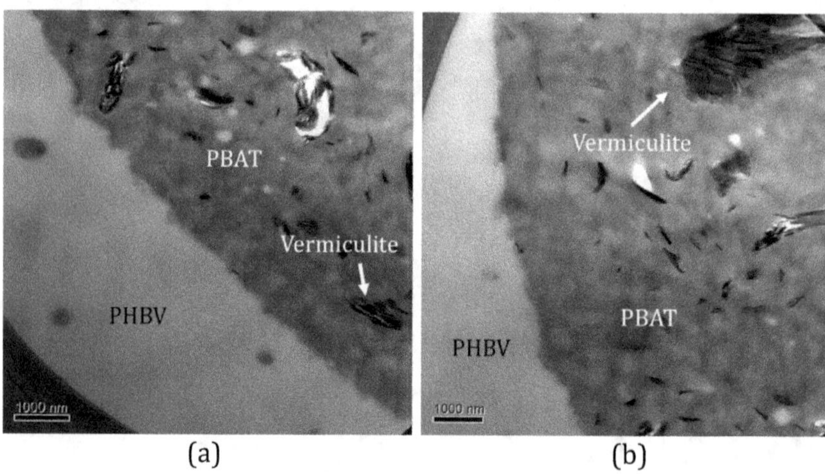

Figure 3.6 (a,b) TEM inages of the PHBV/PBAT/5% vermiculite composite.

The TGA analysis also supported the earlier findings from the morphological analysis about the phase immiscibility in the blends, as shown in Figure 3.7. PHBV polymer exhibited the peak degradation around 280 °C as the cleavage leading to structural degradation occurs at the ester bond by β elimination reaction. On the other hand, the PBAT polymer had much higher peak degradation temperature of around 420 °C. The 60:40 blend exhibited two degradation peaks attributed to the individual polymers, thus, indicating the absence of a

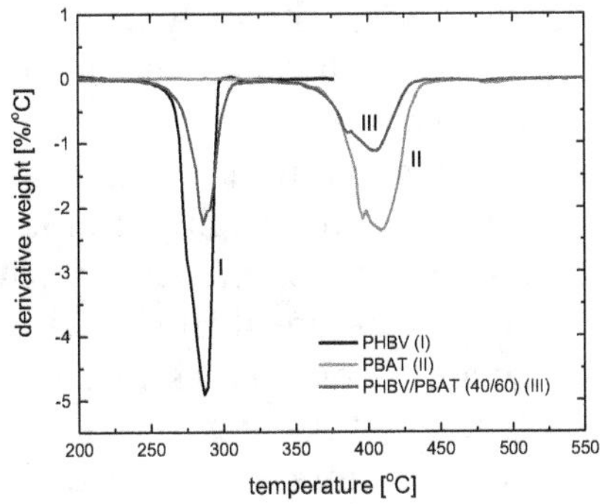

(a)

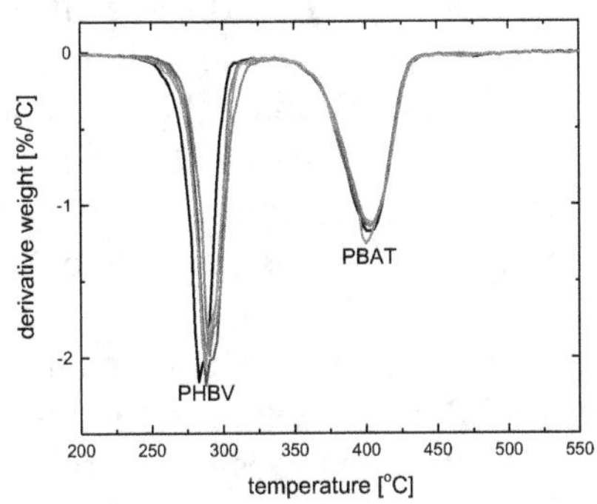

(b)

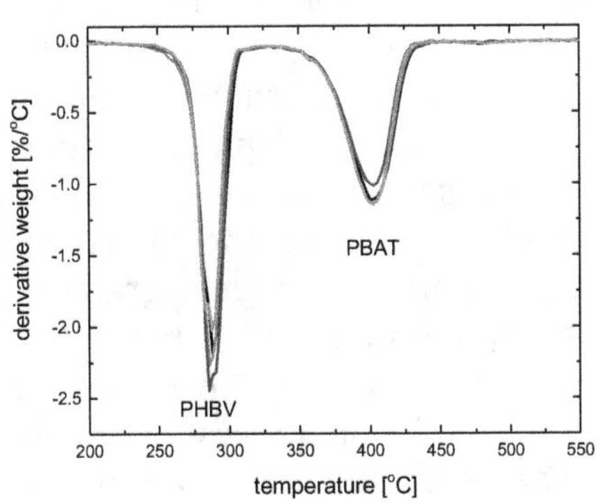

(c)

Figure 3.7 TGA thermograms of (a) PHBV, PBAT and 60/40 PHBV/PBAT blend; (b) PHBV/PBAT composites with varying amounts of graphene and (c) PHBV/PBAT composites with varying amounts of vermiculite.

molecular mixing between the polymers. Even though the degradation of PHBV was observed at significantly lower temperature than PBAT, the blend was still thermally stable under the processing conditions used to generate the nanocomposites. The dynamic TGA carried out for the blend at the compounding temperature as a function of time also revealed no significant weight loss for the overall duration used during compounding, thus, further confirming the thermal stability. Figure 3.7b and c also demonstrate the TGA thermograms of the graphene and vermiculite composites in comparison with the pure blend. As evident from the thermograms, no change in the degradation mechanism of the polymers took place in the presence of the filler. In addition, no appreciable change in the peak degradation temperatures of the PHBV and PBAT phases was observed on the addition of varying amounts of fillers. Thus, the specific localization of the fillers in the PBAT phase didn't impact its thermal degradation behavior. Overall, the fillers did not impact the thermal degradation behavior of the blend, though enhancement in the high temperature performance of the composites can still be expected due to efficient heat transfer from the polymer chains to the filler particles at the interface.

DSC thermograms of the polymers, blend and nanocomposites are demonstrated in Figure 3.8. PHBV exhibited more than one melting peak probably due to the various types of crystalline structures and different thicknesses of lamellar structure [31,43]. The DSC thermogram of the blend exhibited melt transitions of the individual polymers, thus, indicating the phase immiscibility, as observed earlier. No change in the peak melting point was observed in the composites as compared to pure blend, thus, confirming no effect of filler on the thermal transitions of the blend. In graphene nanocomposites, only a single melting peak was observed for PHBV, whereas two main melting endotherms were observed in vermiculite composites. The peak of PBAT was also less obvious in the thermograms of the composites which indicated that the filler in the PBAT phase may have hindered the crystallization of the PBAT phase. However, as the PBAT is relatively amorphous, the reduction in its crystallinity did not impact the overall melt enthalpy and crystallinity of the polymer phase in the composites. The glass transition temperature of the PBAT phase was also earlier reported to decrease in the blend, indicating enhancement in flexibility [31].

The mechanical performance of the nanocomposites was elucidated through DMA and tensile testing. Figure 3.9a demonstrates the

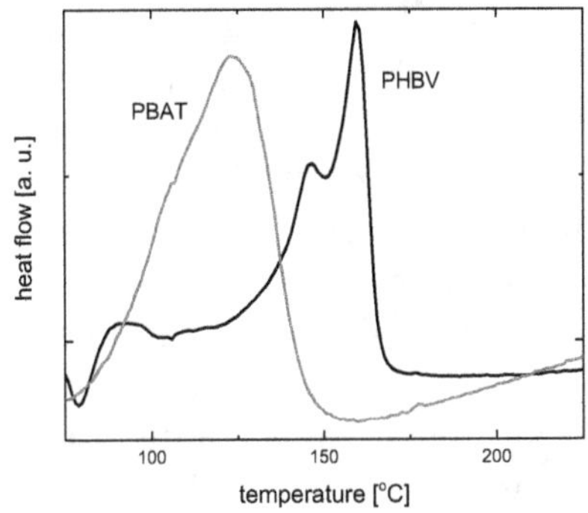

(a)

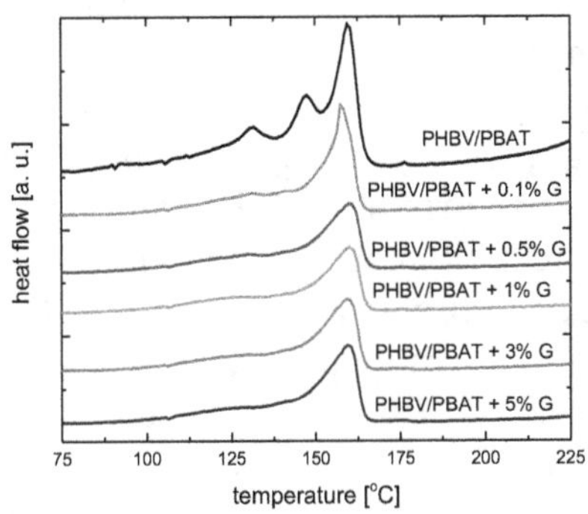

(b)

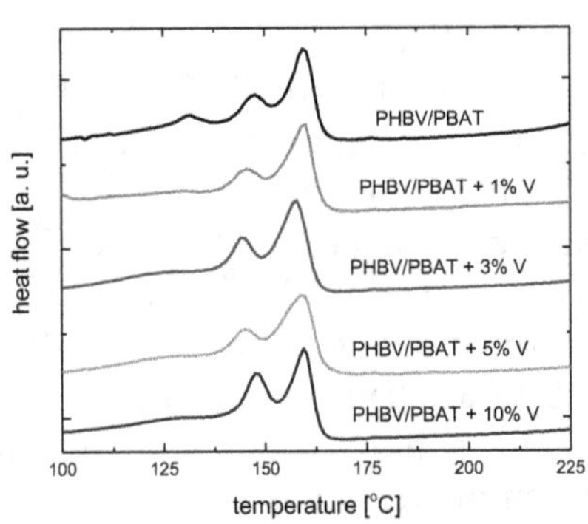

(c)

Figure 3.8 DSC thermograms of (a) PHBV and PBAT, (b) PHBV/PBAT blend and composites with varying amounts of graphene and (c) PHBV/PBAT blend and composites with varying amounts of vermiculite.

storage and loss modulus of PHBV/PBAT blend as a function of temperature. The blend exhibited much higher storage modulus as compared to loss modulus over the measured temperature range, thus, confirming the elastic dominance in the polymer. Figure 3.9b and c represent the storage modulus of PHBV/PBAT composites with graphene and vermiculite respectively. The addition of filler significantly enhanced the storage modulus of the blend, with the increase proportional to the filler content. In the composites too, the elastic dominance was maintained as G' was always higher than G" irrespective of the filler content. The storage modulus of the 0.1% graphene composite was also observed to be lower than the polymer blend, probably due to the plasticization effect at very low filler fraction. In context, the storage modulus of the blend at 50 °C was observed to be 2.79E8 Pa, which was enhanced to 4.15E8 and 4.50E8 Pa for 5% graphene and vermiculite composites, respectively. Further, it was improved to 5.02E8 for the 10% vermiculite composites. Thus, the addition of the filler to the blend significantly enhanced its high temperature performance. The superior performance of the composites was maintained even at 90 °C. The shear rheology of the nanocomposites also exhibited enhanced storage modulus as compared to the blend. The complex viscosity of the nanocomposites was also observed to increase as a function of filler fraction, however, the composites were still processable using the same conditions as the pure blend.

Figure 3.10 demonstrates the tensile properties of PHBV, PBAT, blend and nanocomposites using ASTM D638 tensile test on injection-molded specimens. PHBV had the high tensile modulus of 509 MPa, as compared to 77 MPa for pure PBAT. On the other hand, the elongation at break for PHBV was only 4% as compared to 123% for PBAT (Figure 3.10a and b). The 60:40 blend of PBAT and PHBV exhibited combination of tensile properties of individual polymers, with tensile modulus and elongation at break of 394 MPa and 27% respectively. Comparison of the blend properties with the tensile modulus and elongation at break values of 415 MPa and 22% for the earlier reported 50:50 PBAT/PHBV blend [31] indicated decreased modulus and increased flexibility on enhancing the PBAT content. Figure 3.10c and d demonstrate the tensile properties of graphene and vermiculite nanocomposites as a function of filler fraction. In both the cases, the tensile modulus enhanced with increasing filler fraction. In the case of graphene nanocomposites, a sharp increase in the modulus was observed till 1% filler fraction. On increasing the filler content further, a decreased rate of enhancement in the tensile

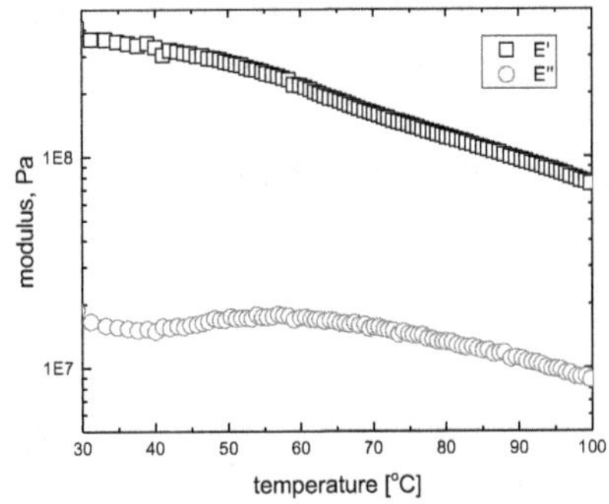

(a)

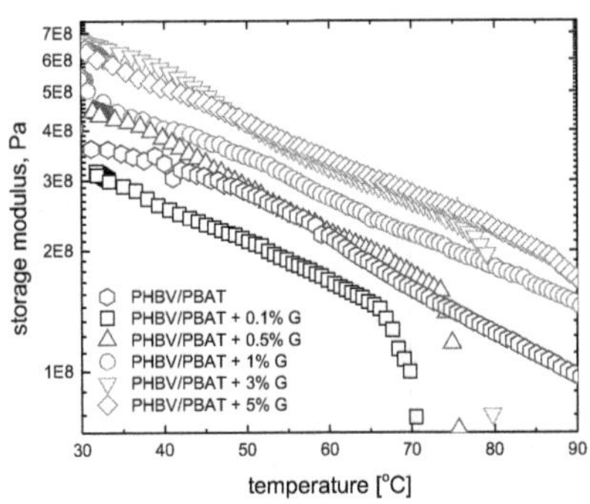

(b)

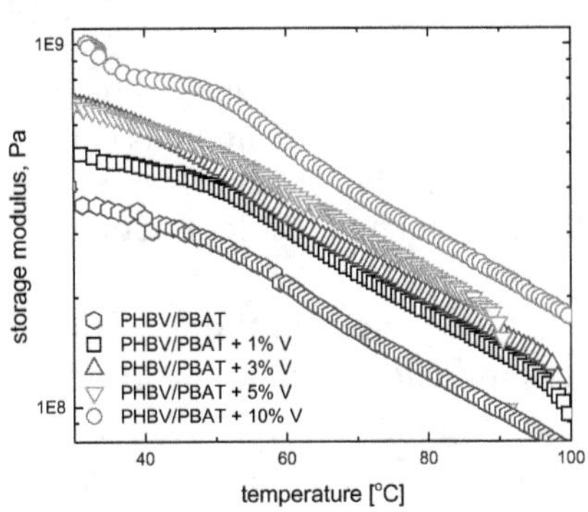

(c)

Figure 3.9 (a) Storage and loss modulus of PHBV/PBAT blend as a function of temperature, (b) storage modulus of PHBV/PBAT composites with graphene and (c) storage modulus of PHBV/PBAT composites with vermiculite.

modulus was observed. It has been reported earlier that the graphene sheets at higher loadings often restack due to van der Waal's interactions, resulting in slippage of the restacked sheets, thereby, leading to less effective stress transfer from the polymer to the graphene sheets [44,45]. The vermiculite filled nanocomposites, on the other hand,

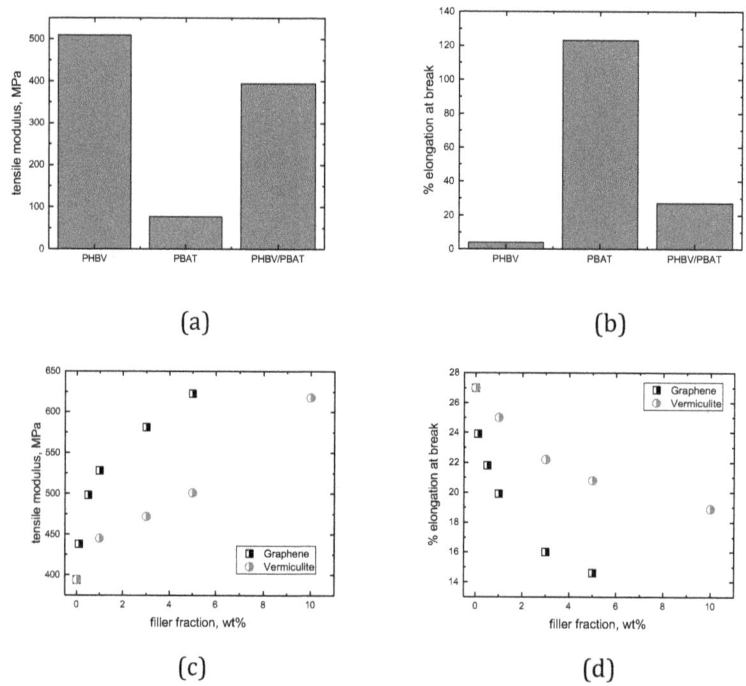

Figure 3.10 (a) Tensile modulus and (b) elongation at break of PHBV, PBAT and PHBV/PBAT blend; (c) tensile modulus of the graphene and vermiculite composites as a function of filler fraction and (d) elongation at break of the composites as a function of filler fraction.

exhibited more gradual increment in the modulus. Further comparison between the fillers indicated that incorporation of graphene was more effective in enhancing the tensile modulus. For instance, the 5% composite with graphene exhibited a tensile modulus of 623 MPa, which was similar to 618 MPa observed for the 10% vermiculite containing composite. Thus, the reinforcement effect of higher aspect ratio graphene filler was much superior in enhancing the tensile properties. Also, as PBAT is relatively amorphous in nature as compared to PHBV, thus, the earlier observed localization of filler phase in the amorphous phase would have also helped to effectively enhance the overall mechanical properties in the composites. Also, the composites of 60:40 PBAT/PHBV blend exhibited reduction in tensile modulus as compared to the 50:50 PBAT/PHBV blend, though the difference between the two blend systems was not significant, due to the dominance of the filler in reinforcing the tensile properties. The elongation of break of all the composites was observed to be much higher than PHBV, thus, signifying simultaneous enhancement of modulus and flexibility in the composites. On comparing the two filler systems, vermiculite nanocomposites exhibited higher degree of flexibility as compared to the graphene based composites. For instance, the elongation of break of 5% graphene and vermiculite nanocomposites was measured as 14.6% and 20.8% respectively. The composites also exhibited higher flexibility than the corresponding composites generated using 50:50 PBAT:PHBV blend [31] due to higher extent of PBAT.

Figure 3.11 demonstrates the biodegradation of the nanocomposites as a function of soil embedding duration. The degradation of the blend was around 2.2% after 120 days. An earlier study reported similar soil embedding of PBAT and approx. 2.1% weight loss was observed after 120 days [30]. However, the earlier study had milder temperature conditions, whereas the current study was carried out in very hot peak summer conditions. Thus, overall comparison of the findings indicated that the addition of PHBV slowed the rate of degradation of PBAT, which was also reported earlier in lipase and phosphate buffered saline media [31]. As PHBV is more crystalline in nature, thus, the addition of PHBV to relatively amorphous PBAT is expected to slow down the degradation as the amorphous part biodegrades faster than the crystalline domains. As evident in Figure 3.11, the incorporation of vermiculite enhanced the rate of degradation as a function of its content. For instance, the composite with 10% vermiculite content had enhanced degradation of > 5% after 120 days. On the other hand, the incorporation of graphene led to a reduction in the rate of degradation. For comparison, the composite with 5% graphene content exhibited a biodegradation of 1% after 120 days in comparison with 2% for the polymer blend and > 4% for the 5% vermiculite composite. Recent studies reported that the addition of vermiculite to the polymer matrices leads to an enhancement in the hydrophilic nature of the composites [46-47], thus, generating conducive pathways for the microbes. This behavior would also have improved due to the localization of vermiculite in the largely amorphous PBAT phase. Similarly, it is well known that graphene is highly hydrophobic in nature [48-49], which would have resulted in the observed reduction in biodegradation. In addition, the rate of weight loss was higher during the initial days, however, it seemed to become slower with time for most of the materials. Similar behaviors have also been

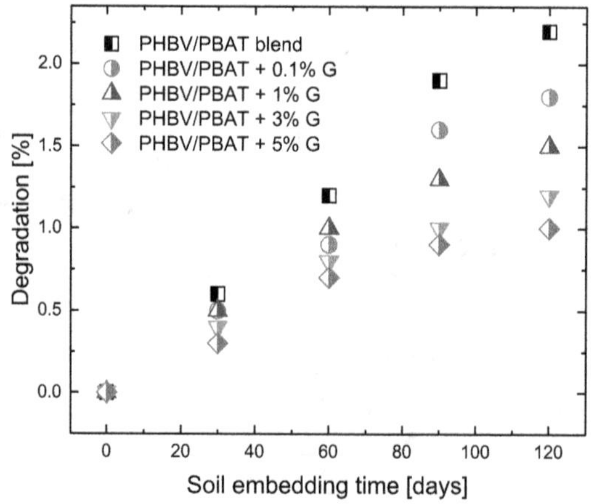

(a)

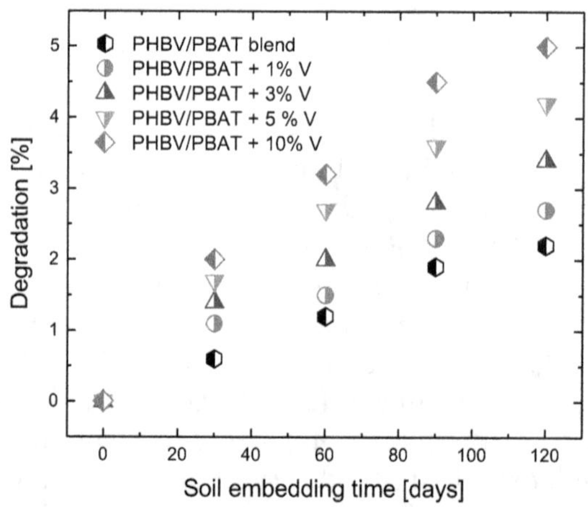

(b)

Figure 3.11 Soil degradation analysis of the (a) graphene and (b) vermiculite composites as a function of embedding time.

observed earlier for PBS/chitosan bio-nanocomposites [39]. The initial faster rate of degradation is generally attributed to the loss of the amorphous phase, which is followed by the degradation of the crystalline phase. This phenomenon can also be studied in the context of increase in crystallinity as a function of soil embedding time, as shown in Figure 3.12. The crystallinity of the samples was observed

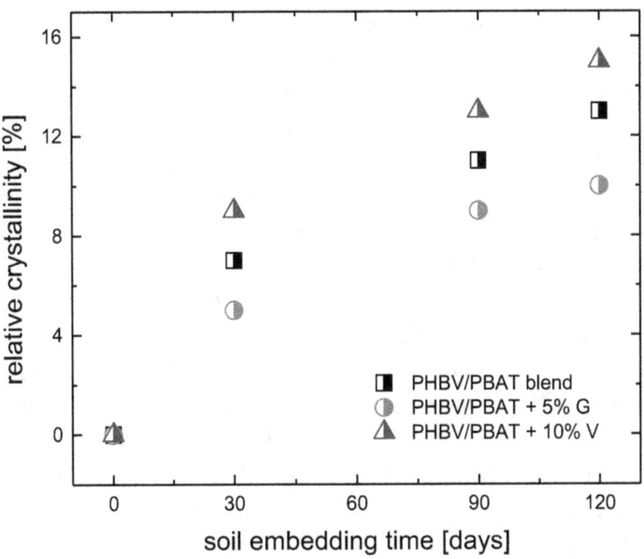

Figure 3.12 Relative crystallinity of the nanocomposites before and after soil burial.

to increase faster in the initial days due to the degradation of the amorphous part (probably in PBAT). However, the rate of increase of crystallinity slowed down (same as slowing down of loss of weight in Figure 3.11) probably due to the slow degradation of the crystalline phase. Obviously, not all the amorphous phase would have degraded at this stage being embedded deep inside the material. Thus, probably the weight loss at this stage resulted from the loss of material from the surface. Earlier studies have also indicated the generation of surface cracks and micro-channels when the degradation was initiated [39].

3.4 Conclusions

In the current study, 60:40 blend of PBAT:PHBV was reinforced with varying amounts of vermiculite and graphene, using conventional melt mixing method, with a goal to achieve increment in both stiffness and flexibility. The blend exhibited immiscible co-continuous phase morphology and the filler was localized in the PBAT phase. The AFM analysis also confirmed the presence of an interfacial layer around the filler particles in the PBAT phase. The layer was confirmed by plasmon energy analysis in TEM to be a mixture of PHBV and PBAT. Both fillers were observed to be uniformly distributed in the PBAT phase, irrespective of the filler content. The thermal degradation behavior of the blend was not affected by the incorporation of the fillers. Similarly, no change was observed in the peak melting point and melt enthalpy in the composites as compared to the blend. The DMA confirmed the improvement in the high temperature performance of the composites as significant increase in the storage modulus of the composites was observed. In addition, the composites also exhibited elastic dominance throughout the measurement range. The shear rheology also confirmed the increase in the storage modulus and viscosity of the composites as a function of increasing filler content, however, the composites could still be processed using the same conditions as the blend. The tensile modulus of the blend enhanced with both graphene and vermiculite, though graphene was more effective due to higher aspect ratio. The composite with 5% graphene had the tensile modulus of 623 MPa in comparison with 603 MPa for the 10% vermiculite composite and 394 MPa for the blend. The modulus was also higher than PHBV which exhibited a value of 509 MPa. The increment of modulus in the graphene composites was also observed to slow down after 1% filler fraction due to restacking of the high aspect ratio graphene sheets. The elongation of break of the composites though decreased with increasing filler content, however, it was still very high as compared to PHBV. For instance, the composite with 10% vermiculite had an elongation at break of 18.9% in comparison with 14.6% for the 5% graphene composite and 4% for PHBV. The vermiculite composites were also observed to have a faster rate of degradation as compared to graphene based composites. In comparison with 2% weight loss for the blend after 120 days, the 5% vermiculite composite exhibited a weight loss of > 4%, which was around 1% for the corresponding graphene composite. Polar nature of vermiculite as well as its localization in the largely amorphous phase in PBAT would have resulted in the enhanced degree of biodegradation. On the other hand, the non-polar nature of graphene hindered the degradation by affecting the transport of degradation agents in the materials. The rate of weight loss during degradation was also observed to decrease with time probably due to initial faster rate of degradation of the amorphous phase, followed by slower rate of degradation of the crystalline phase. This was also corroborated by corresponding increase in the crystallinity in the materials, which similar to the weight loss increased faster in the initial stages and slowed down subsequently. Overall, significant improvements in the properties could be achieved by combining the blending and composite generation strategies. In comparison with the PHBV and PBAT materials, the obtained PHBV/PBAT blend bio-nanocomposites represent functional materials with much higher potential of commercial applications. The controllable morphology of the blend and composites also opens the opportunities for the development of advanced materials for diverse uses by tuning structure-property correlations.

Acknowledgements

The authors are indebted to Aditya Ranjan and Fakhruddin Patwary for experimental support.

References

1. Averous, L., and Boquillon, N. (2004) Biocomposites based on plasticized starch: Thermal and mechanical behaviours. *Carbohydrate Polymers*, **56**, 111-122.
2. Vert, M., Schwarch, G., and Coudane, J. (1995) Present and future of PLA polymers. *Journal of Macromolecular Science A*, **32**, 787-796.
3. Lunt, J. (1998) Large-scale production, properties, and commercial applications of polylactic acid polymers. *Polymer Degradation and Stability*, **59**, 145-152.
4. Doi, Y. (1990) *Microbial Polyesters*, John Wiley and Sons, USA.
5. Chiellini, E., and Solaro, R. (1996) Biodegradable polymeric materials. *Advanced Materials*, **8**, 305-313.
6. Tokiwa, Y., and Suzuki, T. (1977) Hydrolysis of polyesters by lipases. *Nature*, **270**, 76-78.
7. Martin, O., and Averous, L. (2001) Poly(lactic acid): plasticization and properties of biodegradable multiphase systems. *Polymer*, **42**, 6209-6219.
8. Wang, L., Ma, W., Gross, R. A., and McCarthy, S. P. (1998) Reactive compatibilization of biodegradable blends of poly(lactic acid) and poly(e-caprolactone). *Polymer Degradation and Stability*, **59**, 161-168.

9. H. Tsuji, and Ikada, Y. (1996) Blends of aliphatic polyesters. I. Physical properties and morphologies of solution-cast blends of poly(DL-lactide) and poly(e-caprolactone). *Journal of Applied Polymer Science*, **60**, 2367-2375.
10. Koyama, Y., and Doi, Y. (1997) Miscibility of binary blends of poly[(R)-3-hydroxybutyric acid] and poly[(S)-lactic acid]. *Polymer*, **38**, 1589-1593.
11. Ohkoshi, I., Abe, H., and Doi, Y. (2000) Miscibility and solid-state structures for blends of poly[(S)-lactide] with atactic poly[(R,S)-3-hydroxybutyrate]. *Polymer*, **41**, 5985-5992.
12. Jiang, L., Chen, F., Qian, J., Huang, J., Wolcott, M., Liu, L., and Zhang, J. (2010) Reinforcing and toughening effects of bamboo pulp fiber on Poly(3-hydroxybutyrate-co-3-hydroxyvalerate) fiber composites. *Industrial & Engineering Chemistry Research*, **49**, 572-577.
13. Choi, J. S., and Park, W. H. (2004) Effect of biodegradable plasticizers on thermal and mechanical properties of poly(3-hydroxybutyrate). *Polymer Testing*, **23**, 455-460.
14. Kotnis, M. A., O'Brien, G. S., and Willett, J. (1995) Processing and mechanical properties of biodegradable Poly(hydroxybutyrate-co-valerate)-starch compositions. *Journal of Environmental Polymer Degradation*, **3**(2), 97-105.
15. Avella, M., La Rota, G., Martuscelli, E., Raimo, M., Sadocco, P., Elegir, G., and Riva, R. (2000) Poly(3-hydroxybutyrate-co-3-hydroxyvalerate) and wheat straw fibre composites: thermal, mechanical properties and biodegradation behavior. *Journal of Materials Science*, **35**(4), 829-836.
16. Carli, L. N., Crespo, J. S., and Mauler, R. S. (2011) PHBV nanocomposites based on organomodified montmorillonite and halloysite: the effect of clay type on the morphology and thermal and mechanical properties. *Composites Part A: Applied Science and Manufacturing*, **42**, 1601-1608.
17. Zhao, H., Cui, Z., Sun, X., Turng, L.-S., and Peng, X. (2013) Morphology and properties of injection molded solid and microcellular polylactic acid/polyhydroxybutyrate-valerate (PLA/PHBV) blends. *Industrial & Engineering Chemistry Research*, **52**(7), 2569-2581.
18. Nagarajan, V., Mohanty, A. K., and Misra, M. (2013) Sustainable green composites: Value addition to agricultural residues and perennial grasses. *ACS Sustainable Chemistry & Engineering*, **1**, 325-333.
19. Nagarajan, V., Misra, M., and Mohanty, A. K. (2013) New engineered biocomposites from poly(3-hydroxybutyrate-co-3-hydroxyvalerate) (PHBV)/poly(butylene adipate-co-terephthalate)

(PBAT) blends and switchgrass: Fabrication and performance evaluation. *Industrial Crops and Products*, **42**, 461-468.

20. Iannace, S., Ambrosio, L., Huang, S., and Nicolais, L. (1994) Poly(3-hydroxybutyrate)-co-(3-hydroxyvalerate)/Poly-L-lactide blends: Thermal and mechanical properties. *Journal of Applied Polymer Science*, **54**, 1525-1535.
21. Koenig, M., and Huang, S. (1995) Biodegradable blends and composites of polycaprolactone and starch derivatives. *Polymer*, **36**, 1877-1882.
22. Miao, L., Qiu, Z., Yang, W., and Ikehara, T. (2008) Fully biodegradable poly(3-hydroxybutyrate- cohydroxyvalerate)/ poly(ethylene succinate) blends: Phase behavior, crystallization and mechanical properties. *Reactive and Functional Polymers*, **68**, 446-457.
23. Wang, S., Ma, P., Wang, R., Wang, S., Zhang, Y., and Zhang, Y. (2008) Mechanical, thermal and degradation properties of poly (d, l-lactide)/poly (hydroxybutyrate-co-hydroxyvalerate)/poly (ethylene glycol) blend. *Polymer Degradation and Stability*, **93**, 1364-1369.
24. Witt, U., Einig, T., Yamamoto, M., Kleeberg, I., Deckwer, W.-D., and Muller, R. J. (2001) Biodegradation of aliphatic-aromatic copolyesters: evaluation of the final biodegradability and ecotoxicological impact of degradation intermediates. *Chemosphere*, **44**, 289-299.
25. Someya, Y., Sugahara, Y., and Shibata, M. (2005) Nanocomposites based on poly(butylene adipate-co-terephthalate) and montmorillonite. *Journal of Applied Polymer Science*, **95**, 386-392.
26. Chivrac, F., Kadlecova, Z., Pollet, E., and Averous, L. (2006) Aromatic copolyester-based nano-biocomposites: Elaboration, structural characterization and properties. *Journal of Polymers and the Environment*, **14**, 393-401.
27. Chivrac, F., Pollet, E., and Averous, L. (2007) Non-isothermal crystallization behavior of poly(butylene adipate-co-terephthalate)/clay nano-biocomposites. *Journal of Polymer Science, Part B: Polymer Physics*, **45**, 1503-1510.
28. Mittal, V., and Al Zaabi, Khalid (2013) Biodegradable polyester nanocomposites: Phase miscibility and properties. *Journal of Applied Polymer Science*, **130**, 516-525.
29. Mittal, V., Chaudhry, A. U., and Luckachan, G. E. (2014) Biopolymer-thermally reduced graphene nanocomposites: Structural characterization and properties. *Materials Chemistry and Physics*, **147**, 319-332.

30. Mittal, V., Chaudhry, A. U., and Matsko, N. B. (2014) "True" biocomposites with biopolyesters and date seed powder: Mechanical, thermal, and degradation properties. *Journal of Applied Polymer Science*, **131**(19), doi: 10.1002/app.40816.
31. Pawar, S. P., Misra, A., Bose, S., Chatterjee, K., and Mittal, V. (2015) Enzymatically degradable and flexible bio-nanocomposites derived from PHBV and PBAT blend: Assessing thermal, morphological, mechanical, and biodegradation properties. *Colloid and Polymer Science*, **293**, 2921-2930.
32. Wu, C. S. (2005) A comparison of the structure, thermal properties, and biodegradability of polycaprolactone/chitosan and acrylic acid grafted polycaprolactone/chitosan. *Polymer*, **46**, 147-155.
33. Shogren, R. L., Doane, W. M., Garlotta, D., Lawton, J. W., and Willett, J. L. (2003) *Polymer Degradation and Stability*, **79**, 405-411.
34. Rizzarelli, P., Puglisi, C., and Montaudo, G. (2004) Soil burial and enzymatic degradation in solution of aliphatic co-polyesters. *Polymer Degradation and Stability*, **85**(2), 855-863.
35. Kim, H. S., Kim, H. J., Lee, J. W., and Choi, I. G. (2006) Biodegradability of bio-flour filled biodegradable poly(butylene succinate) bio-composites in natural and compost soil. *Polymer Degradation and Stability*, **91**(5), 1117-1127.
36. Jakubowicz, I. (2003) Evaluation of degradability of biodegradable polyethylene (PE). *Polymer Degradation and Stability*, **80**(1), 39-43.
37. Pandey, J. K., and Singh, R. P. (2001) UV-irradiated biodegradability of ethylene--propylene copolymers, LDPE, and I-PP in composting and culture environments. *Biomacromolecules*, **2**(3), 880-885.
38. Pagga, U. (1997) Testing biodegradability with standardized methods. *Chemosphere*, **35**(12), 2953-2972.
39. Patwary, F., Matsko, N., and Mittal, V. (2016) Biodegradation properties of melt processed PBS/chitosan bio-nanocomposites with silica, silicate and thermally reduced graphene. *Polymer Composites*, doi: 10.1002/pc.23947.
40. McAllister, M. J., Li, J. L., Adamson, D. H., Schniepp, H. C., Abdala, A. A., Liu, J., Herrera-Alonso, M., Milius, D. L., Car, R., Prud'homme, R. K., and Aksay, I. A. (2007) Single sheet functionalized graphene by oxidation and thermal expansion of graphite. *Chemistry of Materials*, **19**(18), 4396-4404.
41. Hummers, W. S., and Offeman, R. E. (1958) Preparation of graphitic oxide. *Journal of the American Chemical Society*, **80**(6), 1339-1339.

42. Chaudhry, A. U., and Mittal, V. (2013) High-density polyethylene nanocomposites using masterbatches of chlorinated polyethylene/graphene oxide. *Polymer Engineering and Science*, **53**(1), 78-88.
43. Javadi, A., Srithep, Y., Lee, J., Pilla, S., Clemons, C., Gong, S., and Turng, L.-S. (2010) Processing and characterization of solid and microcellular PHBV/PBAT blend and its RWF/nanoclay composites. *Composites Part A: Applied Science and Manufacturing*, **41**, 982-990.
44. Li, D., Mueller, M. B., Gilje, S., Kaner, R. B., and Wallace, G. G. (2008) Processable aqueous dispersions of graphene nanosheets. *Nature Nanotechnology*, **3**, 101-105.
45. Yang, S.-Y., Chang, K.-H., Tien, H.-W., Lee, Y.-F., Li, S.-M., Wang, Y.-S., Wang, J.-Y., Ma, C.-C. M., and Hu, C.-C. (2011) Design and tailoring of a hierarchical graphene-carbon nanotube architecture for supercapacitors. *Journal of Materials Chemistry*, **21**, 2374-2380.
46. Zheng, Y., Li, P., Zhang, J., and Wang, A. (2007) Study on superabsorbent composite XVI. Synthesis, characterization and swelling behaviors of poly(sodium acrylate)/vermiculite super absorbent composites. *European Polymer Journal*, **43**, 1691-1698.
47. Tjong, S., Meng, Y., and Hay, A. (2002) Novel preparation and properties of polypropylene-vermiculite nanocomposites. *Chemistry of Materials*, **14**, 44-51.
48. Leenaerts, O., Partoens, B., and Peeters, F. (2009) Water on graphene: Hydrophobicity and dipole moment using density functional theory. *Physical Review B*, **79**, 235440.
49. Nguyen, D. D., Tai, N.-H., Lee, S.-B., and Kuo, W.-S. (2012) Superhydrophobic and superoleophilic properties of graphene-based sponges fabricated using a facile dip coating method. *Energy & Environmental Science*, **5**, 7908-7912.

4

Ternary Blends of PCL, PBAT and PBS: Morphology and Properties

4.1 Introduction

The physical polymer reinforcement means such as blending or filler incorporation have gained favorable position for acquiring new polymeric materials with altered properties [1-5]. The properties of the resultant materials depend on features like interaction between the phases, compatibilization, processing conditions, composition, etc. The generation of polymer blends specially offers advantages such as property profile combinations, aid in processing, low capital costs as compared to the development of new monomers and polymers with required properties. Numerous different blends have been reported in the literature by combining a large number of polymers. For instance, in a recent study, flexible blends of high-density polyethylene (HDPE) with chlorinated polyethylene (CPE) of different chlorination levels were generated using melt mixing [6]. The blends with CPE with lower level of chlorination exhibited superior mechanical properties than the blends with the corresponding CPE with higher degree of chlorination.

In the case of biopolymers, blending is also a common strategy to combine the properties of individual polymers so as to make them competitive to the traditional non-degradable polymers. Correlo *et al.* [7] reported the properties of melt processed blends of chitosan with biodegradable polyesters like poly-ε-caprolactone (PCL), polybutylene succinate (PBS), polylactic acid (PLA), poly(butylene terephthalate adipate) and poly(butylene succinate adipate). In another study, blending of PBS with chitosan helped to enhance the properties and processability of the individual polymers [8]. In addition, the

Vikas Mittal[a,], Yuddish Manna[a,**] and Nadejda Matsko[b]*
[a]The Petroleum Institute (part of Khalifa University of Science and Technology), Abu Dhabi, UAE; [b]Graz Centre of Electron Microscopy, Graz, Austria
**Current address: Bletchington, Wellington County, Australia; **Current address: University of Waterloo, Canada*

blends were incorporated with fillers to further enhance the properties [8]. Similarly, in another recent study, in order to enhance flexibility of poly(3-hydroxybutyrate-co-3-hydroxyvalerate) (PHBV), blending was carried out with poly(butylene adipate-co-terephthalate) (PBAT) to generate PHBV/PBAT blend (50/50 wt%/wt%) using conventionally accepted melt blending process [9]. Zhao *et al.* [10] also studied blends of PHBV and PLA over wide range of compositions. No significant increment in the modulus was observed by blending, however, a small increment in the strain was observed. Nagarajan *et al.* [11] similarly reported some improvement in structural properties of PHBV/PBAT blend composites with lignocellulosic fibers. To overcome the limitations of thermoplastic starch (TPS), biodegradable blends of TPS with polyesters PCL and PBAT have also been industrially generated [12-15]. Similarly, to overcome the limitations of low flexibility and low resistance to heat and humidity of PLA, blends of PLA with PBAT have been reported, which combined the toughness and flexibility characteristics of PBAT with the mechanical properties of PLA [16-18]. The resulting blends were observed to have ductile failure and reduced brittleness depending on the amount of PBAT. Similarly, blends of PLA with PCL have also been reported to complement the weaknesses of both polymers [19,20]. Martin and Averous [21] also reported the physical blends of PLA and TPS. It was observed that the mechanical properties of the blends decreased significantly, and the properties were lower than even the individual components due to phase incompatibility and transformation of TPS as very brittle matrix when added with PLA as dispersed phase. Thus, this particular blend combination of TPS and PLA required compatibilization with a flexible and tough material in order to synergistically combine the properties of the phases. In another study, Sarazin *et al.* [22] reported the binary and ternary blends of PLA, PCL and TPS using two different types of starch in a single-step extrusion process and demonstrated their superior physical properties. TPS grades with glycerol contents of 24 and 36 % were used, and the obtained blends alleviated the weaknesses of each individual polymer.

Mittal *et al.* [23] also reported the use of one-step melt mixing procedure to generate binary and ternary blends containing different compositions of PLA, TPS and PCL polymers. $PLA_{50}TPS_{50}$ binary blend was observed to be highly immiscible, however, the addition of PCL in the blends resulted in decreased domain sizes and better dispersion. TPS was observed in the ternary blends to be present in or around the PCL phase, especially at >10 % content. Contraction of PLA and TPS phases and subsequent expansion of PCL phase due to physical aging also resulted in the loss of ultrastructure in PCL as well as phase separation [24]. For instance, Figure 4.1 shows the time-dependent variations in the phase of $PLA_{50}TPS_{10}PCL_{40}$ ternary blend.

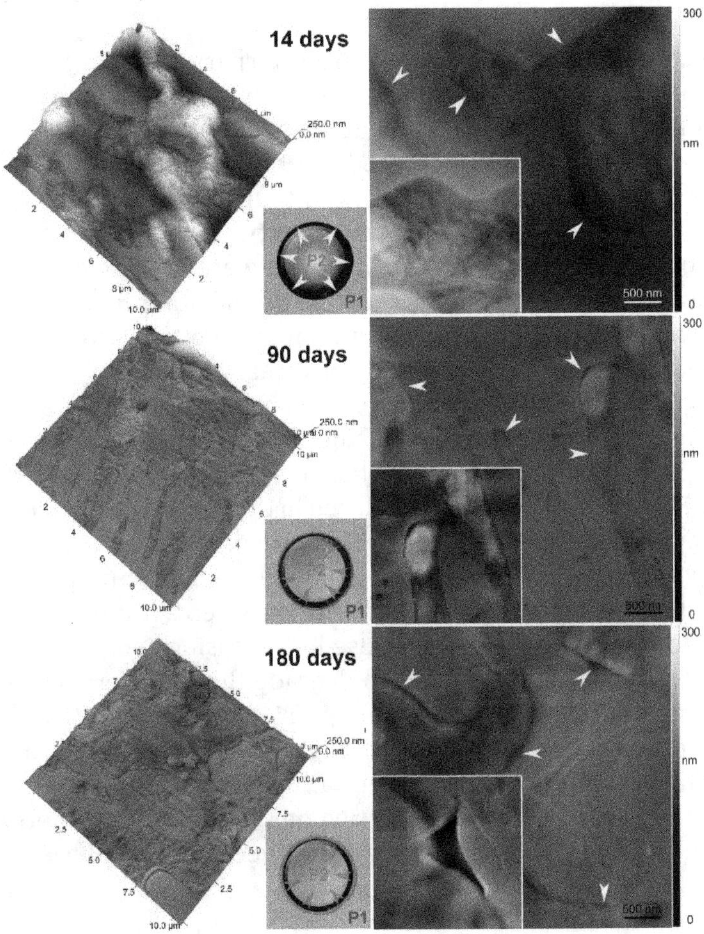

Figure 4.1 The effect of phase separation between different polymer phases as a function of time for the PLA-PCL-TPS blend. Room temperature AFM topographical images of freshly prepared cross-sections of the sample blocks with the appropriate age: (A and B) 14 days after production, (C and D) 60 days after production and (E and F) 180 days after production. Yellow arrows indicate areas of interface between PCL and PLA. Insets: P1 - polymer 1; P2 - polymer 2. Reproduced from Reference 24 with permission from Royal Society of Chemistry.

Organization of PCL domains was partially or completely lost. In contrast, the ultrastructural organization of PLA within polymer bulk did not change significantly, but the phase separation between the phases was observed with aging. As PCL exists in a rubbery state at room temperature, therefore, the expansion of PCL phase most probably was a consequence of a volumetric contraction of PLA and TPS phases.

Ravati and Favis [25] also reported ternary blends of PBS with PLA, PBAT and PCL in order to improve the limitation of low elongation of PBS. PCL was chosen due to its flexibility, whereas PLA was incorporated to enhance bio-based content in the blends. PBAT was chosen due its capability to enhanced toughness and elongation at break. The authors observed partial wetting for the ternary blend based on PBS/PLA/PCL, as PLA droplets were observed to self-assemble at the interface of PBS and PCL. On the other hand, the blend based on PBS/PLA/PBAT exhibited complete wetting with tri-continuous morphology. In comparison, the ternary blend PBS/PBAT/PCL had combined morphology, where both partial and complete wetting were observed. In order to gain further insights into the design of microstructure and physical properties of such ternary blends, blends of PCL, PBAT and PBS are also explored in the current study. The composition of the ternary blends was varied in order to study the impact of individual components on the blend morphology and properties. Conventional melt mixing method with one-step addition of the polymers to the compounder was used. Processing temperature and time were optimized so as not to degrade the polymers, while still ensuring uniform mixing. Morphological analysis was carried out using atomic force and optical microscopy, along with X-ray diffraction and infra-red spectroscopy. Thermal, mechanical and rheological properties of the blends were carried out by thermogravimetry, differential scanning calorimetry, dynamic mechanical analysis, tensile testing, and shear rheometry.

4.2 Experimental

4.2.1 Materials

PBAT biopolyester with trade name Ecoflex® F Blend C1200 (density ~1 g.cm^{-3}, peak melting point of 123 °C) was supplied by BASF SE, Germany. As mentioned earlier, it is a biodegradable aliphatic-aromatic copolyester based on the monomers 1,4-butanediol, adipic acid and terephthalic acid in the polymer chain, which biodegrades to the basic monomers and eventually to carbon dioxide, water and biomass when metabolized in the soil or compost under standard conditions. PCL with a trade name of CAPA 6506 S was procured from Perstorp Polyols, USA. It had weight average molecular weight (M_w) of 50,000 g/mol and peak melting temperature (T_m) of 60 °C (supplier information). PBS used was Bionelle 1050 grade from Showa Highpolymer Co., Japan. The materials were used as-received, without any further treatment.

4.2.2 Generation of Blends

Polymer blends were prepared by melt mixing of PCL, PBAT and PBS using mini twin conical screw extruder (MiniLab HAAKE Rheomex CTW5, Germany), as shown in Figure 4.2. The screw length and screw diameter were 109.5 mm and 5/14 mm conical respectively. The

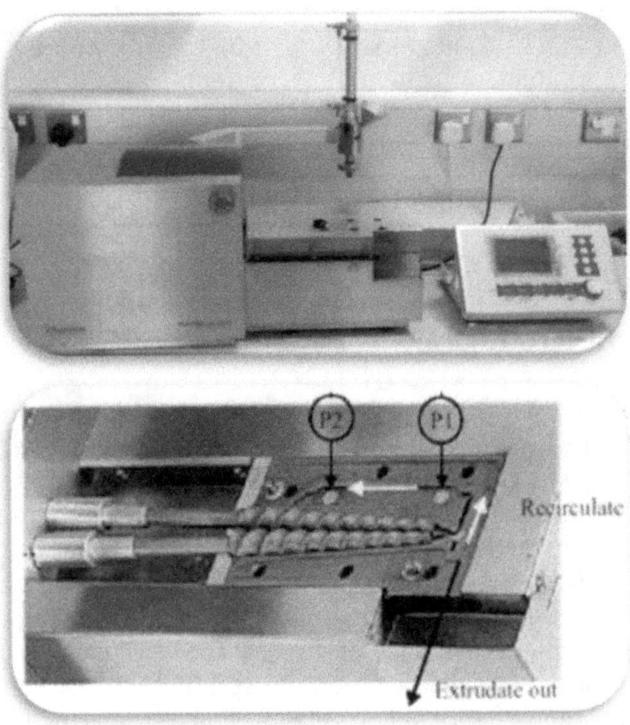

Figure 4.2 Mini twin conical screw extruder used for the generation of the blends.

composition of the blends is shown in Table 4.1. As evident from the table, the blends had varying amounts of PBS from 10-40%, PCL from 33-50% and PBAT 10-40%. A mixing temperature of 185 °C, rotational speed of 70 rpm and mixing time of 8 min were used and batches 4 g size were prepared. The mixing temperature was chosen on the basis of the crystalline melting phenomena exhibited by the polymers. Pure polymers were also extruded similarly for effective comparison with the blends, though different compounding temperatures were used (145 °C for PBAT and PBS, 90 °C for PCL). An average mass of 12 g of each sample was extruded. The disc, dumbbell and bar-shaped test samples were prepared by compression molding using hot press MTP-10 from Tetrahedron Associates. The pressure was kept at of 345 bar (5 psi) for 3 min at 185 °C (365 °F) using the pneumatic press. A controlled rate of cooling of the samples was used so as not to alter the microstructure due to uncontrolled cooling.

Table 4.1 Composition of the blends

Sample	PCL content, %	PBS content, %	PBAT content, %	Abbreviation
1	50	10	40	$PCL_{50}PBS_{10}PBAT_{40}$
2	50	25	25	$PCL_{50}PBS_{25}PBAT_{25}$
3	33	33	33	$PCL_{33}PBS_{33}PBAT_{33}$
4	50	40	10	$PCL_{50}PBS_{40}PBAT_{10}$

4.2.3 Characterization Methods

The wide-angle X-ray diffraction (WAXRD) analysis of the materials was performed using a Panalytical powder diffractometer (X'Pert PRO) using Cu Kα radiation (λ = 1.5406 Å) in reflection mode. A zero-background holder was used to minimize the noise. The samples were step-scanned from 2Θ = 5-35 ° at room temperature using a step size of 2Θ = 0.02 ° and a step time of 10 s. IR spectra were collected on Nicolet iS10 spectrometer equipped with SmartiTR diamond ATR accessory (angle of incidence of 45 °), DTGS KBr detector and KBr beam splitter. It had a diamond ATR crystal (index of refraction = 2.4 at 1000 cm^{-1}) and a depth penetration of 2 μm at 1000 cm^{-1} for sample with refractive index of 1.5. Spectra were recorded by the OMNIC software in the 4000-525 cm^{-1} region with a resolution of 4 cm^{-1} from 16 scans.

For the microscopy analysis, the polymer samples were mounted in special holders which at the same time fit in the microtome and were suitable for the examination of the block face by light microscopy and atomic force microscopy. The block faces of specimens after cryo ultramicrotomy were investigated at ambient conditions using ZEISS Axioplan light microscope (ZEISS, Germany) equipped with ZEISS Axio Cam ICc 1 CCD camera in reflected polarized light, and by Dimension 3100 AFM/SPM (Veeco, USA) atomic force microscope. AFM images were collected in tapping mode using silicon nitride cantilevers with natural frequencies in the 300 kHz range (force constant 20 N/m, tip radius 10 nm (NTMDT, Russia)). AFM image processing was performed using Nanoscope v720 software (Veeco, USA).

Thermal properties of the pure polymers and blends were analyzed using Netzsch thermogravimetric analyzer (TGA). Nitrogen was used as a carrier gas and the scans were obtained from 50 to 700 °C at a heating rate of 20 °C/min. Differential scanning calorimetric measurements on the polymers and blends were performed using a Netzsch DSC under nitrogen atmosphere. Compression molded samples were used for the analysis. The scans were obtained from 30 to 200 to 30 °C using heating and cooling rates of 15 °C/min and 5 °C/min, respectively.

Tensile testing of pure polymers and blends was performed on universal testing machine (Testometric, UK) using standard ASTM D638. The dumbbell shaped samples with 53 mm length, 4 mm width, and 2 mm thickness were used. A loading rate of 5 mm/min was employed and the tests were carried out at room temperature. WinTest Analysis software was used for the calculation of tensile modulus and other tensile properties. An average of five values was calculated. AR 2000 rheometer from TA Instruments was used to characterize the rheological properties such as storage modulus (G'), loss modulus (G''), viscosity (η'), elasticity (η''), as well as complex viscosity (η^*). Disc-shaped samples of 25 mm diameter and 2 mm thickness were measured at 185 ° C using a gap opening of 1.2 mm. Strain sweep scans were recorded at ω = 1 rad/s from 0.1% to 100% strain and the samples were observed to be shear stable up to 10% strain. For the comparison, frequency sweep scans (dynamic testing) of all the samples were recorded at 4% strain from ω = 0.1 to 100 rad/s. DMA (dynamic mechanical analyzer) RSA3 of TA Instruments was used to study the change in modulus with temperature (25-100 °C). Dynamic temperature ramp test was performed at a frequency of 5 rad/s, ramp rate 3 °C/min and 1% strain for all samples.

4.3 Results and Discussion

PCL, PBAT and PBS are among the most commonly used biopolymers, thus, alleviating their inherent structural limitations through blending would result in further enhancement of property profile as well as application potential. It is, however, challenging to optimize the blend compositions which provide most effective performance enhancement. Keeping this in perspective, the fraction of the individual polymers was varied in the current study and corresponding impact on morphology and physical properties was studied. It is also important to note that the morphology and properties are affected not only by the composition, but also by the processing conditions, especially compounding and post-processing conditions. Different morphology and physical properties may result depending on the variations in the processing conditions, which is also one of the factors for differences between the different literature studies.

Figure 4.3 shows the light microscopy images of the blends. At the micro-scale, the surface of the blends was uniform, irrespective of the

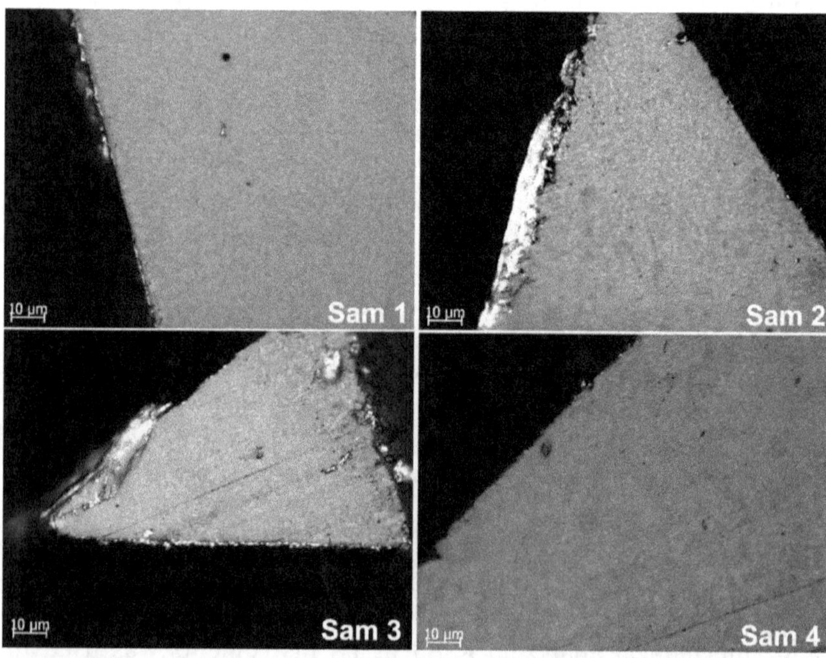

Figure 4.3 Light micorscopy images of $PCL_{50}PBAT_{40}PBS_{10}$ (Sample 1), $PCL_{50}PBAT_{25}PBS_{25}$ (Sample 2), $PCL_{33}PBAT_{33}PBS_{33}$ (Sample 3) and $PCL_{50}PBAT_{10}PBS_{40}$ (Sample 4).

blend composition. Though the component distribution was visible at the surface indicating phase immiscibility, however, the surface did not exhibit any large aggregates and phase domains. The microstructure of the blends also remained unchanged with time, in contrast to the earlier findings of phase separation in the PLA/PCL/TPS blends [23,24]. As all three components of the blend are in rubbery state at room temperature, it may have avoided the phase separation observed earlier.

Figure 4.4-4.7 demonstrate the AFM height and phase images of $PCL_{50}PBAT_{40}PBS_{10}$, $PCL_{50}PBAT_{25}PBS_{25}$, $PCL_{33}PBAT_{33}PBS_{33}$ and $PCL_{50}PBAT_{10}PBS_{40}$ blends. In the images, the dark phase corresponds to PBAT (largely amorphous), whereas the lighter phase represents PCL (semi-crystalline with good degree of crystallinity). PBS is observed as the bright/white phase (also contains high degree of crystallinity). The morphology of $PCL_{50}PBAT_{40}PBS_{10}$ blend was co-continuous in nature with uniform intermixing of PCL and PBAT phases. PBS was observed to be present as spherical dispersed phase. Interestingly, PBS domains were either present in PBAT or localized at the interface of PCL and PBAT. No mixing of PBS in PCL phase was observed. As PBS and PBAT have similar melting temperatures, thus, the observed morphology of specific distribution of PBS in and

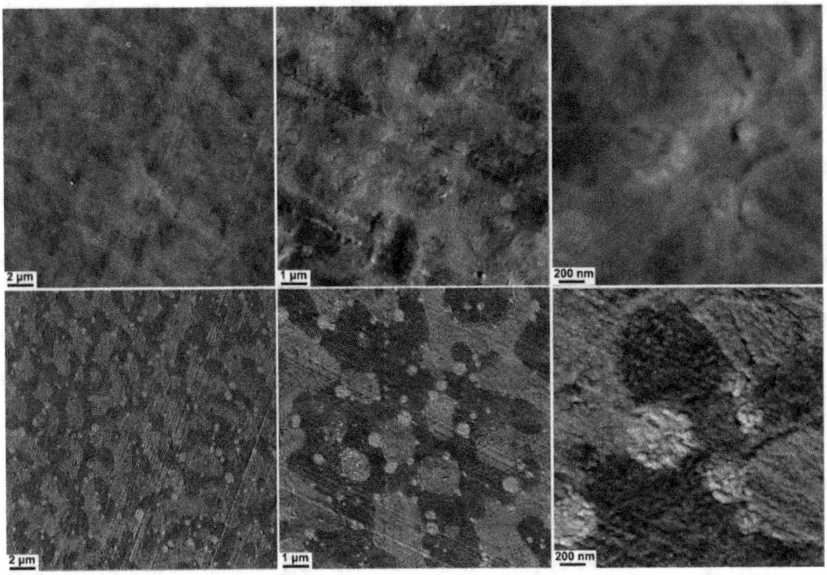

Figure 4.4 AFM height (top row) and phase (bottom row) images of $PCL_{50}PBAT_{40}PBS_{10}$.

around PBAT can be probably a kinetic phenomenon introduced during the compounding process. Also, as PBS is relatively highly crystalline, thus, intermixing of the PBS with PBAT may also affect its capability to crystallize or degree of crystallinity. Similar effect may also be observed for PCL due to its constrained environment inside the co-continuous morphology with PBAT. As also observed in the optical images, the polymer domains were smaller in size, thus, indicating good intermixing of the polymers. Enhancing the PBS content (and correspondingly decreasing the PBAT content) in $PCL_{50}PBAT_{25}PBS_{25}$ resulted in corresponding increase in the domain size of PBS, as shown in Figure 4.5. The co-continuous morphology exhibited in the earlier case by PCL and PBAT phases was also affected as more discrete domains of PBAT could be observed, which was expected due to the reduced fraction of PBAT in the blend. However, both PCL and PBAT polymers still exhibited uniform intermixing. Similar to the $PCL_{50}PBAT_{40}PBS_{10}$ blend in Figure 4.4, PBS phase was present as spherical domains predominantly inside the PBAT phase. The interface between the PBS and PBAT phases was also stable, as no pullout of the PBS phase was observed during block face preparation at cryo conditions. Largely amorphous nature of PBAT may have acted as a glue for the PCL phase. Similar observations were observed for the

Figure 4.5 AFM height (top row) and phase (bottom row) images of $PCL_{50}PBAT_{25}PBS_{25}$.

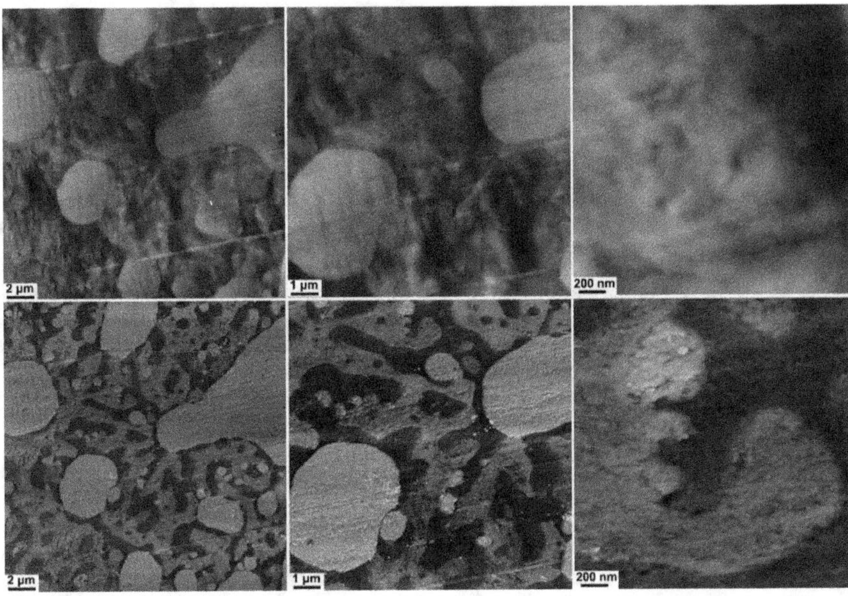

Figure 4.6 AFM height (top row) and phase (bottom row) images of $PCL_{33}PBAT_{33}PBS_{33}$.

ternary blend $PCL_{33}PBAT_{33}PBS_{33}$, as shown in Figure 4.6. However, the domain size of PBS was observed to be significantly enhanced. Similar to earlier cases, PBS phase was surrounded by PBAT. Among all the polymers, PCL appeared to form a continuous phase and was well mixed with PBAT. Further reducing the PBAT amount (or increasing the PBS amount) in the ternary blend $PCL_{50}PBAT_{10}PBS_{40}$ led to large domains of PBS surrounded by PCL, as evident in Figure 4.7. The PBAT phase was well mixed as dispersed spherical droplets inside the PCL phase due to its lower fraction. Interestingly, the smaller domains of PBS were still observed to be present with the PBAT phase, clearly confirming the effect of kinetic mixing during the compounding process. Thus, it can be inferred from the morphological analysis that specific distribution of the polymers took place in the ternary blend, which was dependent on the blend composition. Due to this phenomenon, some changes in the crystal structure or crystallinity of the polymers might have also been induced.

Further insights into the structure and morphology were gained through IR spectroscopy and X-ray diffraction. Figure 4.8 demonstrates the IR spectra of pure polymers and blends. The pure polymers (Figure 4.8a) exhibited the characteristic C=O stretching bands

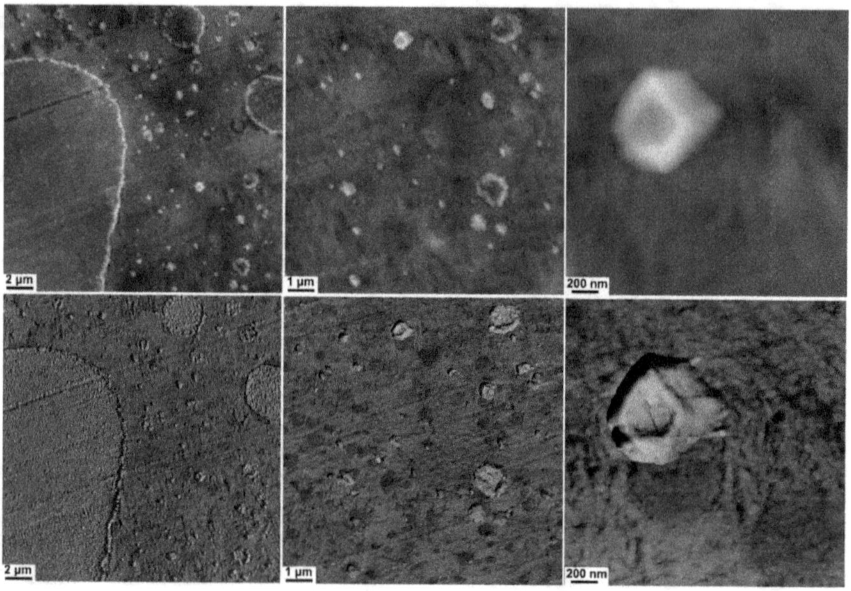

Figure 4.7 AFM height (top row) and phase (bottom row) images of $PCL_{50}PBAT_{10}PBS_{40}$.

at 1754 cm^{-1} along with a shoulder at slightly lower wavenumber. Differences in the absorption bands of the three polymers between 1500-600 cm^{-1} were observed. In comparison to the pure polymers, the IR spectra of the blends also exhibited the C=O stretching band at 1754 cm^{-1}. The blends exhibited no change in the peaks corresponding to pure PCL, indicating no change in the crystal structure of PCL. A small reduction in the corresponding peak intensity was though observed as the amount of PBS was increased (while keeping the content of PCL constant at 50%), which could have been resulted due to the reduction in the crystallinity of PCL in the blend. On the other hand, peaks corresponding to PBS in the 1500-750 cm^{-1} range were observed to be shifted or completely absent, indicating a change in the crystal structure of the polymer in the blends. These findings also supported the earlier opinion from the AFM analysis. Figure 4.9 also shows the X-ray diffractograms of the pure polymers and blends. X-ray diffractogram of pure PCL contained two peaks at 21.8° and 24.1° 2Θ corresponding to (111) and (211) crystal plane indices, respectively [26]. PBS also exhibited two diffraction peaks at 20.5° and 23° 2Θ. On the other hand, PBAT did not exhibit any defined diffraction peak owing to its largely amorphous nature. In the blends, the peaks

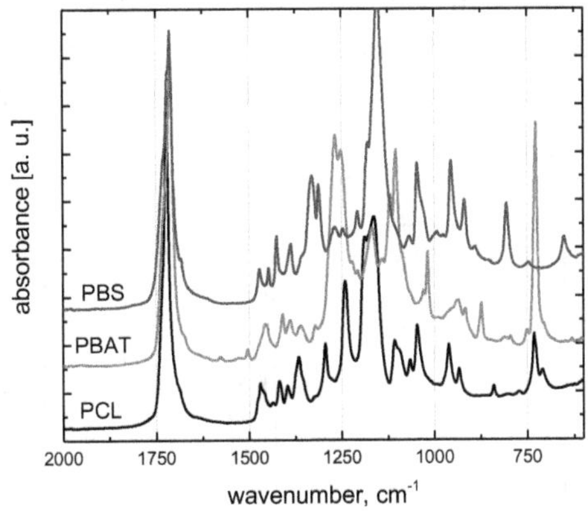

(a)

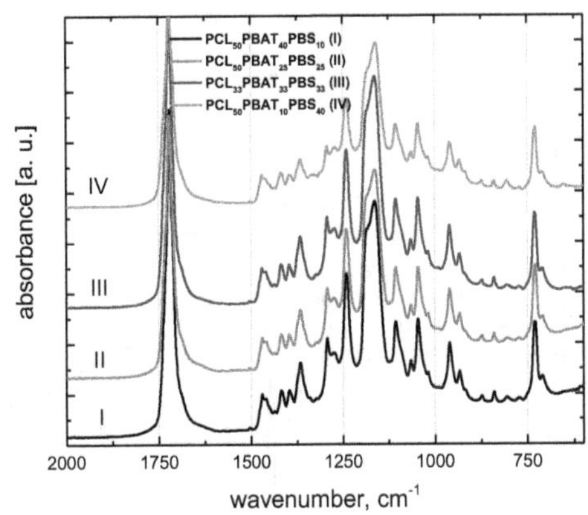

(b)

Figure 4.8 IR spectra of (a) pure polymers and (b) ternary blends.

corresponding to PCL did not shift their positions, also confirming the earlier finding from IR abut no change in the crystal structure of PCL in the blends. However, the melt enthalpy analysis of the blends exhibited a decrease in the melt enthalpy (normalized) for the PCL melting peak as the amount of PBS was increased in the blends ($PCL_{50}PBAT_{40}PBS_{10}$, $PCL_{50}PBAT_{25}PBS_{25}$, and $PCL_{50}PBAT_{10}PBS_{40}$). Probably, the presence of PBS in the blend hindered the crystallization of PCL. On the other hand, the diffraction peak positions corresponding to PBS were affected, thus, indicating changes in the crystal structure. In fact, the first diffraction peak at 20.5° was largely absent in the blends. In addition, the ternary blend $PCL_{50}PBAT_{40}PBS_{10}$ also did not exhibit 23° 2Θ peak. As the content of PBS in this blend was 10%, it could have suffered extensive hindrance to crystallization in the presence of PCL and PBAT. On comparing the $PCL_{50}PBAT_{40}PBS_{10}$,

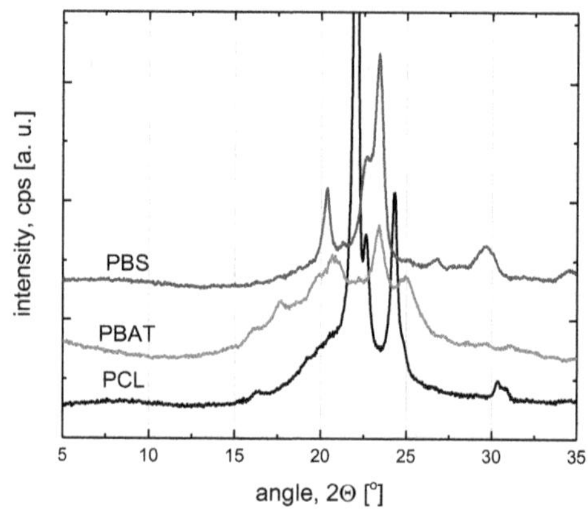

(a)

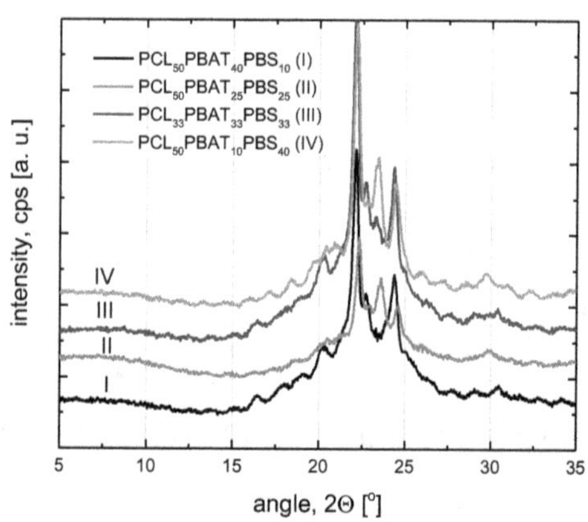

(b)

Figure 4.9 X-ray diffractograms of (a) pure polymers and (b) ternary blends.

$PCL_{50}PBAT_{25}PBS_{25}$, and $PCL_{50}PBAT_{10}PBS_{40}$ ternary blends, where the amount of PBS was increased (while keeping the PCL content constant at 50 wt%), it was observed that a relative increase in the peak

stability. The onset of degradation temperature was also observed to reduce as a function of PBS content. Especially, the ternary blend $PCL_{50}PBAT_{10}PBS_{40}$ exhibited onset of degradation temperature around 310 °C. Also, $PCL_{50}PBAT_{40}PBS_{10}$ and $PCL_{50}PBAT_{10}PBS_{40}$ blends exhibited a weight loss of 2% and 10% respectively after 15 min during measurement. In addition, owing to similar degradation profiles of the pure polymers with narrow differences in temperatures, the TGA thermograms of the blends also exhibited a single degradation pattern.

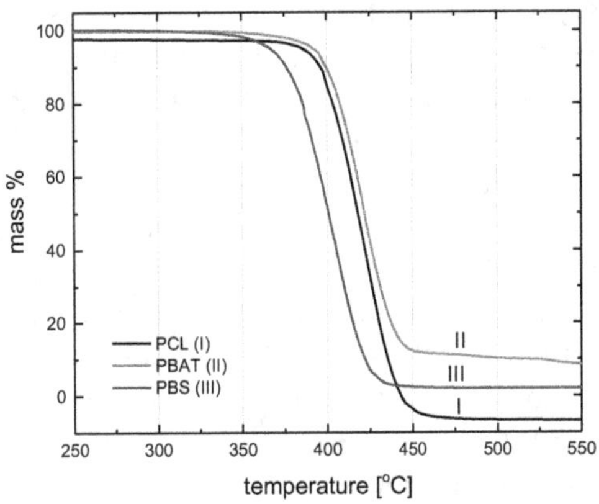

(a)

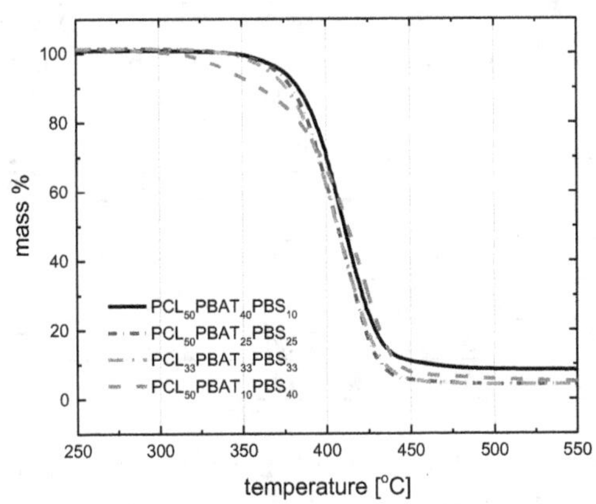

(b)

Figure 4.11 TGA thermograms of the (a) pure polymers and (b) ternary blends.

Table 4.2 and Figure 4.12 demonstrate the tensile properties of the blends in comparison with the pure polymers. PBS had the highest tensile modulus of 500 MPa, followed by 392 MPa for PCL. PCL exhibited the highest strain at break of 813%, whereas a poor elongation of 3% was observed for PBS. PBAT has moderate tensile modulus of 78 MPa and significant elongation at break of 123%. Increasing the PBS concentration in the blends resulted in enhancement of the tensile modulus from 280 MPa for $PCL_{50}PBAT_{40}PBS_{10}$ to 408 MPa for $PCL_{50}PBAT_{10}PBS_{40}$ (Figure 4.12a). Owing to high extent of flexibility of PCL and PBAT, the ternary blends $PCL_{50}PBAT_{40}PBS_{10}$ and $PCL_{50}PBAT_{25}PBS_{25}$ had elongation at break of 215% and 194% respectively (Figure 4.12b). Further increasing the PBS content, however, negatively impacted the flexibility of the blends. Probably much larger domain size of PBS in $PCL_{33}PBAT_{33}PBS_{33}$ and $PCL_{50}PBAT_{10}PBS_{40}$ ternary blends resulted in strain hardening of the chains. The enhanced modulus and elongation was also accompanied with a decrease in peak stress and stress at break in the blends. Overall, by the generation of blends of varying compositions, the flexibility of PBS could be tuned which is of high commercial relevance due to its wide spectrum of applications. Especially, the blends $PCL_{50}PBAT_{40}PBS_{10}$ and $PCL_{50}PBAT_{25}PBS_{25}$ are optimum due to enhanced flexibility and high modulus.

Table 4.2 Tensile properties of the materials

Polymer/blend	Tensile modulus, MPa	Peak stress, MPs	Peak strain, %	Stress at break, MPa	Strain at break, %
PCL	392	38.5	652	35.5	813
PBAT	78	27.9	99	25.8	123
PBS	500	37.0	3.0	13.4	3
$PCL_{50}PBAT_{40}PBS_{10}$	280	10.5	15.2	1.4	215
$PCL_{50}PBAT_{25}PBS_{25}$	345	13	13.3	5.8	194
$PCL_{33}PBAT_{33}PBS_{33}$	344	12.4	8.8	5.1	11
$PCL_{50}PBAT_{10}PBS_{40}$	408	14.5	4.4	6.8	6

DMA analysis of the blends as a function temperature is also presented in Figure 4.13. Similar to the tensile properties, the storage modulus of the blends was observed to increase as a function of PBS content. Though the modulus decreased on enhancing the temperature, the trends were still maintained. The blends $PCL_{50}PBAT_{25}PBS_{25}$, $PCL_{33}PBAT_{33}PBS_{33}$ and $PCL_{50}PBAT_{10}PBS_{40}$ were observed to exhibit a

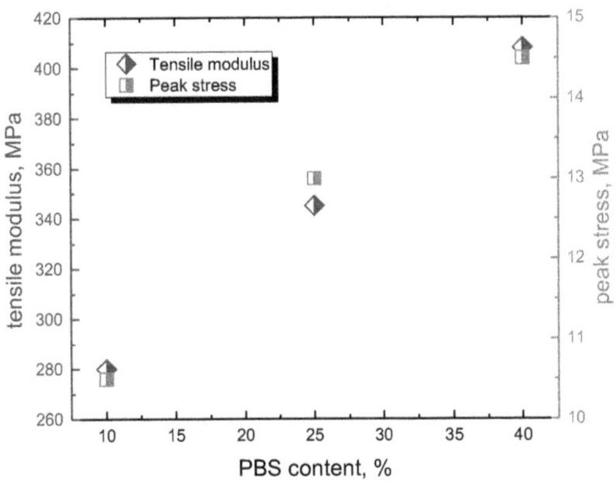

(a)

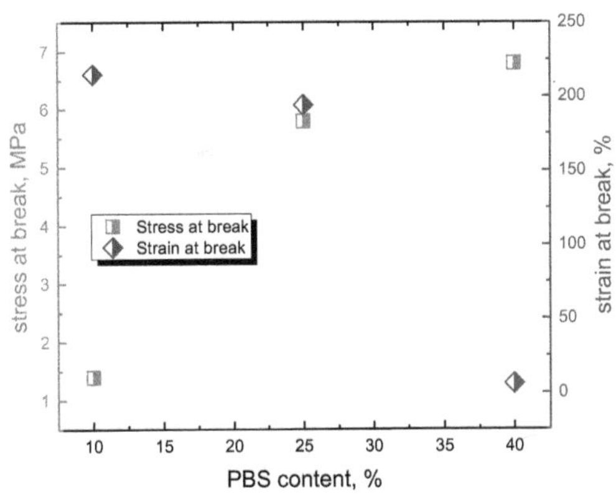

(b)

Figure 4.12 (a) Tensile modulus and peak stress; (b) stress at break and strain at break of the ternary blends as a function of increasing PBS content. The PCL content was 50%, whereas the PBAT content varied inversely to PBS content.

significantly high modulus even at 90 °C, thus, indicating the high temperature performance. These blends were also observed to exhibit a sharp drop in the modulus around the peak melting temperature of PCL, however, the modulus was still appreciable due to the presence of high temperature melting PBS and PBAT. The ternary blend $PCL_{50}PBAT_{40}PBS_{10}$ also exhibited a sharp drop near the melting temperature of PCL, however, its performance was lost around 65 °C.

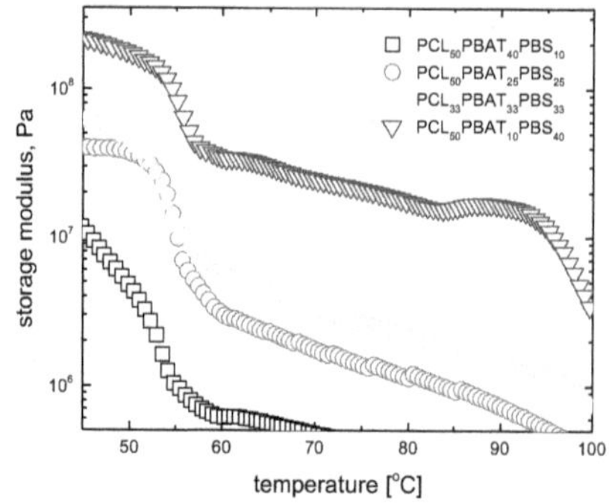

(a)

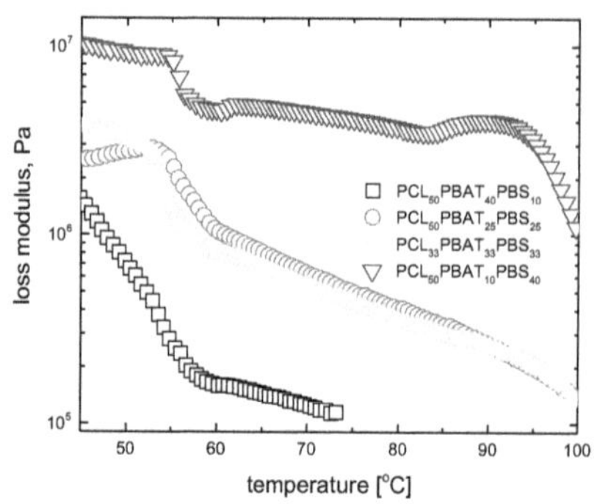

(b)

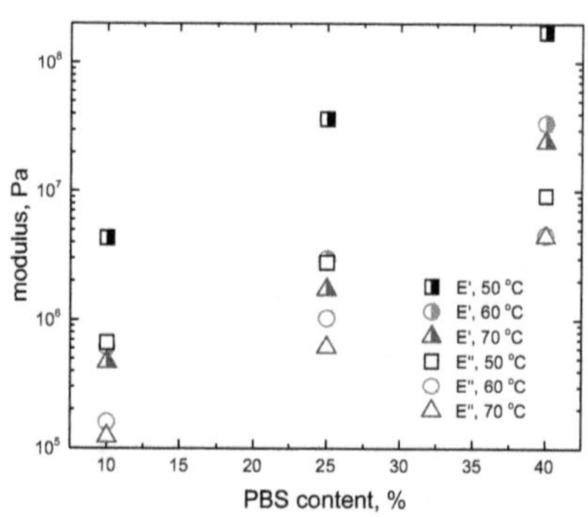

(c)

Figure 4.13 (a) Storage and (b) loss moduli of the ternary blends as a function of temperature from the DMA analysis; (c) variation of the storage and loss modulus as a function of PBS content at 50, 60 and 70 °C

As shown in Figure 4.13b, the loss modulus of the blends also exhibited similar behavior with temperature as the storage modulus. In addition, the values of loss modulus were lower than the storage modulus for the whole temperature range, indicating the elastic dominance in the blends. Figure 4.13c also presents the variation of the storage and loss moduli as a function of PBS content at 50, 60 and 70 °C. For instance, the storage modulus for the $PCL_{50}PBAT_{10}PBS_{40}$ blend at 50 °C was observed to be 1.7E8 Pa, as compared to 2.4E7 Pa at 70 °C. In comparison, the loss modulus at 50 °C was 9.2E6 Pa, which was reduced to 4.4E6 Pa at 70 °C. Other blends also exhibited similar differences as a function of temperature as well as PBS content, thus, confirming the positive effect of increased PBS content on the mechanical performance of the blends. Overall comparison of the morphological, thermal and mechanical properties revealed that the blend $PCL_{50}PBAT_{25}PBS_{25}$ exhibited optimum properties due to uniform component distribution, enhanced flexibility, moderate tensile modulus, good thermal stability as well as significant storage modulus which was also retained at high temperatures. Further insights can be gained by generating blends with the compositions in the vicinity of this blend.

Figure 4.14 demonstrates the rheological analysis of the blends as a function of angular frequency. The shear rheological analysis was carried out at a high temperature of 185 °C, which led to the formation of low viscosity melt during measurement. The storage modulus increased for all the blends as a function of angular frequency. Comparing the behavior of different blends, though no significant difference in the storage modulus was observed with angular frequency, small differences could still be observed. Highest storage modulus was observed for $PCL_{50}PBAT_{40}PBS_{10}$ and $PCL_{50}PBAT_{25}PBS_{25}$ blends. Probably, the inclusion of smaller PBS phase in largely amorphous PBAT acted as a reinforcing filler. Somewhat lower storage modulus was observed for $PCL_{33}PBAT_{33}PBS_{33}$ and $PCL_{50}PBAT_{10}PBS_{40}$ due to larger PBS phase which was not effectively mixed with PBAT. Thus, although the measurements were made in the melt form, differences in the corresponding stress relaxation led to different behaviors in the blends. It is also expected due the absence of molecular mixing among the polymer phases. Figure 4.14b also shows the loss modulus of the blends as a function of angular frequency. Similar trends as storage modulus were observed. In addition, the loss modulus of the blends was observed to be higher than storage modulus at all frequency values as well as for all blend compositions, which indicated

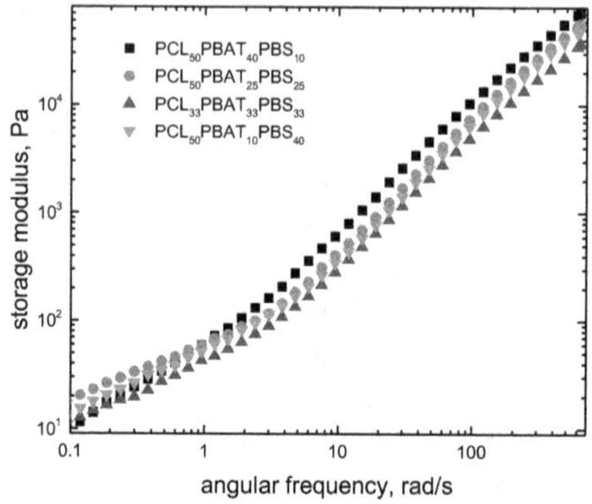

(a)

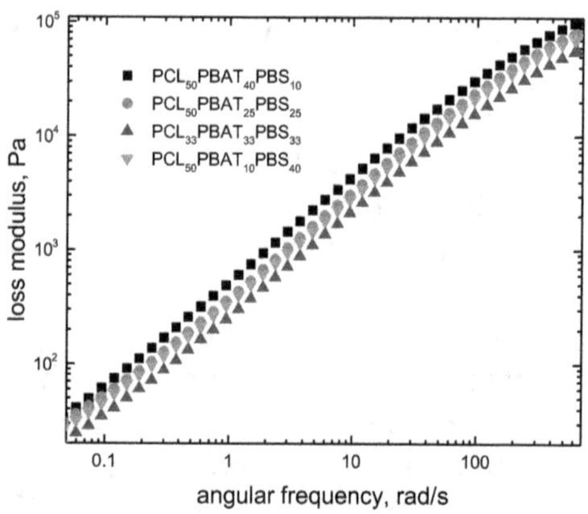

(b)

Figure 4.14 (a) Storage and (b) loss moduli of the ternary blends as a function of angular frequency.

dominance of the viscous character during melt shearing. Similar behavior was also observed in the case of complex viscosity, as shown in Figure 4.15. The viscosity decreased for all the blends with angular frequency due to shear thinning nature. Comparison of the blends exhibited higher viscosity for the $PCL_{50}PBAT_{40}PBS_{10}$ blend, followed by $PCL_{50}PBAT_{25}PBS_{25}$, due to the specific blend morphology and composition, as also observed for the modulus in Figure 4.14. The lowest viscosity values were observed for the $PCL_{33}PBAT_{33}PBS_{33}$ blend. However, the observed changes in the viscosity did not pose any hindrance to the processing of the blends.

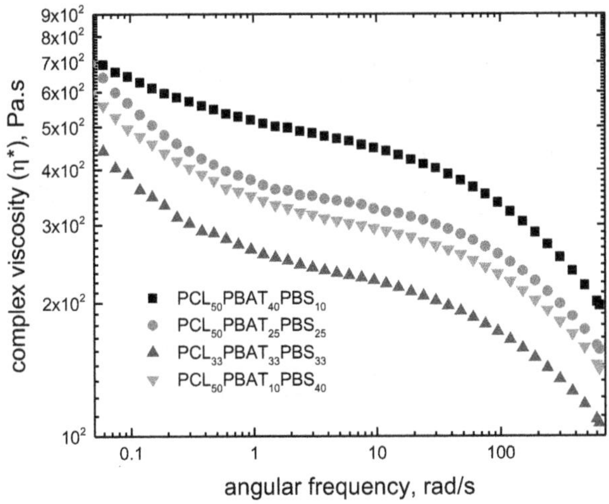

Figure 4.15 Complex viscosity of the blends as a function of angular frequency.

Figure 4.16 shows the phase miscibility analysis of the blends using commonly used Cole-Cole, van Gurp and Han Chuang methods [27-33]. In Cole-Cole analysis, a smooth, semi-circular shape of the graph between real (η') and imaginary (η'') parts of complex viscosity generally suggests miscible blends with homogenous phase. As observed in Figure 4.16a, the curves of the blends exhibited semi-circular shape at lower viscosity values, especially for $PCL_{50}PBAT_{40}PBS_{10}$ and $PCL_{50}PBAT_{25}PBS_{25}$ blends. However, a deviation from this behavior was observed at higher viscosity. This indicated some degree of intermixing in the phases, as also observed in AFM images due to good intermixing between PCL and PBAT along with dispersion of PBS in PBAT for $PCL_{50}PBAT_{40}PBS_{10}$ and $PCL_{50}PBAT_{25}PBS_{25}$ blends. Figure 4.16b also demonstrates the analysis of the rheological data based on van Gurp plots representing the relationship between complex modulus (G*) and phase angle, delta (δ). The van Gurp analysis for the blends confirmed the findings from Cole-Cole analysis. In the blends, the time-temperature superposition principle was observed to hold (indicated by the merging of the curves into a common curve) for all the blends at higher G* values. At lower G* values, deviations

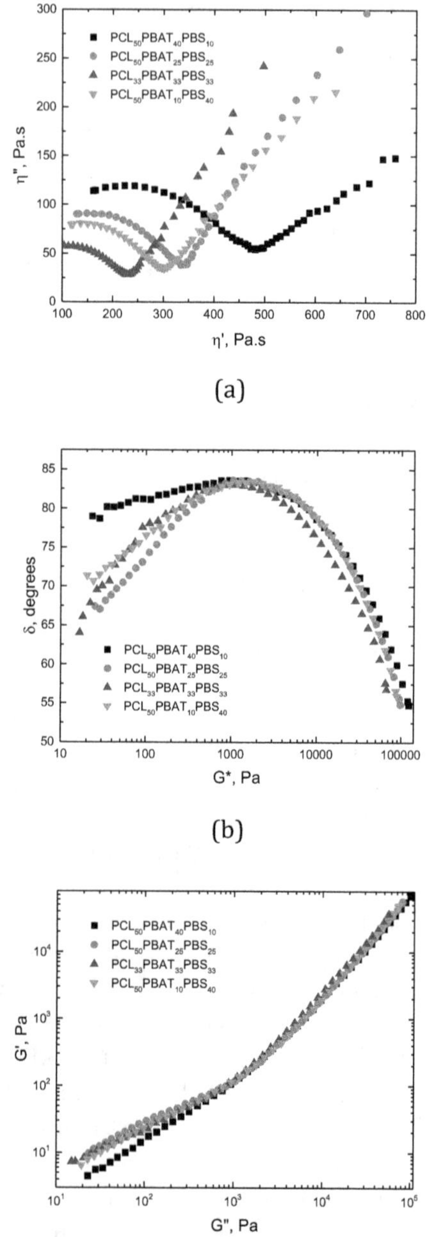

Figure 4.16 (a) Cole-Cole, (b) van Gurp and (c) Han Chuang analysis of the ternary blends.

were observed, which confirmed the absence of molecular phase miscibility in the blends. Similarly, analysis of the compatibility of polymer blends by using the criteria reported by Han and Chuang (where G' is plotted vs. G") in Figure 4.16c revealed composition dependent correlation at lower modulus values. On the other hand, the behavior was composition independent at the higher modulus values, thus, confirming the earlier observation of presence of some intermixing among the blend components.

4.4 Conclusions

In the current study, one-step melt mixing method was successfully employed to generate the ternary blends of PCL, PBAT and PBS. The fraction of the blend constituents was varied in order to study the effect on morphology and physical properties. The morphology analysis of the blends indicated absence of complete phase miscibility, however, good level of intermixing in the blends was observed. The domain sizes of the individual phases were relatively smaller and the distribution was retained with time. PBS was observed to be dispersed in PBAT or at PBAT/PCL interface, probably due to the kinetic effects introduced during the compounding process. The IR and X-ray analysis also confirmed no change in the crystalline peak positions of PCL, though a reduction in its degree of crystallinity was observed with increasing PBS content in the blends. On the other hand, structural changes were observed in PBS in the blends. Also, increasing the content of PBS in the blends exhibited higher intensity crystalline signals for PBS, indicating increment in its crystallinity, though the crystal structure was modified. It was also indicated by the small increment in the peak melting temperature of PBS as a function of PBS content in the blends. Though the thermal stability of the blends was negatively impacted by the addition of PBS, the blends were still stable for high temperature processing conditions. The tensile modulus of the blends was observed to increase as the PBS content was enhanced. Specifically, the $PCL_{50}PBAT_{10}PBS_{40}$ blend exhibited highest modulus of 408 MPa, however, its elongation was only 6% as compared to 3% for pure PBS. On the other hand, the blend $PCL_{50}PBAT_{25}PBS_{25}$ exhibited more optimal performance with a strain at break of 194% and tensile modulus of 345 MPa. DMA analysis also exhibited improvement in the high temperature performance of the blends with increasing PBS content in the blends. In shear rheology, blends exhibited only minor differences in the storage modulus.

Probably, the observed differences resulted from the dispersed morphology where PBS may have acted as reinforcement for PBAT phase, especially for $PCL_{50}PBAT_{40}PBS_{10}$ and $PCL_{50}PBAT_{25}PBS_{25}$ blends. As a result, differences in the relaxation of the polymers in the blends would have caused the differences in the modulus, as a function of blend composition. The loss modulus of the blends was observed to be higher than storage modulus for complete frequency range, indicating viscous dominance. The phase miscibility analysis also signaled some degree of phase intermixing, but absence of molecular mixing among the blend components.

References

1. Hamad, K., Kaseem, M., and Deri, F. (2011) Rheological and mechanical characterization of poly(lactic acid)/polypropylene polymer blends. *Journal of Polymer Research*, **18**, 1799-1806.
2. Jancar, J. (1999) *Mineral Fillers in Thermoplastics I: Raw Materials and Processing*, Springer, Gemany.
3. Wang, B., Zhang, Y, Gou, Z., Cheng, J., and Fang, Z. (2011) Biodegradable aliphatic/aromatic copoly(ester-ether)s: The effect of poly(ethylene glycol) on physical properties and degradation behavior. *Journal of Polymer Research*, **18**, 187-196.
4. Mittal, V. (2012) *Functional Polymer Blends: Synthesis, Properties and Performance*, CRC Press, USA.
5. Wang, L., Zhang, Z., Chen, H., Zhang, S., and Xiong, C. (2010) Preparation and characterization of biodegradable thermoplastic Elastomers (PLCA/PLGA blends). *Journal of Polymer Research*, **17**, 77-82.
6. Chaudhry, A. U., and Mittal, V. (2014) Blends of high-density polyethylene with chlorinated polyethylene: Morphology, thermal, rheological, and mechanical properties. *Polymer Engineering and Science*, **54**(1), 85-95.
7. Correlo, V. M., Boesel, L. F., Bhattacharya, M., Mano, J. F., Neves, N. M., and Reis, R. L. (2005) Properties of melt processed chitosan and aliphatic polyester blends. *Materials Science and Engineering A*, **403**, 57-68.
8. Patwary, F., Matsko, N., and Mittal, V. (2016) Biodegradation properties of melt processed PBS/chitosan bio-nanocomposites with silica, silicate and thermally reduced graphene. *Polymer Composites*, doi: 10.1002/pc.23947.
9. Pawar, S. P., Misra, A., Bose, S., Chatterjee, K., and Mittal, V. (2015) Enzymatically degradable and flexible bio-nanocomposites derived

from PHBV and PBAT blend: Assessing thermal, morphological, mechanical, and biodegradation properties. *Colloid and Polymer Science*, **293**, 2921-2930.

10. Zhao, H., Cui, Z., Sun, X., Turng, L.-S., and Peng, X. (2013) Morphology and properties of injection molded solid and microcellular polylactic acid/polyhydroxybutyrate-valerate (PLA/PHBV) blends. *Industrial & Engineering Chemistry Research*, **52**, 2569-2581.
11. Nagarajan, V., Mohanty, A. K., and Misra, M. (2013) Sustainable green composites: Value addition to agricultural residues and perennial grasses. *ACS Sustainable Chemistry & Engineering*, **1**, 325-333.
12. Averous, L., Moro, L., Dole, P., and Fringant, C. (2000) Properties of thermoplastic blends: starch–polycaprolactone. *Polymer*, **41**, 4157-4167.
13. Vikman, M., Hulleman, S. H. D., van der Zee, M., Myllarinen, P., and Feil, H. (1999) Morphology and enzymatic degradation of thermoplastic starch–polycaprolactone blends. *Journal of Applied Polymer Science*, **74**, 2594-2604.
14. Singh, R. P., Pandey, J. K., Rutot, D., Degee, P., and Dubois, P. (2003) Biodegradation of poly(epsilon-caprolactone)/starch blends and composites in composting and culture environments: the effect of compatibilization on the inherent biodegradability of the host polymer. *Carbohydrate Research*, **338**, 1759-1769.
15. Nayak, S. K. (2010) Biodegradable PBAT/Starch nanocomposites. *Polymer - Plastics Technology & Engineering*, **49**, 1406-1418.
16. Xu, Y. X., and Hanna, M. A. (2005) Preparation and properties of biodegradable foams from starch acetate and poly (tetramethylene adipate-co-terephthalate). *Carbohydrate Polymers*, **59**, 521-529.
17. Quero, E., Muller, A. J., Signori, F., Coltelli, M. B., and Bronco, S. (2012) Isothermal cold-crystallization of PLA/PBAT blends with and without the addition of acetyl tributyl citrate. *Macromolecular Chemistry and Physics*, **213**, 36-48.
18. Bhatia, A., Gupta, R. K., Bhattacharya, S. N., and Choi, H. J. (2007) An investigation of melt rheology and thermal stability of poly(lactic acid)/ poly(butylene succinate) nanocomposites. *Journal of Applied Polymer Science*, **114**(5), 2837-2847.
19. Dell'Erba, R., Groeninckx, G., Maglio, G., Malinconico, M., and Migliozzi, A. (2001) Immiscible polymer blends of semicrystalline biocompatible components: thermal properties and phase morphology analysis of PLLA/PCL blends. *Polymer*, **42**, 7831-7840.
20. Maglio, G., Malinconico, M., Migliozzi, A., and Groeninckx, G. (2004) Immiscible poly(L-lactide)/poly(ε-caprolactone) blends: Influence of the addition of a poly(L-lactide)-poly(oxyethylene) block copolymer on thermal behavior and morphology. *Macromolecular Chemistry and Physics*, **205**(7), 946-950.

21. Martin, O., and Averous, L. (2001) Poly(lactic acid): plasticization and properties of biodegradable multiphase systems. *Polymer*, **42**, 6209-6219.
22. Sarazin, P., Li, G., Orts, W. J., and Favis, B. D. (2008) Binary and ternary blends of polylactide, polycaprolactone and thermoplastic starch. *Polymer*, **49**, 599-609.
23. Mittal, V., Akhtar, T., Luckachan, G., and Matsko, N. (2015) PLA, TPS and PCL binary and ternary blends: structural characterization and time-dependent morphological changes. *Colloid and Polymer Science*, **293**(2), 573-585.
24. Matsko, N. B., and Mittal, V. (2014) Thermotropic dynamic processes in multiphase polymer systems by (cryo-)AFM. *Soft Matter*, **10**, 5478-5488.
25. Ravati, S., and Favis, B. D. (2013) Tunable morphologies for ternary blends with poly(butylene succinate): Partial and complete wetting phenomena. *Polymer*, **54**(13), 3271-3281.
26. Liao, H.-T., and Wu, C.-S. (2009) Preparation and characterization of ternary blends composed of polylactide, poly(ε-caprolactone) and starch. *Materials Science and Engineering A*, **515**, 207-214.
27. Cho, K., Lee, B. H., Hwang, K. M., Lee, H., and Choe, S. (1998) Rheological and mechanical properties in polyethylene blends. *Polymer Engineering and Science*, **38**(12), 1969-1975.
28. Kim, H. K., Rana, D., Kwag, H., and Choe, S. (2001) Melt rheology of ethylene 1-octene copolymer blends synthesized by Ziegler-Natta and metallocene catalysts. *Korea Polymer Journal*, **8**, 34-43.
29. Kwag, H., Rana, D., Choe, K., Rhee, J., Woo, T., Lee, B. H., and Choe, S. (2000) Binary blends of metallocene polyethylene with conventional polyolefins: Rheological and morphological properties. *Polymer Engineering and Science*, **40**(7), 1672-1681.
30. Utracki, L. A. (1989) *Rheology of Polymer Alloys and Blends*, Hanser Publishers, Germany.
31. van Gurp, M., and Palmen, J. (1998) Time-temperature superposition for polymeric blends. *Rheology Bulletin*, **67**(1), 5-8.
32. Chuang, H. K., and Han, C. D. (1984) Rheological behavior of polymer blends. *Journal of Applied Polymer Science*, **29**(6), 2205-2229.
33. Han, C. D., and Chuang, H. K. (1985) Criteria for rheological compatibility of polymer blends. *Journal of Applied Polymer Science*, **30**(11), 4431-4454.

5

Morphology and Properties of PCL, PLA and PBS Ternary Blends

5.1 Introduction

Due to the exponential growth in the consumption of plastic materials, a large amount of non-biodegradable plastics is being disposed into the nature annually. Environmental pollution has become a global issue because of the negative impact of the plastic waste on our ecological systems. Recycling is one of the routes to alleviate the challenges posed by the plastic waste, however, it consumes a lot of energy in washing and processing the plastics. One of the other alternatives is to partially replace the commodity synthetic polymers with biopolymers, which are readily susceptible to microbial attack [1]. For instance, introduction of green bags in the Australian markets has resulted in PE bag usage to fall from around 6 billion in 2002 to 3.9 billion in 2007 [2]. In Europe, despite the economic crisis, market of biodegradable polymers grew 5-10% (depending on products and applications) in 2009 compared with 2008. A report from Business Communications Company (BCC) Research estimated a compound annual growth rate (CAGR) of 22% for the biodegradable polymers for the 5-year period starting from 2012 [3].

Major drawbacks of biodegradable polymers hindering their widespread application are their high cost and moderate mechanical properties as compared to conventional thermoplastics [4]. For instance, poly(lactic acid) (PLA), a thermoplastic polymer, has been used to produce packaging due to its high modulus and strength. However, its high brittleness, low melt viscosity, and low toughness lead to its restricted applications [5]. Similarly, poly(butylene succinate) (PBS) is another commonly employed biopolymer with good

Vikas Mittal[a,], Yuddish Manna[a,**] and Nadejda Matsko[b]*
[a]The Petroleum Institute (part of Khalifa University of Science and Technology), Abu Dhabi, UAE; [b]Graz Centre of Electron Microscopy, Graz, Austria
**Current address: Bletchington, Wellington County, Australia; **Current address: University of Waterloo, Canada*

thermal properties and modulus, however, it suffers from poor flexibility [6]. On the other hand, biopolymers like polycaprolactone (PCL) and poly(butylene adipate-co-terephthalate) (PBAT) exhibit high degree of flexibility, but require further enhancements in the modulus to expand their spectrum of applications. Thus, extensive efforts have been made in order to improve the property profiles of the biopolymers. Commonly employed methodologies to improve the weak physical properties include copolymerization, plasticization and polymer blending. Among these methods, blending of biopolymers has evolved into a useful and cost-effective strategy to combine the beneficial properties of individual polymers, thereby, also eliminating their inherent limitations. In addition, the blending is completely compatible with the melt mixing processes commonly employed commercially for polymer processing. In a recent study, in order to enhance flexibility of poly(3-hydroxybutyrate-co-3-hydroxyvalerate) (PHBV), blending was carried out with PBAT using conventionally accepted melt blending process [7]. Similarly, to overcome the limitations of PLA, the blends of PLA with PBAT and PCL have been reported, which combined the toughness and flexibility characteristics of PBAT and PCL with the mechanical properties of PLA [8-12]. Some literature studies have also reported the binary blends of PLA and PBS [13-14].

In another study, Sarazin *et al.* [15] reported the blends of PLA, PCL and thermoplastic starch (TPS). The authors observed that the mechanical response of the blends exhibited a synergistic effect which was superior to their binary blends. Similarly Liao *et al.* [16] also reported the ternary blends of PLA with PCL and starch. It was reported that PLA could be transformed from rigid to ductile by the addition of PCL. The study also revealed that the price of PLA/PCL blends was reduced by the addition of starch, thus, indicating the generation of cost-effective blends with superior properties. Mittal *et al.* [17,18] also reported successful generation of binary and ternary blends of PLA, TPS and PCL using melt mixing method. Raman analysis confirmed that PLA in $PLA_{50}PCL_{50}$ blend exhibited amorphous character, whereas addition of amorphous TPS did not hinder the crystallization of PLA in $PLA_{50}TPS_{50}$ blend. Morphology of the blends was observed to exhibit physical aging phenomenon as a function of time, as shown in Figure 5.1 for PLA/PCL/TPS blend [19]. Three main factors were observed for the blend as a function of time: densification of the PCL bulk (e. g. the distance between polymer nanofibers was reduced (Figure 5.1(A), (B), (D) and (E))), deformation of the PCL ultrastructure 180 days from the production (e. g. the polycrystalline organization of PCL fibers was partially or sometimes completely lost (Figure 5.1(C) and (F)) and in contrast, the ultrastructural organization of PLA within polymer bulk did not change much within 180 days of observation [19]. In order to gain further insights into the distribution of three components in the ternary blends, application of Raman mapping was also reported [17,20]. Raman mapping/imaging, which

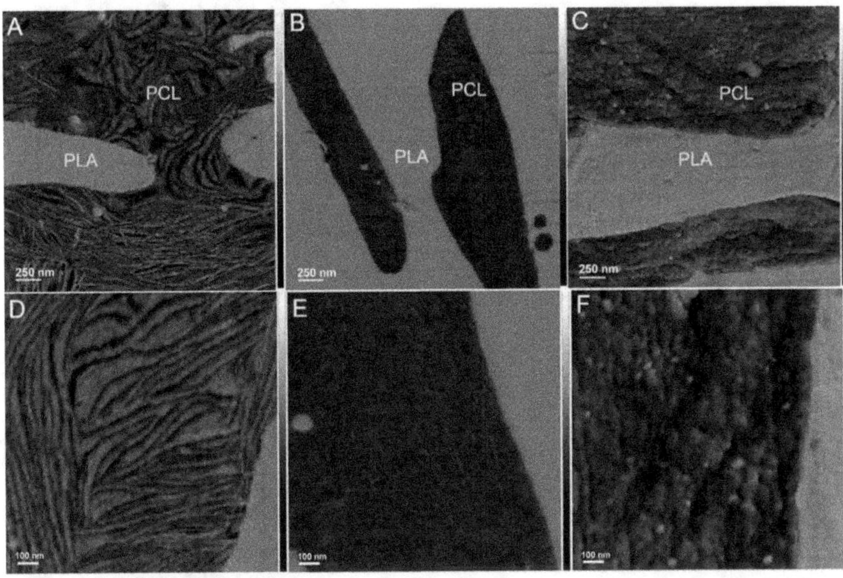

Figure 5.1 Evolution of ultrastructural organization of polymer phases as a function of time for the PLA/PCL/TPS blend. Room temperature AFM phase images of freshly prepared cross-sections of the sample blocks with the appropriate age: (A and D) 14 days after production, (B and E) 60 days after production and (C and F) 180 days after production. Reproduced from Reference 19 with permission from Royal Society of Chemistry.

is a spatially resolved technique, allows to characterize the distribution of components as well as intermolecular interactions in the multiphase systems [20-22]. For instance, Figure 5.2 demonstrates the Raman mapping performed for $PLA_{50}TPS_{30}PCL_{20}$ sample [17]. The C=O stretching bands in the regions of 1744 to 1785 and 1711 to 1732 cm^{-1} corresponding, respectively, to PLA and PCL were used for the mapping analysis and are denoted in the mapping images as violet and green colors, respectively. As TPS did not generate any significant band in its Raman spectra due to its largely amorphous nature, the vacant spaces in between violet and green domains in the mapping image could be assigned to TPS phase. Mittal *et al.* [23] further studied the ternary blends of PCL, PBAT and PBS generated using one-step melt mixing method. PBS was observed to be dispersed in PBAT or at PBAT/PCL interface. In addition, structural changes were observed in PBS in the blends.

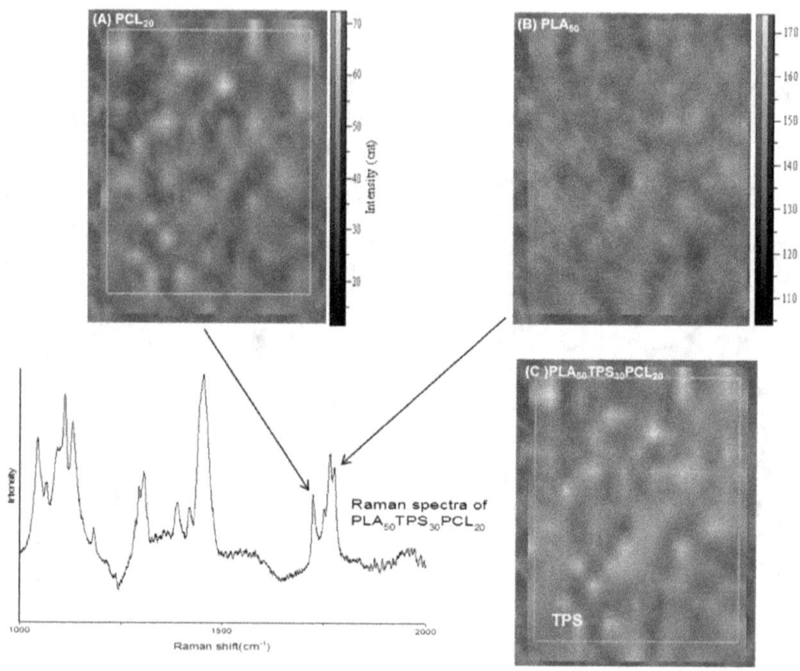

Figure 5.2 Raman mapping of $PLA_{50}TPS_{30}PCL_{20}$ ternary blend presented as superimposition on the area of the sample where mapping was conducted. (A) & (B) represent individual images of PCL and PLA components in the ternary blend, which were obtained by using the intensity of the bands indicated by the arrows and (C) represents the combined image. Reproduced from Reference 17 with permission from Springer.

In a recent study, Ravati and Favis [24] also studied the morphology of PBS, PLA, and PCL blends. The authors observed partial wetting for the ternary blend based on PBS/PLA/PCL system, as PLA droplets were observed to self-assemble at the interface of PBS and PCL. The current study further explores this useful category of blends for morphological and physical properties as a function of blend composition.

5.2 Experimental

5.2.1 Materials

PLA, procured from Biomer (Germany), had a density of 1.25 g/cm^3, glass transition temperature (T_g) of 55 °C and peak melting temperature (T_m) of 168 °C. PCL with a trade name of CAPA 6506 S was procured from Perstorp Polyols, USA. It had weight average molecular weight (M_w) of 50,000 g/mol and peak melting temperature (T_m) of 60 °C (supplier information). PBS used was Bionelle 1050 grade from Showa Highpolymer Co., Japan. The materials were used as-received, without any further treatment.

5.2.2 Preparation of Blends

Mini twin conical screw extruder (MiniLab HAAKE Rheomex CTW5, Germany) was used for the melt mixing of PCL, PLA and PBS. The screw length and screw diameter were 109.5 mm and 5/14 mm conical respectively. The composition of the blends is shown in Table 5.1. As evident from the table, the blends had varying amounts of PBS from 5-45%, PCL from 33-50% and PLA 5-45%. A mixing temperature of 185 °C, rotational speed of 70 rpm and mixing time of 8 min were used. Pure polymers were also extruded similarly for effective comparison with the blends, though different compounding temperatures were used (185 °C for PLA, 145 °C for PBS, and 90 °C for PCL). An average mass of 12 g of each sample was extruded. For the subsequent characterization, the disc, dumbbell and bar-shaped test samples were prepared by compression molding using hot press MTP-10 from Tetrahedron Associates. The pressure was kept at of 345 bar (5 psi) for 3 min at 185 °C (365 °F) using the pneumatic press. A controlled rate of cooling of the samples was used in order to avoid unwanted changes in the crystallinity of the polymers.

5.2.3 Blend Characterization

Thermal degradation behavior of the pure polymers and blends was analyzed using Netzsch thermogravimetric analyzer (TGA). Nitrogen was used as a carrier gas and the scans were obtained from 50 to 700 °C at a heating rate of 20 °C.min^{-1}. The calorimetric measurements on the polymers and blends were performed using a Netzsch DSC under nitrogen atmosphere. Compression molded samples were used for

Table 5.1 Weight fractions of PCL, PLA and PBS used to generate ternary blends

Sample	PCL content, %	PLA content, %	PBS content, %	Abbreviation
1	50	45	5	$PCL_{50}PLA_{45}PBS_5$
2	50	20	30	$PCL_{50}PLA_{20}PBS_{30}$
3	50	5	45	$PCL_{50}PLA_5PBS_{45}$
4	33	33	33	$PCL_{33}PLA_{33}PBS_{33}$

the analysis. The scans were obtained from 30 to 200 to 30 °C using heating and cooling rates of 15 °C.min^{-1} and 5 °C.min^{-1}, respectively. The samples were dried in vacuum oven overnight before carrying out the thermal analysis.

IR spectra were collected on Nicolet iS10 spectrometer equipped with SmartiTR diamond ATR accessory (angle of incidence of 45°), DTGS KBr detector and KBr beam splitter. It had a diamond ATR crystal (index of refraction = 2.4 at 1000 cm^{-1}) and a depth penetration of 2 μm at 1000 cm^{-1} for sample with refractive index of 1.5. Spectra were recorded by the OMNIC software in the 4000-525 cm^{-1} region with a resolution of 4 cm^{-1} from 16 scans. Wide-angle X-ray diffraction (WAXRD) analysis of the materials was performed using a Panalytical powder diffractometer (X'Pert PRO) using CuKα radiation (λ = 1.5406 Å) in reflection mode. The samples were step-scanned from 2Θ = 5-35° at room temperature using a step size of 2Θ = 0.02° and a step time of 10 s.

The rheological properties such as storage modulus (G′), loss modulus (G″), viscosity (η′), elasticity (η″), as well as complex viscosity (η*) were characterized using AR 2000 rheometer from TA Instruments. Disc-shaped samples of 25 mm diameter and 2 mm thickness were measured at 185 °C using a gap opening of 1.2 mm. Strain sweep scans were recorded at ω = 1 rad.s^{-1} from 0.1% to 100% strain and the samples were observed to be shear stable up to 10% strain. For the comparison, frequency sweep scans (dynamic testing) of all the polymer and blend samples were recorded at 4% strain from ω = 0.1 to 100 rad.s^{-1}.

The block faces of specimens after cryo ultramicrotomy were investigated at ambient conditions using ZEISS Axioplan light microscope (ZEISS, Germany) equipped with ZEISS Axio Cam ICc 1 CCD camera in reflected polarized light, and by Dimension 3100 AFM/SPM (Veeco, USA) atomic force microscope. AFM images were collected in tapping mode using silicon nitride cantilevers with natural frequencies in the 300 kHz range (force constant 20 N/m and tip radius 10 nm (NTMDT, Russia)). AFM image processing was performed using Nanoscope v720 software (Veeco, USA).

Universal testing machine (Testometric, UK) using standard ASTM D638 was employed for the tensile testing of pure polymers and blends. The dumbbell shaped samples with 53 mm length, 4 mm width, and 2 mm thickness were used. A loading rate of 5 mm.min^{-1} was employed and the tests were carried out at room temperature. WinTest Analysis software was used for the calculation of tensile modulus and other tensile properties. An average of five values was calculated. DMA (dynamic mechanical analyzer) RSA3 of TA Instruments was used to study the change in modulus with temperature (25-100 °C). Dynamic temperature ramp test was performed at a frequency of 5 rad.s^{-1}, ramp rate 3 °C.min^{-1} and 1% strain for all samples.

5.3 Results and Discussion

Owing to their commercial relevance, the biopolymers PCL, PLA and PBS have been blended in the current study in different proportions so as to study the effect of blend composition on the morphology and physical properties of the blends. Specifically, the content of PLA and PBS was varied in the PCL phase. In addition, another equal weight blend was also generated. Conventional one-step melt mixing process was used for the generation of the ternary blends. As the processing conditions have a significant influence on the resulting morphology and properties, same conditions were employed for all the blends. In addition, the compounding temperature and time were chosen so as to achieve uniform mixing, however, also avoiding thermal degradation of the polymers at high temperature. Also, due to above mentioned changes in the morphology as a function of time, the characterization of the blends was carried out immediately after the blend generation.

Figure 5.3 demonstrates the TGA thermograms of the pure polymers and blends. As evident from Figure 5.3a, PCL was observed to be the most stable with peak degradation temperature of 423 °C. On the other hand, the peak degradation temperatures of PBS and PLA were 400 and 378 °C. Similarly, the onset of degradation of the PCL, PBS and PLA was observed at 380, 360 and 330 °C respectively. The TGA thermograms of the blends (Figure 5.3b and c) exhibited early degradation owing to the presence of PLA phase. In fact, the onset of degradation temperature was observed to decrease proportionally with increase in the PLA content in the blends. In addition, the thermograms exhibited two degradation steps, indicating phase immiscibility in the blends. The first lower temperature degradation signal corresponded to PLA, whereas the second higher temperature degradation could be attributed to PBS and PCL. Though PBS and PCL are not known to be miscible with each other, however, a single degradation signal was observed probably because of their degradation temperature range being close to each other. The peak position of this high temperature degradation step was also observed to depend on the blend composition. Figure 5.3d also quantifies the effect of PLA or

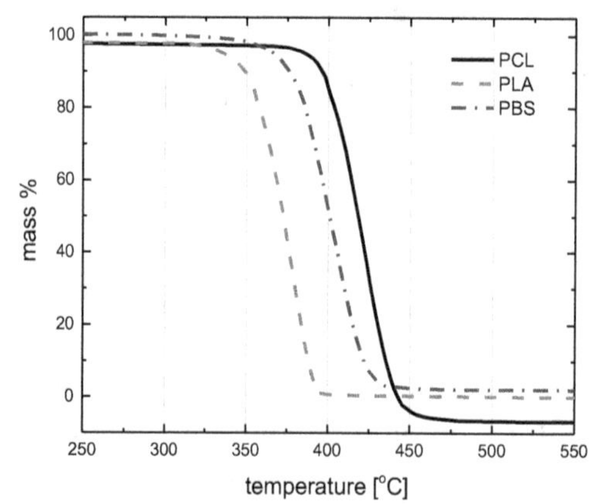

(a)

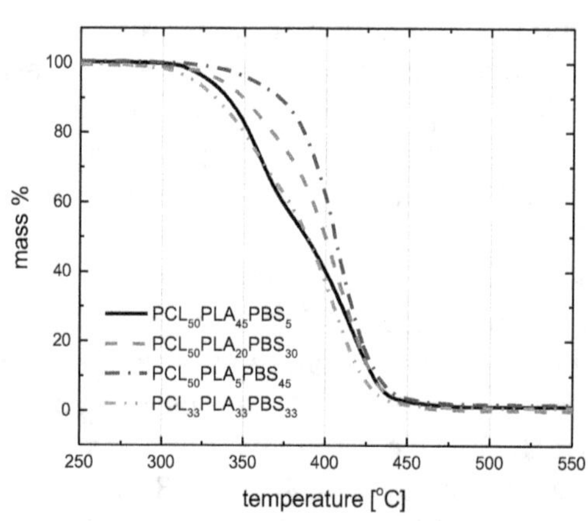

(b)

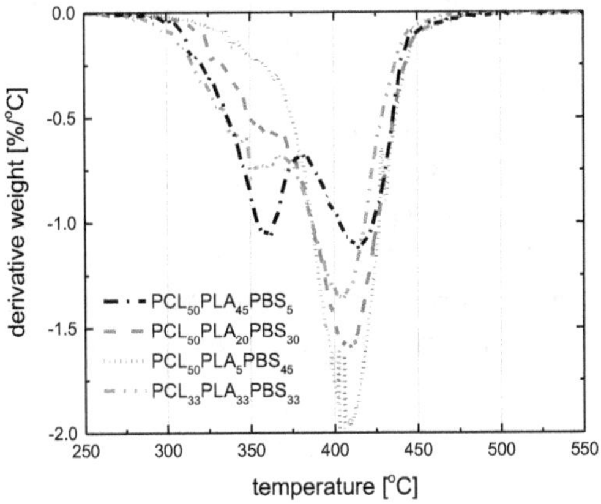

(c)

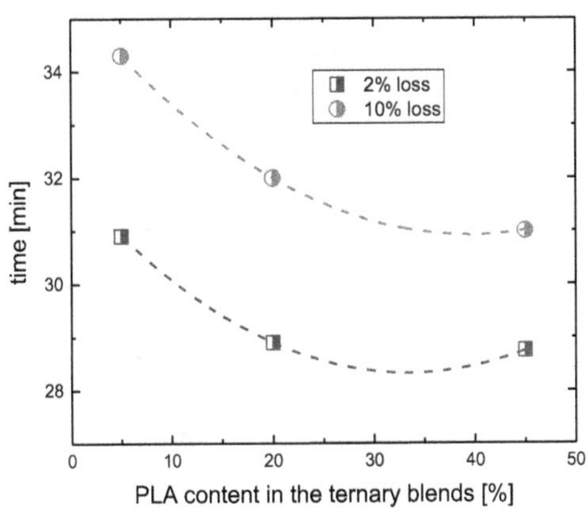

(d)

Figure 5.3 (a) Cumulative TGA thermograms of pure polymers; (b) and (c) cumulative and derivative TGA thermograms of the blends (the first degradation peak in the derivative plots corresponds to PLA, whereas the second peak corresponds to PCL and PBS), and (d) time taken by the blends to lose 2% and 10% weight as a function of PLA content.

or PBS on the thermal stability of the blends, as indicated through the time taken by the blends to lose 2% and 10% weight as a function of PLA content. Even in the case of the blend with the highest PLA content, the time taken to lose 2% weight was nearly 29 min. Thus, even though the thermal stability of the blends was negatively impacted by the presence of PLA, however, the blends remained stable for the high temperate processing. Dynamic TGA carried out at the compounding temperature of 185 °C also exhibited thermal stability of the blends as no significant weight loss was observed with time (15 min).

The calorimetric properties of the pure polymers and blends are shown in Figure 5.4 (second heating curves). As evident in Figure 5.4a, all polymers exhibited semi-crystalline nature with peak melting points of 59.1, 114.1 and 171.3 °C for PCL, PBS and PLA respectively. In the DSC thermograms of the blends in Figure 5.4b, the individual transitions of all the polymers were observed, confirming the

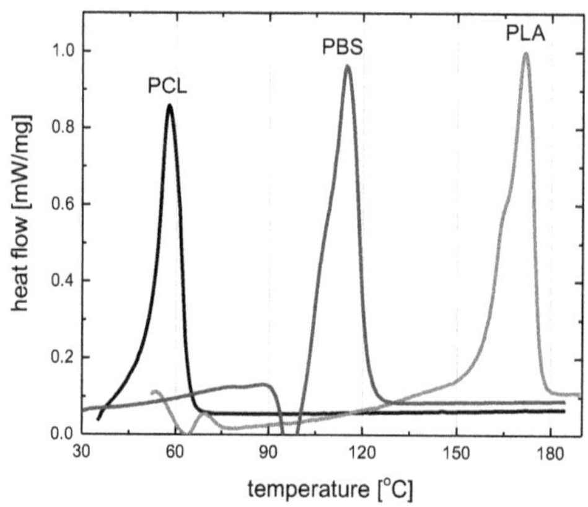

(a)

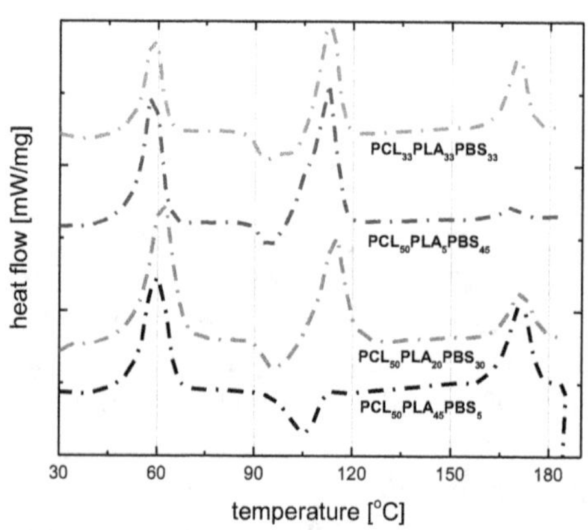

(b)

Figure 5.4 DSC thermograms of (a) pure polymers and (b) ternary blends earlier findings from TGA about phase immiscibility. The melt enthalpy of the PLA phase was reduced in the blends as compared to the pure polymer, thus, confirming the reduction in its extent of crystallinity in the blends. Overall comparison of the magnitude of the crystalline melting peaks also revealed significantly smaller peak for PLA. For the blend with 5% PLA, slight reduction in its peak melting point was also observed (168.7 °C). The smaller amount of PLA in the blend would have resulted in its constrained morphology inside the other phases, thus, further enhancing the reduction in crystallinity. On the other hand, PCL and PBS did not exhibit any significant shift in the peak melting point in the blends. In the ternary blend $PCL_{50}PLA_{45}PBS_5$, the melting transition corresponding to PBS was absent, probably due to the similar constrained morphology of PBS, as opined earlier for the blend with 5% PLA.

The IR spectra of the pure polymers and blends are demonstrated in Figure 5.5. The IR spectra of pure polymers exhibited characteristic C=O bands at 1721 cm^{-1} and 1754 cm^{-1}, along with other absorption peaks in the fingerprint region. In the blends, the PLA absorbance bands were almost absent, probably due to reduced crystallinity, as also observed in DSC. No change in the PCL absorbance bands was observed in the blends, which along with DSC analysis suggests no change in its crystallinity. On the other hand, PBS exhibited absence/shift of the absorbance peaks, indicating some changes in the crystal nature or conformations. Also, the observed behavior of PBS was consistent in all the blends. Thus, in summary, PLA could not effectively crystallize in the presence of PBA and PCL, whereas the PBS phase had some modification in the crystal structure.

Further insights into the morphology of the blends were also gained through X-ray diffraction, as shown in Figure 5.6. The diffractogram of pure PLA had two diffraction peaks at 2θ values of 16.8° and 19°. The peak recorded at 2θ 16.8° was recognized as the (200, 110) reflection of the α-form of crystals, and the other peak appearing at 19° corresponded to the (010) reflection [25]. X-ray diffractogram of pure PCL contained two peaks at 21.8° and 24.1° 2θ corresponding to (111) and (211) crystal plane indices, respectively [16]. PBS also exhibited two diffraction peaks at 20.5° and 23° 2θ. PLA did not exhibit any diffraction peak in the blends, indicating largely amorphous nature or reduced crystallinity, thus, confirming the earlier findings from DSC and IR. In addition, the blend with 5% PBS did not exhibit any crystalline diffraction peak attributed to PBS, due to hindrance to its crystallinity in the blends, which also confirmed the

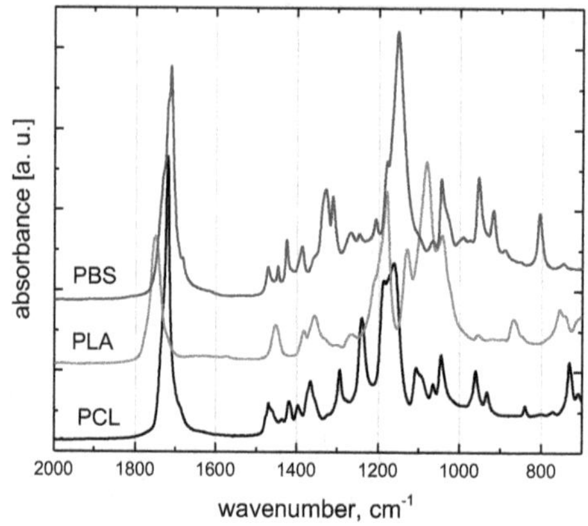

(a)

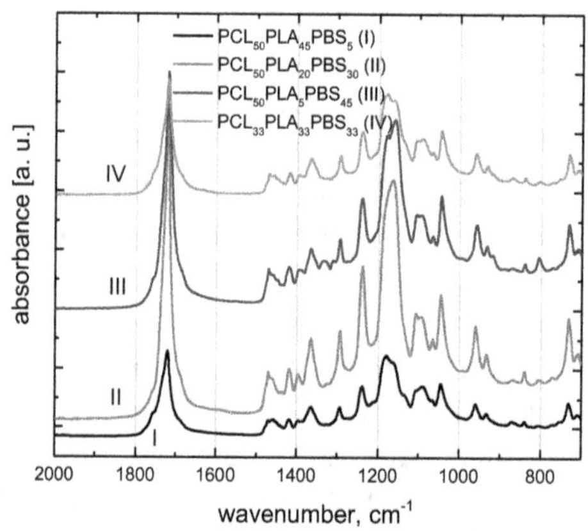

(b)

Figure 5.5 IR spectra of (a) pure polymers and (b) ternary blends.

finding from DSC. Also, in other blends, the first characteristic diffraction peak of PBS was observed as a diffuse signal, whereas the intensity of the second diffraction peak was qualitatively enhanced as a function of PBS content. These findings, also supported by IR, indicated some changes in the crystalline nature of PBS in the blends. On the other hand, PCL did not exhibit any change in the corresponding crystalline peaks in the blends, indicating no hindrance or alteration in its ability to crystallize.

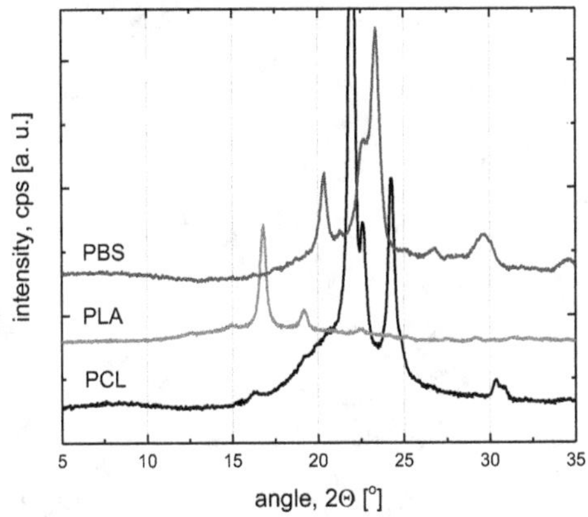

(a)

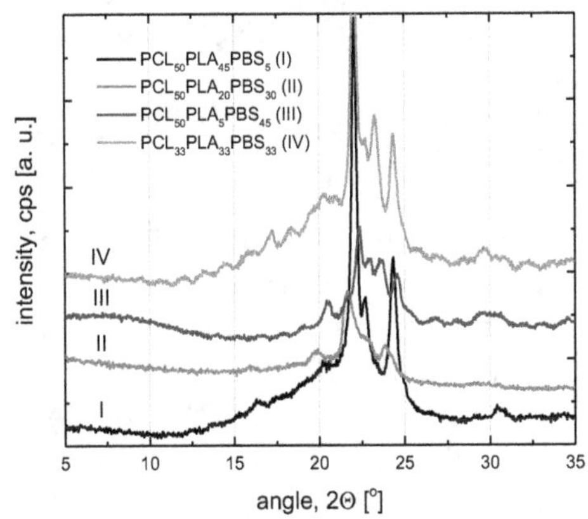

(b)

Figure 5.6 X-ray diffractograms of (a) pure polymers and (b) ternary blends.

Figures 5.7-5.10 demonstrate the AFM height and phase images of the blends $PCL_{50}PLA_{45}PBS_5$, $PCL_{50}PLA_{20}PBS_{30}$, $PCL_{50}PLA_5PBS_{45}$, and $PCL_{33}PLA_{33}PBS_{33}$. The morphological analysis was carried out at ambient conditions so as to retain the true microstructure. In the mages, the bright phase represents PLA, dark phase is PCL, whereas lighter phase is PBS. For the sample $PCL_{50}PLA_{45}PBS_5$ (Figure 5.7), 5% PBS phase was observed as spherical droplets at the interface of PLA and PCL. Such stabilization effect has also been observed earlier by Ravati

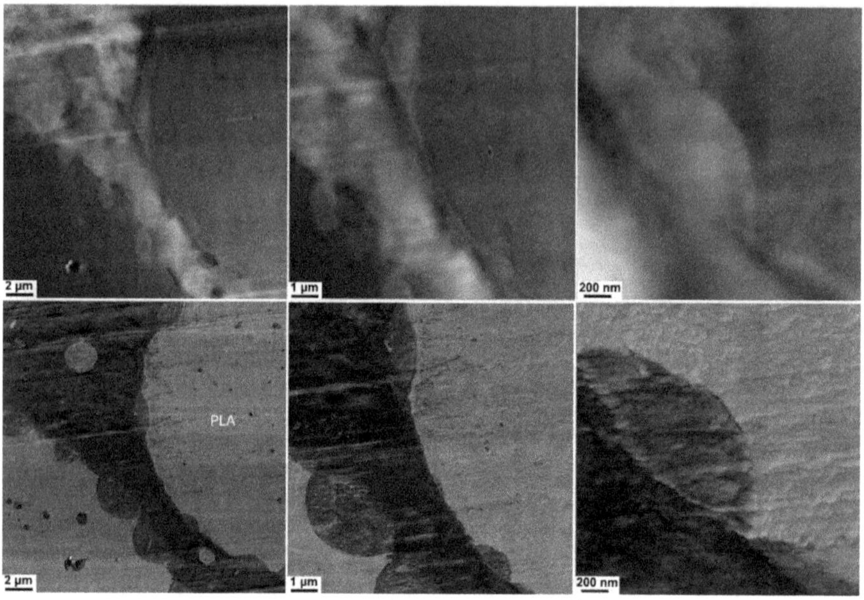

Figure 5.7 AFM height (top row) and phase (bottom row) image of the ternary blend $PCL_{50}PLA_{45}PBS_5$.

and Favis [24]. The observed morphology of PBS would have resulted in hindered crystallization, as observed earlier. The domain size of PLA was large, indicating large extent of phase immiscibility. However, small domains of PLA could be observed in PCL, probably trapped kinetically during melt mixing in the compounder. The interface between the PLA and PCL, not covered with PBS, did not exhibit significant phase separation. However, it has to be noted that the analysis was performed on a fresh sample, whereas, extensive phase separation has been reported in the PCL/PLA blends as a function of time due to contraction/expansion phenomenon exhibited by combination of polymers with above and below glass transition temperature at ambient conditions [17,19]. Decreasing the PLA content to 20% in the blend $PCL_{50}PLA_{20}PBS_{30}$ led to significant changes in the morphology as compared to $PCL_{50}PLA_{45}PBS_5$, as evident in Figure 5.8. Larger PBS domains were observed, whereas PLA domains were observed to become smaller. A fraction of PBS could still be observed at the interface of PLA and PCL. The interface between PCL and PLA as well as PBS and PCL was clearly separated with the presence of voids and cracks, whereas these were absent at the interface between PLA and PBS. As also observed in Figure 5.7, a small fraction of PLA was

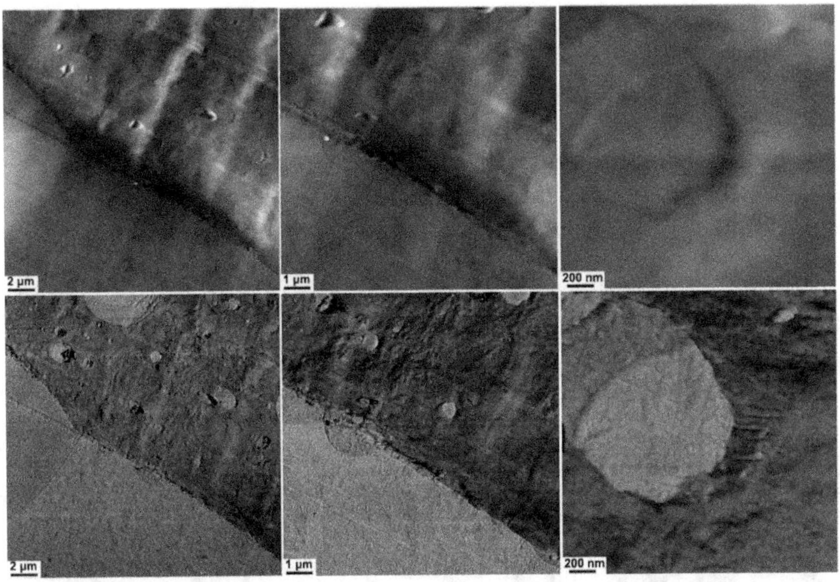

Figure 5.8 AFM height (top row) and phase (bottom row) image of the blend $PCL_{50}PLA_{20}PBS_{30}$.

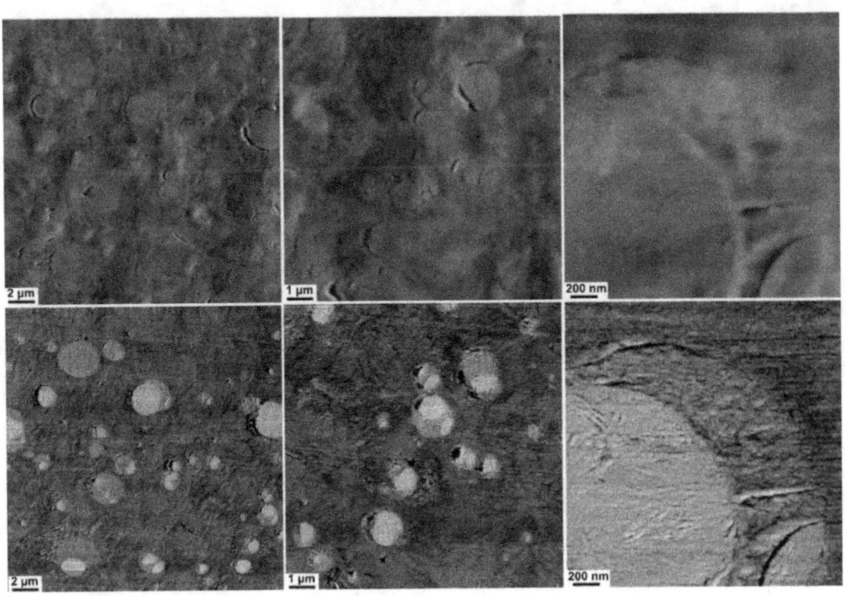

Figure 5.9 AFM height (top row) and phase (bottom row) image of the blend $PCL_{50}PLA_{5}PBS_{45}$.

observed to be scattered in the PCL phase as spherical domains. Decreasing the PLA content further in the blend $PCL_{50}PLA_{5}PBS_{45}$ exhibited opposite morphology as compared to $PCL_{50}PLA_{45}PBS_{5}$, as evident in Figure 5.9. Due to lower fraction of PLA in the blend, the PLA phase was dispersed as spherical domains and present at the interface of PCL and PBS. Though the PBS and PCL are not miscible with each other, however, the morphology was observed to be uniform with good intermixing of PBS and PCL. In addition, some of the PLA domains were also observed to be surrounded with PBS. The blend with equal fraction of three polymers had a similar morphology as the $PCL_{50}PLA_{20}PBS_{30}$ blend, as shown in Figure 5.10. The PLA and PCL phases were observed to contain large domains of PBS at the interface. As also observed earlier, the interface between PLA and PCL as well as PBS and PCL also exhibited lack of adhesion, though the interface between PBS and PLA did not show any crack or void.

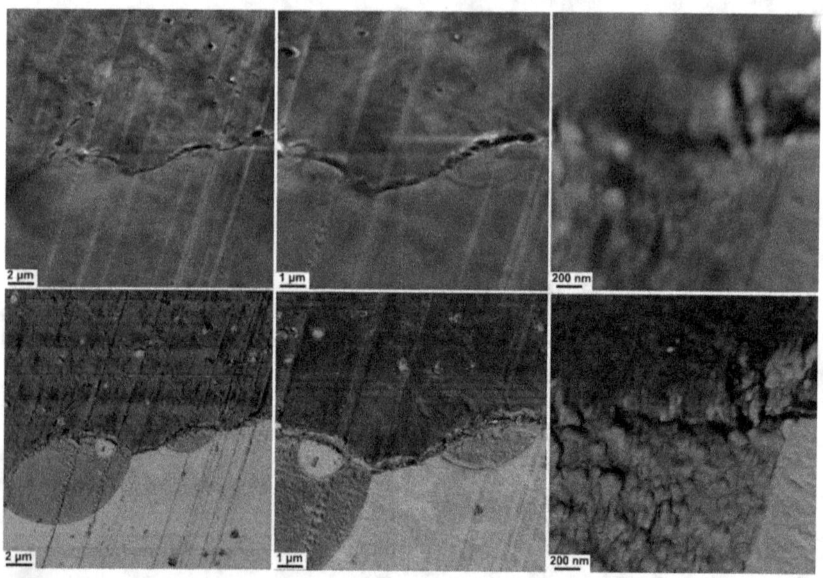

Figure 5.10 AFM height (top row) and phase (bottom row) image of the ternary blend $PCL_{33}PLA_{33}PBS_{33}$.

Figure 5.11 also shows the schematic of the specific distribution of PLA or PBS as a function of blend composition for $PCL_{50}PLA_{45}PBS_{5}$ and $PCL_{50}PLA_{5}PBS_{45}$, also termed as partial wetting behavior [24]. Such kind of component distribution at interface generally acts as a compatibilizer between the two phases. Also, at these particular blend compositions, better control on morphology was observed, which provides opportunities for tuning the behavior of the blends. Though only partial wetting of the components by dispersed PBS or PLA phases was achieved, and more efficient wetting by the formation of a continuous domain of the dispersed phase at the interface between the larger components (instead of spherical domains) would

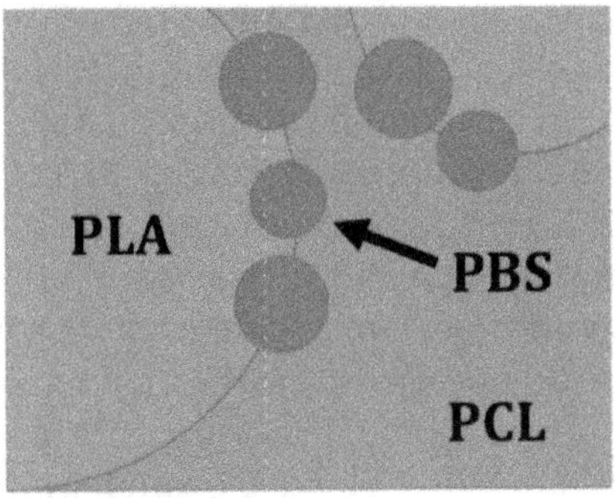

$PCL_{50}PLA_{45}PBS_5$

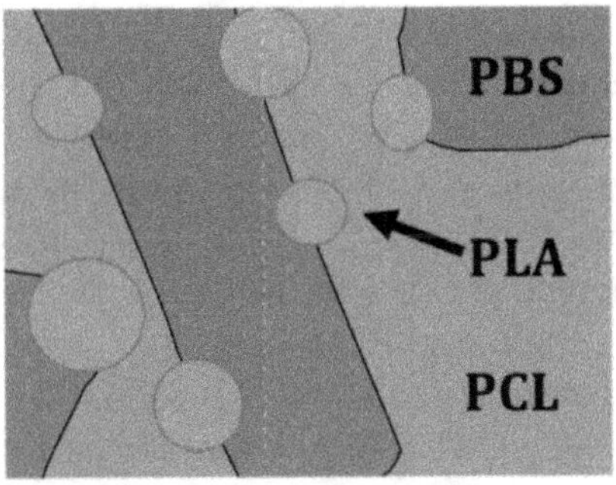

$PCL_{50}PLA_5PBS_{45}$

Figure 5.11 Schematic of the distribution of polymers in the ternary blends $PCL_{50}PLA_{45}PBS_5$ and $PCL_{50}PLA_5PBS_{45}$.

be preferred, however, partial wetting is still beneficial to generate synergy in the properties of the blends consisting of completely immiscible components. Overall, from the microscopy analysis, it could be observed that the $PCL_{50}PLA_5PBS_{45}$ blend had the most uniform morphology with absence of large domain of PLA as well as good intermixing of PCL and PBS. Though the $PCL_{50}PLA_{45}PBS_5$ blend also exhibited similar morphology as the $PCL_{50}PLA_5PBS_{45}$ blend, however, large domains of PLA did not lead to effective intermixing. To study the effect of time on the morphology of the blends, the AFM analysis on the $PCL_{50}PLA_{45}PBS_5$ and $PCL_{50}PLA_5PBS_{45}$ blends was also carried out after 120 days of blend generation. The interface between PLA and PCL in $PCL_{50}PLA_{45}PBS_5$ was exhibited separation, and significant changes in the microstructure of PCL were observed, as reported earlier [17,19]. No significant modification in the microstructure of the PLA and PBS phases was observed, probably due to the polymers below their glass transition temperature at ambient conditions. The morphology of the $PCL_{50}PLA_5PBS_{45}$ blend did not exhibit enhanced phase separation with time. Some changes in the microstructure of PLA were observed, however, overall, the blend morphology was not impacted significantly with time. This also confirmed the relative stability of the $PCL_{50}PLA_5PBS_{45}$ blend morphology towards dynamic changes.

Further insights into the morphology of the blends were also gained through the light microscopy analysis of the block face surface, as shown in Figure 5.12. The light microscopy analysis confirmed the earlier findings from AFM. In the blend $PCL_{50}PLA_{45}PBS_5$, large PLA domains were observed, whereas PCL and PBS phases had good intermixing. Increasing of the PBS content (and reduction of PLA content) in the blend $PCL_{50}PLA_{20}PBS_{30}$ led to reduction in the size of the PLA domains and more uniform distribution. Large PBS domains were also observed at the PLA-PCL interface. Further reducing the PLA content in the blend $PCL_{50}PLA_5PBS_{45}$ resulted in a uniform morphology, as also observed earlier in AFM. The phases were observed to be well mixed with PLA dispersed in the form of smaller spherical domains. Similar to the $PCL_{50}PLA_{20}PBS_{30}$ blend, $PCL_{33}PLA_{33}PBS_{33}$ also exhibited three distinct phases with PBS present as large domains at the interface of PCL and PLA, thus, supporting the earlier microscopy observations. Thus, the optical microscopy also confirmed that the blend $PCL_{50}PLA_5PBS_{45}$ had the optimal morphology and good phase intermixing. It has to be noted that even though good intermixing among the phases was observed, the phases were still molecularly immiscible, as observed in DSC.

Figure 5.13 also demonstrates the phase miscibility analysis of the blends from rheology, as a function of PLA/PBS content in the blends (while keeping the PCL content constant). The methods like Cole-Cole

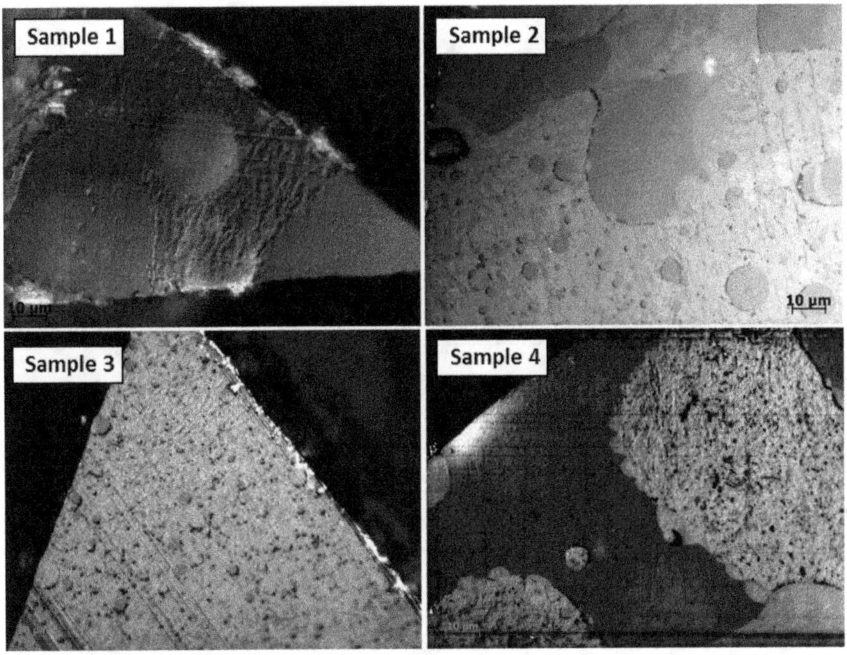

Figure 5.12 Optical images of the blends. Sample 1: $PCL_{50}PLA_{45}PBS_5$, Sample 2: $PCL_{50}PLA_{20}PBS_{30}$, Sample 3: $PCL_{50}PLA_5PBS_{45}$, and Sample 4: $PCL_{33}PLA_{33}PBS_{33}$.

and van Gurp (as well as Han-Chuang) have been commonly used to gain insights into the phase miscibility of the polymer blends. Cole-Cole analysis develops relationships between real (η') and imaginary (η") parts of complex viscosity [26-28]. A smooth, semi-circular shape of the graph would suggest miscible system with homogenous phase. The deviation from this behavior indicates phase segregation due to immiscibility of the components. The van Gurp plots represent the relationship between complex modulus (G*) and phase angle delta (δ) [29,30]. In Han and Chuang analysis, G' is plotted vs. G", and such analysis generates composition independent correlation for compatible systems, whereas the correlation is composition dependent for incompatible cases [31-32]. As seen in Figure 5.13, the Cole-Cole analysis of the blends with varying amount of PLA (and PBS) did not follow semi-circular path suggesting phase immiscibility in the blends, which was also observed earlier. However, it is necessary to further support the findings from Cole-Cole plots, as these findings could sometimes be misleading [33]. The van Gurp analysis also confirmed the findings from the Cole-Cole analysis. In all the blends, the

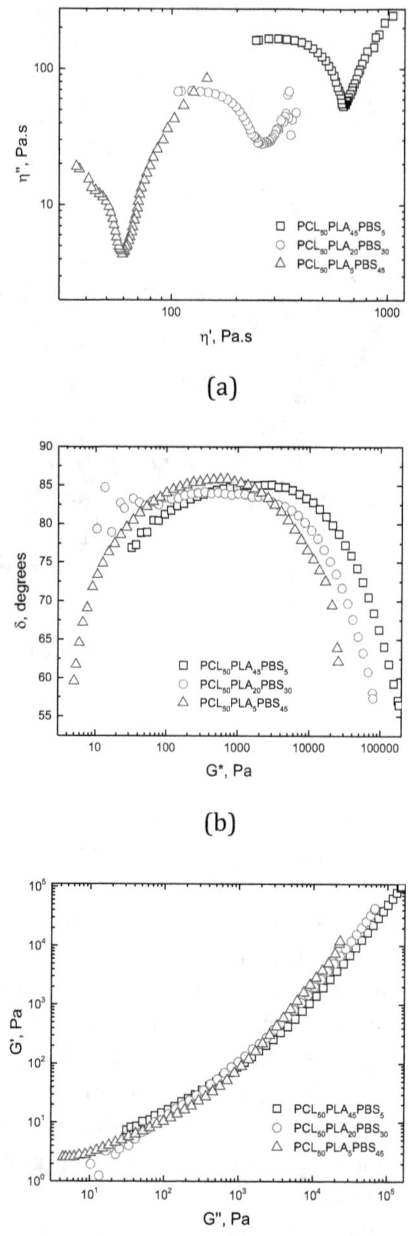

Figure 5.13 (a) Cole-Cole, (b) van Gurp, and (c) Han Chuang analysis of the blends.

time-temperature superposition principle did not hold (indicated by non-merging of the curves into a common curve). The deviation was observed for the whole range of G*, thus, confirming immiscibility among the phases. Similar to other analyses, concentration dependent correlation was also observed for the blends in the Han Chuang analysis.

Tensile properties of the blends are demonstrated in Table 5.2, in comparison with the pure polymers. Figure 5.14 also shows the effect of PLA content in the blends on the mechanical properties. PLA exhibited highest tensile modulus of 1926 MPa, whereas moderate modulus values of 500 and 392 MPa were observed for PBS and PCL respectively. PCL had significantly high elongation of 813%, in comparison to 5% and 3% for PLA and PBS respectively. As observed in Figure 5.14a, the tensile modulus of the blends increased as a function of PLA content. For instance, the blend with 5% PLA exhibited the modulus of 471 MPa, which was enhanced to 921 MPa for the blend with 45% PLA. Thus, depending on the blend composition, the blends exhibited moderate to high modulus values. The peak stress of the blends was also observed to increase with the PLA content (Figure 5.14a), though the values were smaller than any of the polymers. For instance, the blend with 5% PLA exhibited a peak stress of 9.6 MPa, which enhanced to 19.6 MPa at 45% PLA content. The strain at break of the blends was observed to be significantly low, which was similar to PBS and PLA. Thus, it was evident that good intermixing of PCL with PBS led to significant reduction in its elongation. The strain at break was also observed to decrease with the PLA content. Similar to the peak stress, the stress at break in the blends also increased with the PLA content. Thus, the mechanical properties of the blends

Table 5.2 Tensile properties of the materials

Polymer/blend	Tensile modulus, MPa	Peak stress, MPa	Peak strain, %	Stress at break, MPa	Strain at break, %
PCL	392	38.5	652	35.5	813
PLA	1926	57.7	3.4	50.8	5
PBS	500	37.0	3.0	13.4	3
$PCL_{50}PLA_{45}PBS_5$	921	19.6	3.9	18.6	5.1
$PCL_{50}PLA_{20}PBS_{30}$	592	12.0	4.3	11.7	6.3
$PCL_{50}PLA_5PBS_{45}$	471	9.6	5.1	7.9	7.7
$PCL_{33}PLA_{33}PBS_{33}$	852	14.1	4.1	13.5	5.9

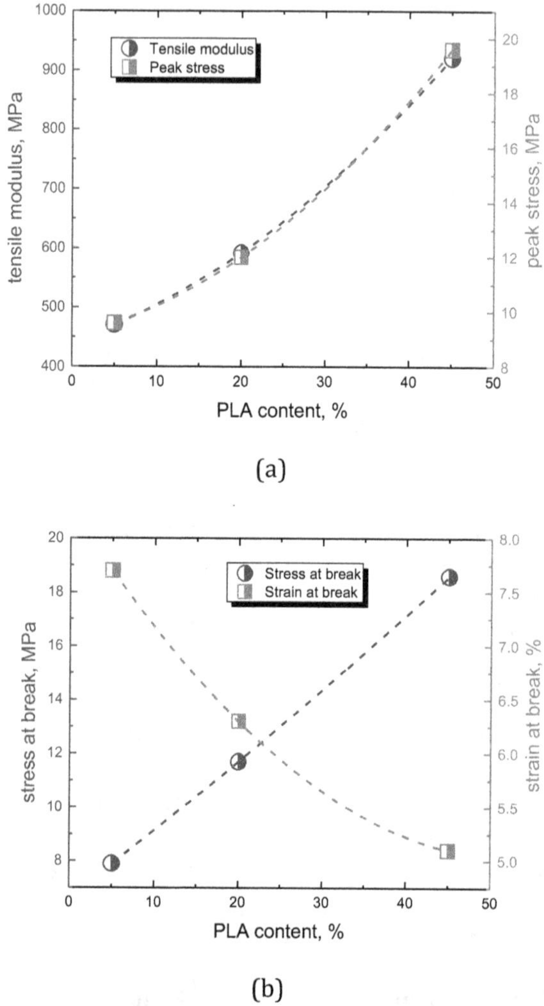

Figure 5.14 (a) Tensile modulus and peak stress, and (b) stress at break and strain at break of the blends as a function of increasing PLA content (or decreasing PBS content).

exhibited effective combination of the component properties, however, enhanced elongation would have increased the potential of their commercial application further. Thus, further effort is required to explore possibilities to enhance the elongation, while retaining the modulus.

Dynamic mechanical analysis of the blends as a function of temperature is demonstrated in Figure 5.15. As evident in Figure 5.15a, the storage modulus exhibited similar behavior as the tensile modulus. Till 50 °C, the storage modulus of the blends increased with increasing PLA content in the blends. There was a declination in the storage modulus afterwards owing to the melting of the PCL component. The drop was significant in the case of $PCL_{50}PLA_{45}PBS_5$ blend. As the PLA phase was separately present in the blend and as no intermixing of PCL with PBS could be achieved due to lower fraction of PBS, the melting of PCL content led to sharp reduction in the modulus. In other cases, owing to the good intermixing between the PCL and PBS phases, the reduction in modulus was moderate. Especially, the minimum reduction was observed for the $PCL_{50}PLA_5PBS_{45}$ blend, where the morphology analysis had indicated most optimal phase intermixing between PBS and PCL. The storage modulus of the blends was also observed to increase in the temperature range of 65-85 °C (except $PCL_{50}PLA_5PBS_{45}$). Such a phenomenon is attributed to PLA and has been reported due to the crystallization of the PLA phase occurring during the DMA measurements [34]. In addition, the extent of this observed increase was a function of the amount of PLA in the blends. For this reason, the $PCL_{50}PLA_5PBS_{45}$ blend with the lowest amount of PLA did not exhibit any increase in the storage modulus. Overall, the blends exhibited high modulus values even at 90 °C, especially for $PCL_{50}PLA_{20}PBS_{30}$, $PCL_{50}PLA_5PBS_{45}$, and $PCL_{33}PLA_{33}PBS_{33}$. The loss modulus of the blends exhibited similar behavior as the storage modulus, however, its magnitude was smaller than the storage modulus over the complete temperature range, thus, indicating dominance of elastic character in the blends. Figure 5.15c also demonstrates the transition temperature of PLA in the blends. As compared to 55 °C for the pure PLA, the transition in the blends was observed to occur at higher temperatures as the amount of PLA was enhanced. For instance, the blend with 45% PLA exhibited the transition at 60 °C, in comparison with 56 °C for the blends with 20% and 33% PLA. On the other hand, the blend with 5% PLA did not clearly exhibit the thermal transition, owing to its constrained morphology. These findings indicated the enhancement of thermal performance of PLA in the presence of PCL and PBS. It should also be noted that the though the high temperature stability was reduced on adding PLA, but thermal performance in the temperature range around 25-100 °C was, in fact, improved with the incorporation of PLA, which represents an advantage for the commercial applications in this range.

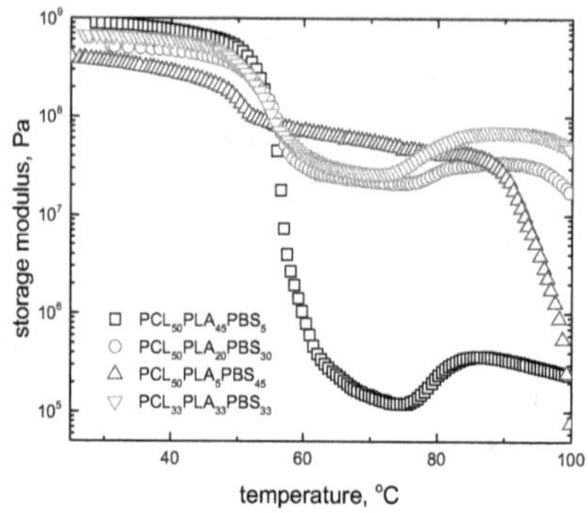

(a)

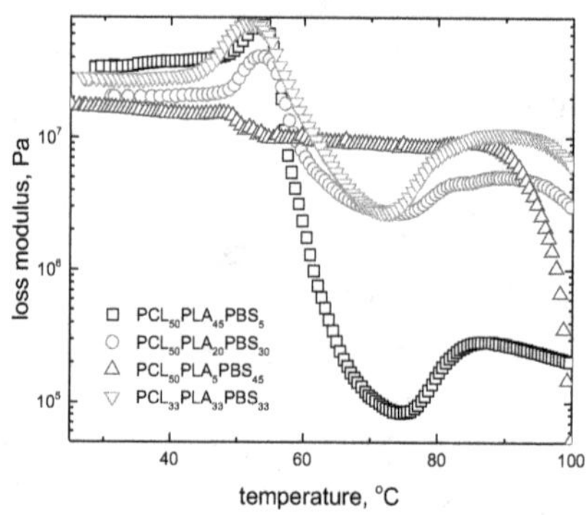

(b)

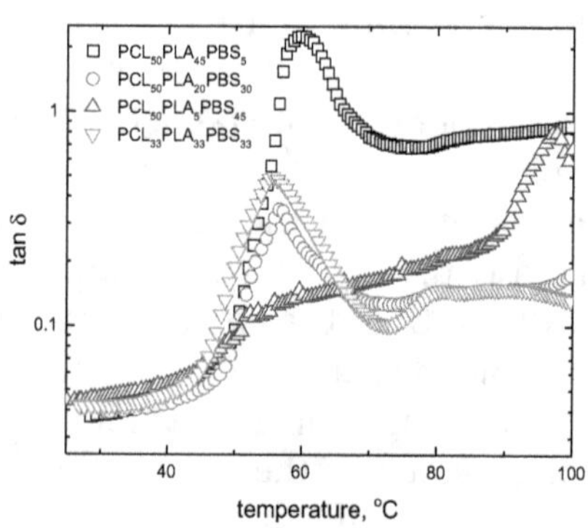

(c)

Figure 5.15 (a) Storage and (b) loss modulus of the blends as a function of temperature; (c) tan δ curves of the blends as a function of temperature.

The rheological performance of the blends as a function of PLA/PBS content (while keeping the PCL content constant at 50%) is also shown in Figure 5.16. As also observed for storage modulus in DMA, the modulus was dependent on PLA/PBS content in the blends. The blend with 45% PLA exhibited highest storage modulus over the whole angular frequency range. Nearly an order of magnitude reduction was observed in the blend with 5% PLA content. The loss modulus of the blends also exhibited similar behavior, however, it was observed to be higher in magnitude than the storage modulus at shear

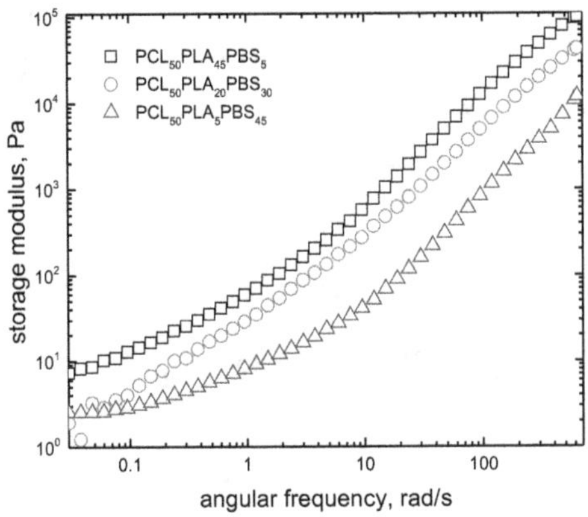

(a)

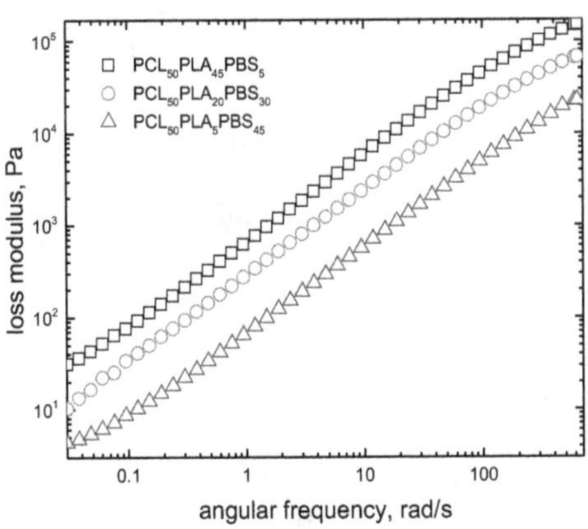

(b)

Figure 5.16 (a) Storage and (b) loss modulus of the ternary blends as a function of angular frequency.

rheology conditions. The higher magnitude of loss modulus was retained over the whole angular frequency range, indicating dominance of viscous nature. As also observed earlier, though all the polymer components were present as melt at the shear rheology conditions, however, the non-uniform nature due to phase immiscibility and different relaxation of the component polymers resulted in different modulus values in the blends [23].

Figure 5.17 also demonstrates the complex viscosity of the blends as a function of angular frequency. Similar to the modulus, it was evident that the viscosity of the blends was affected by the composition. The blend with 45% fraction of PLA had the highest viscosity, followed by its decrease as the PLA content was decreased (or PBS content was increased). Even though the viscosity was modified in the blends, the blends could still be processed using the same processing conditions.

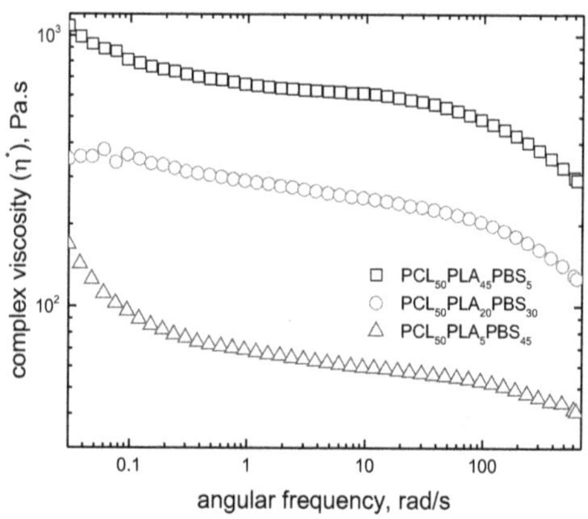

Figure 5.17 Complex viscosity of the blends as a function of angular frequency.

5.4 Conclusions

In the current study, ternary blends of PLA, PBS and PCL were generated using one-step melt mixing method. The blend composition was varied in order to gain more insights into the dependence of morphology and physical properties on composition. The incorporation of PLA led to slight reduction in the thermal stability of the blends, however, the blends were still stable for high temperature processing.

The DSC analysis also revealed phase immiscibility as well as reduced extent of PLA crystallinity in the blends. These findings were also supported by the IR and X-ray analyses. Changes in the crystal nature/structure of PBS were also observed in the blends, irrespective of the blend composition. The morphological analysis revealed the distribution of PBS at the interface of PCL and PLA as spherical domains, when the PBS content was 5%. Similar morphology was also observed in the case of the blend with 5% PLA content, where PLA was dispersed as spherical domains at the interface of PCL and PBS. Other blends exhibited larger domains of all three polymers. The blend with 5% PLA exhibited optimal morphology with good intermixing between PCL and PBS phases, which was also retained with time. Optical microscopy analysis of the blends also supported the AFM findings. In addition, the phase miscibility analysis using rheological insights also revealed the absence of any phase mixing at molecular level, even though some blends exhibited micro-scale intermixing among the components.

The tensile modulus was observed to be modified corresponding to the blend composition. The tensile modulus was 921 MPa for the blend with 45% PLA, which was reduced to 471 MPa for the blend with 5% PLA. The blends exhibited reduced elongation at break, which was similar to pure PBS or PLA, thus, indicating the requirement of evolving further strategies to enhance the flexibility, while retaining the modulus. The PLA also enhanced the thermal performance of the blends in the 25-90 °C temperature range, as the modulus was observed to enhance as a function of PLA content in the blends. In addition, the thermal transition temperature of PLA was also enhanced with its content in the blends. The blend with 5% PLA content had a diffuse transition signal, while the blend with 45% PLA exhibited the transition at 60 °C, in comparison with 55 °C for the pure polymer. The storage modulus as a function of angular frequency was also enhanced as a function of PLA content. The blends also exhibited dominance of viscous nature at the shear rheology conditions. The viscosity of the blends had similar enhancement with PLA fraction, though the blends could still be processed using same processing conditions. The generated ternary blends, especially $PCL_{50}PLA_{5}PBS_{45}$ represent eco-friendly materials with optimal morphology and mechanical properties which can be useful for a number of commercial applications. In addition, the obtained morphology also reveals opportunities for tuning of the component distribution, thus, optimizing the blend properties.

References

1. Patwary, F., Matsko, N., and Mittal, V. (2016) Biodegradation properties of melt processed PBS/chitosan bio-nanocomposites with silica, silicate and thermally reduced graphene. *Polymer Composites*, doi: 10.1002/pc.23947.
2. Plastic Bags (2009). *Australian Government Web Archive*, Department of Sustainability, Environment, Water, Population and Communities, Australia. Online: http://webarchive.nla.gov.au/gov/20130904224818/http://www.environment.gov.au/settlements/waste/plastic-bags/index.html [Assessed 19th April 2017].
3. Global Biodegradable Polymers Market to Amount to 2.5 Billion Pounds in 2016 (2011). *Market Research Reports*, BCC Research. Online: http://www.bccresearch.com/pressroom/report/code/PLS025D [Assessed 21st April 2017].
4. Bordes, P., Pollet, E., and Averous, L. (2009) Nano-biocomposites: Biodegradable polyester/nanoclay systems. *Progress in Polymer Science*, **34**, 125-155.
5. Yuan, H., Liu, Z., and Ren, J. (2009) Preparation, characterization, and foaming behavior of poly(lactic acid)/poly(butylene adipate-co-butylene terephthalate) blend. *Polymer Engineering and Science*, **49**(5), 1004-1012.
6. Jiang, L., and Zhang, J. (2016) Biodegradable and biobased polymers. In: *Applied Plastics Engineering Handbook*, Kutz, M. (ed.), 2nd edition, William Andrew Publishing, USA, pp. 127-143.
7. Pawar, S. P., Misra, A., Bose, S., Chatterjee, K., and Mittal, V. (2015) Enzymatically degradable and flexible bio-nanocomposites derived from PHBV and PBAT blend: Assessing thermal, morphological, mechanical, and biodegradation properties. *Colloid and Polymer Science*, **293**, 2921-2930.
8. Xu, Y. X., and Hanna, M. A. (2005) Preparation and properties of biodegradable foams from starch acetate and poly (tetramethylene adipate-co-terephthalate). *Carbohydrate Polymers*, **59**, 521-529.
9. Quero, E., Muller, A. J., Signori, F., Coltelli, M. B., and Bronco, S. (2012) Isothermal cold-crystallization of PLA/PBAT blends with and without the addition of acetyl tributyl citrate. *Macromolecular Chemistry and Physics*, **213**, 36-48.
10. Bhatia, A., Gupta, R. K., Bhattacharya, S. N., and Choi, H. J. (2007) An investigation of melt rheology and thermal stability of poly(lactic acid)/ poly(butylene succinate) nanocomposites. *Journal of Applied Polymer Science*, **114**(5), 2837-2847.
11. Dell'Erba, R., Groeninckx, G., Maglio, G., Malinconico, M., and Migliozzi, A. (2001) Immiscible polymer blends of semicrystalline

biocompatible components: thermal properties and phase morphology analysis of PLLA/PCL blends. *Polymer*, **42**, 7831-7840.

12. Maglio, G., Malinconico, M., Migliozzi, A., and Groeninckx, G. (2004) Immiscible poly(L-lactide)/poly(ε-caprolactone) blends: Influence of the addition of a poly(L-lactide)-poly(oxyethylene) block copolymer on thermal behavior and morphology. *Macromolecular Chemistry and Physics*, **205**(7), 946-950.
13. Park, J. W., and Im, S. S. (2002) Phase behavior and morphology in blends of poly(L-lactic acid) and poly(butylene succinate). *Journal of Applied Polymer Science*, **86**(3), 647-655.
14. Shibata, M., Inoue, Y., and Miyoshi, M. (2006) Mechanical properties, morphology, and crystallization behavior of blends of poly(L-lactide) with poly(butylene succinate-co-L-lactate) and poly(butylene succinate). *Polymer*, **47**(10), 3557-3364.
15. Sarazin, P., Li, G., Orts, W. J., and Favis, B. D. (2008) Binary and ternary blends of polylactide, polycaprolactone and thermoplastic starch. *Polymer*, **49**, 599-609.
16. Liao, H.-T., and Wu, C.-S. (2009) Preparation and characterization of ternary blends composed of polylactide, poly(ε-caprolactone) and starch. *Materials Science and Engineering A*, **515**, 207-214.
17. Mittal, V., Akhtar, T., Luckachan, G., and Matsko, N. (2015) PLA, TPS and PCL binary and ternary blends: structural characterization and time-dependent morphological changes. *Colloid and Polymer Science*, **293**(2), 573-585.
18. Mittal, V., Akhtar, T., and Matsko, N. (2015) Mechanical, thermal, rheological and morphological properties of binary and ternary blends of PLA, TPS and PCL. *Macromolecular Materials and Engineering*, **300**, 423-435.
19. Matsko, N. B., and Mittal, V. (2014) Thermotropic dynamic processes in multiphase polymer systems by (cryo-)AFM. *Soft Matter*, **10**, 5478-5488.
20. Luckachan, G. E., and Mittal, V. (2015) Evaluation of crystallinity variation and phase dispersion in polymer blends and nanocomposites by Raman mapping. *Journal of Polymer Research*, **22**:237, doi: 10.1007/s10965-015-0878-0.
21. Adar, F., Lee, E., and Whitley, A. (2009) Molecular imaging by confocal Raman mapping: Enabling technologies for speed, multivariate analysis, and convenience. *SPIE Proceedings Series*, **7319**, doi: 10.1117/12.817728.
22. Pastorczak, M., Wiatrowski, M., Kozanecki, M., Lodzinski, M., and Ulanski, J. (2005) Confocal Raman microscopy in 3-dimensional shape and composition determination of heterogeneous systems. *Journal of Molecular Structure*, **744**, 997-1003.
23. Mittal, V., Manna, Y., and Matsko, N. B. (2018) Ternary blends of PCL, PBAT and PBS: Morphology and properties. In: *Biopolymers Based*

Advanced Materials, Mittal, V. (ed.), Central West Publishing, Australia, pp. 87-114.

24. Ravati, S., and Favis, B. D. (2013) Tunable morphologies for ternary blends with poly(butylene succinate): Partial and complete wetting phenomena. *Polymer*, **54**(13), 3271-3281.
25. Wan, Y., Wu, H., Yu, A., and Wen, D. (2006) Biodegradable polylactide/chitosan blend membranes. *Biomacromolecules*, **7**, 1362-1372.
26. Cho, K., Lee, B. H., Hwang, K. M., Lee, H., and Choe, S. (1998) Rheological and mechanical properties in polyethylene blends. *Polymer Engineering and Science*, **38**(12), 1969-1975.
27. Kim, H. K., Rana, D., Kwag, H., and Choe, S. (2001) Melt rheology of ethylene 1-octene copoly- mer blends synthesized by Ziegler-Natta and metallocene catalysts. *Korea Polymer Journal*, **8**, 34-43.
28. Kwag, H., Rana, D., Choe, K., Rhee, J., Woo, T., Lee, B. H., and Choe, S. (2000) Binary blends of metallocene polyethylene with conventional polyolefins: Rheological and morphological properties. *Polymer Engineering and Science*, **40**(7), 1672-1681.
29. Joshi, M., Butola, B. S., Simon, G., and Kukaleva, N. (2006) Rheological and viscoelastic behav- ior of HDPE/octamethyl-POSS nanocomposites. *Macromolecules*, **39**(5), 1839-1849.
30. van Gurp, M., and Palmen, J. (1998) Time-temperature superposition for polymeric blends. *Rheology Bulletin*, **67**(1), 5-8.
31. Chuang, H. K., and Han, C. D. (1984) Rheological behavior of polymer blends. *Journal of Ap- plied Polymer Science*, **29**(6), 2205-2229.
32. Han, C. D., and Chuang, H. K. (1985) Criteria for rheological compatibility of polymer blends. *Journal of Applied Polymer Science*, **30**(11), 4431-4454.
33. Utracki, L. A. (1989) *Rheology of Polymer Alloys and Blends*, Hanser Publishers, Germany.
34. Martin, O., and Averous, L. (2001) Poly(lactic acid): Plasticization and properties of biodegradable multiphase systems. *Polymer*, **42**(14), 6209-6219.

6

Structural Modification of PLA

6.1 Introduction

Biopolymers are classified based on the biosynthesis or chemical synthesis processes as well as the origin of the monomers [1]. One such category corresponds to polymers where both monomers and polymers are obtained by chemical synthesis from fossil resources, however, the obtained polymer is biodegradable. Prominent members of this category are poly(butylene adipate-co-terephthalate) (PBAT) polymers, or aromatic copolyesters as well as biodegradable polyamides (PA) [2,3]. Another category where the monomers are bio-derived and follow the chemical synthesis path to generate biodegradable polymers includes polymers like poly-L-lactide (PLA). PLA is commercially important bio-polyester derived from corn starch or other crop sources [4,5]. In comparison with the common petroleum-based polymers, PLA exhibits superior mechanical properties (phase uniformity shown in Figure 6.1) along with its functionality of biodegradability. However, it also suffers from low flexibility and low resistance to heat and humidity, which need to be improved for enhancing the spectrum of its applications.

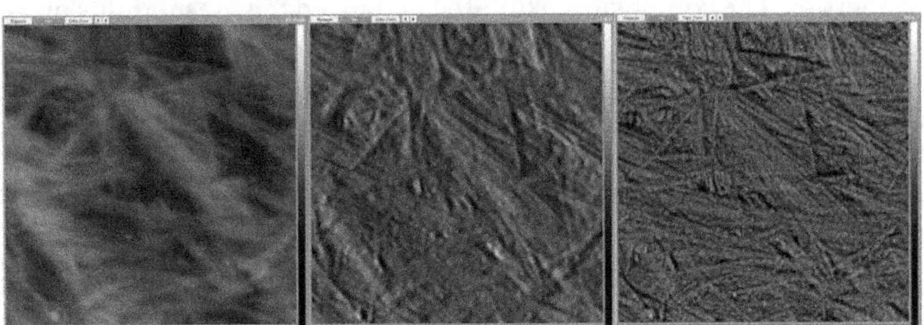

Figure 6.1 Height, amplitude and phase images of PLA depicting its structural uniformity.

Vikas Mittal and Gisha Luckachan, The Petroleum Institute (part of Khalifa University of Science and Technology), Abu Dhabi, UAE*
**Current address: Bletchington, Wellington County, Australia*

To overcome the inherent issues of PLA, different strategies have been reported in the literature. One of the common strategies is the incorporation of fillers. For instance, in a recent study, Mittal *et al.* [6] reported the nanocomposites of PLA with varying amounts of thermally reduced graphene, generated by melt mixing. The findings confirmed the use of graphene as a functional filler to enhance the properties of biopolymers and also indicated the role of filler-matrix interactions in achieving filler dispersion and superior nanocomposite properties. Use of conventional fillers have also been reported to decrease the rate of biodegradation of the biopolymers [7], thus, negating the positive effect of filler on other properties. In this respect, various PLA composites involving biofillers have also been reported. For instance, Chun *et al.* [8] reported the PLA eco-composites with coconut shell powder. The composites with untreated coconut shell powder were observed to have decreased tensile strength and elongation at break. Chun and Husseinsyah [9] also reported PLA composites with corn cob as biofiller. The addition of corn cob resulted in decreased tensile strength and elongation at break. The authors reported that coconut oil coupling agent was needed to improve the mechanical performance of the composites. Wang *et al.* [10] reported the PLA composites with wood flour and the composites had a decrease in tensile strength from 55 to 27 MPa at 30% filler content. Mittal *et al.* [11,12] also reported biodegradable PLA composites with varying amount of date seed powder (DSP) with an aim to utilize the date seed waste as well as to achieve composites with improved properties. The composites exhibited enhanced tensile modulus marginally till 20% filler content, however, it was still significant due to the very high modulus of pure PLA. The storage and loss modulus of the composites also enhanced with filler fraction, though aggregates were observed at higher filler fractions. Figure 6.2 depicts the storage and loss modulus as well as biodegradation performance of PLA composites as a function of filler content. Thus, generation of composites leads to overall enhancement in PLA properties, however, the elongation of the polymer generally remains low.

Another very common approach to improve the properties of PLA is blending with other biopolymers. For instance, blends of PLA with PBAT have been reported, which combined the toughness and flexibility characteristics of PBAT with the mechanical properties of PLA [13-15]. The resulting blends were observed to have ductile failure and reduced brittleness depending on the amount of PBAT. Similarly, blends of PLA with PCL have also been reported to complement the

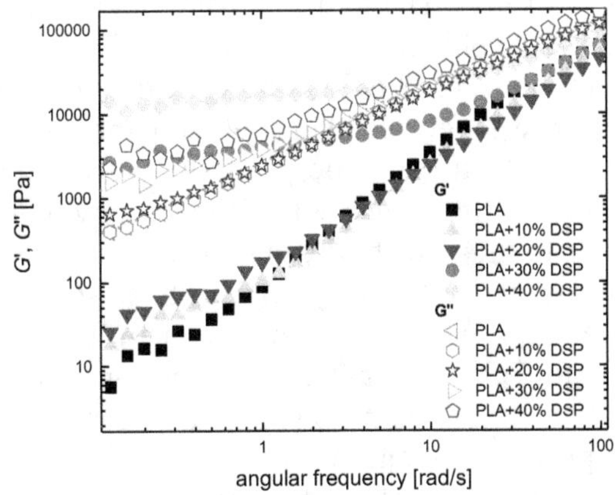

(a)

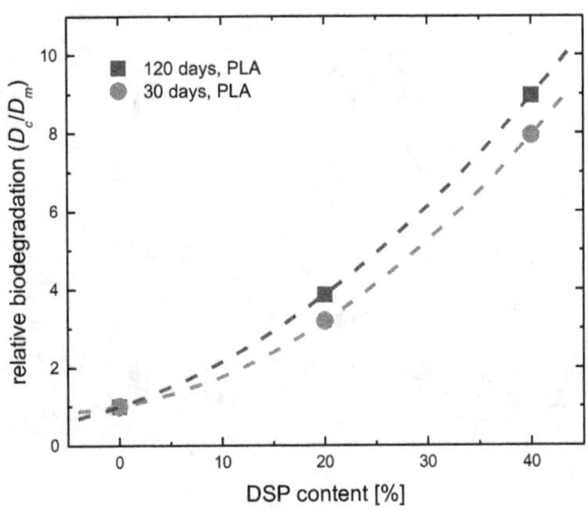

(b)

Figure 6.2 (a) Storage and loss modulus; (b) biodegradation of PLA composites as a function of date seed powder content. Reproduced from Reference 11 with permission from Wiley.

weaknesses of both polymers [16,17]. Sarazin *et al.* [18] reported the binary and ternary blends of PLA, PCL and TPS using two different types of starch in a single-step extrusion process and demonstrated their superior physical properties. Similarly Liao *et al.* [19] also reported the ternary blends of PLA with PCL and starch. It was reported that PLA could be transformed from rigid to ductile by the addition of

PCL. Mittal *et al.* [20,21] also reported successful generation of binary and ternary blends of PLA, TPS and PCL using melt mixing method. The strain at break of the blends was observed to be significantly enhanced as compared to pure PLA. In another study, ternary blends of PLA, PBS and PCL were generated using one-step melt mixing method [22]. The thermal performance and modulus were enhanced as a function of PLA, though the elongation of the ternary blends was observed to be similar to PLA, due to the presence of low extension PBS. In addition, Mittal *et al.* [20] also reported phase separation in the PLA-PCL blend as a function of time. As can be observed for $PLA_{50}PCL_{50}$ blend in Figure 6.3a-c, interfacial connection between PCL and PLA phases exhibited an excellent adhesion, when analyzed shortly after the production, and no phase-separated regions could be observed in the sample. After 90 days, the volumetric contraction of PLA phase (or alternatively the volumetric expansion of PCL

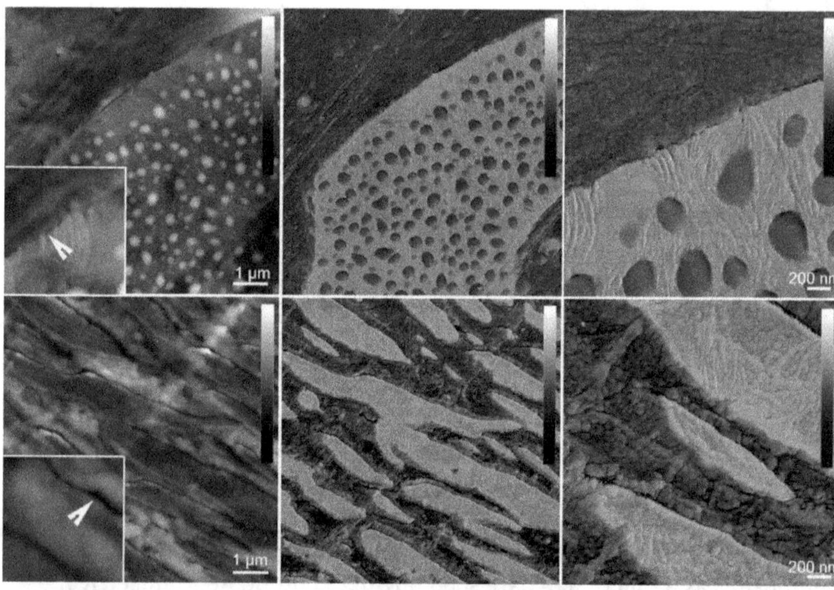

Figure 6.3 (a) Height and (b, c) phase images of $PLA_{50}PCL_{50}$blend; (d) height and (e, f) phase images of the blend after aging of the sample at ambient conditions for 90 days. Reproduced from Reference 20 with permission from Springer.

phase) resulted in almost complete relaxation. As a result, the crystalline structure of the phases exhibited a significant change (Figure 6.3d, f).

Another fundamental strategy to tune the properties of PLA is through structural modification of PLA using molecular reactions, including chain extension, hyperbranching, etc. Different chain extenders have been reported in literature, such as multi-functional epoxides, polycarbodiimide, and diisocyanates. Among these, multi-functional epoxides have found extensive use owing to their ability to react with -OH and -COOH groups of PLA. For instance, Zhou *et al.* [23] reported the chain extension and branching of PLA by reaction with a DGEBA-based epoxy resin. In another study, Elhassan *et al.* [24] used oligomer Joncryl ADR (multifunctional epoxide) to extend the PLA chains using concentrations of 0.5 and 0.9 wt%. Al-Itry *et al.* [25] used reactive extrusion with functionalized epoxy to improve thermal, rheological and mechanical properties of PLA. Positive impact was observed on viscosity and thermal stability in the chain extended polymer. Mesias and Murillo [26] also reported the reaction of hyperbranched polyester polyol with PLA, which also opens the ways to tune the properties of the polymers based on the intended application.

The current study aims to gain further insights into the impact of chain extension on the structure and properties of PLA. The chain extension of PLA has been carried out using reaction extrusion mode with multifunctional epoxide as extender. A comparison with the impact of hyperbranching on the properties has also been presented briefly.

6.2 Experimental

6.2.1 Materials and Structural Modification of PLA

Two grades of PLA were obtained from Nature Works and were denoted as PLA1 and PLA2. Multi-functional epoxide based chain extender was procured from Sigma Aldrich. The reactive extrusion was carried out in a twin screw extruder (MiniLab HAAKE Rheomex CTW5, Germany) at a temperature of 185 °C, following the procedures reported earlier [25]. The screw length and screw diameter were 109.5 mm and 5/14 mm conical respectively. The amount of the chain extender used during the reactive extrusion was 0.5 wt% of the PLA amount. The PLA samples were dried at 80 °C under vacuum overnight so as to remove any moisture which could interfere with the reaction in the extruder. A low rotational speed of 40 rpm and a moderate residence time of 5 min was used so as to avoid any thermal degradation of the polymers. The modified polymers so obtained were designated as PLA1-CE and PLA2-CE. In addition, PLA1 was also similarly reacted with a commercially procured hyperbranched polyester polyol at 200 °C under nitrogen [26]. Similar to the chain extender, the amount of the hyperbranching agent was 0.5 wt% of the PLA amount. The modified PLA was denoted as PLA1-HB. After extrusion, the extrudate was cooled quickly in a cold water bath so as to stop any further reaction. The extrudate was also subsequently dried at 80 °C under vacuum overnight. The unmodified polymers were also processed in the extruder using same processing conditions to maintain similar thermal history for all the samples. For subsequent characterization, the disc and dumbbell shaped test samples of both original as well as modified PLA samples were prepared by compression molding using hot press MTP-10 from Tetrahedron Associates. The pressure was kept at 345 bar (5 psi) for 3 min at 185 °C (365 °F) using the pneumatic press. A controlled rate of cooling of the samples was used in order to avoid unwanted changes in the crystallinity of the polymers.

6.2.2 Characterization Techniques

The PLA materials were characterized by FT-IR spectroscopy using a Bruker VERTEX 70 instrument equipped with a Pike MIRacle ZnSe ATR, DT-DLaTGS detector and KBr beam-splitter. IR acquisitions between 4000 cm^{-1} and 500 cm^{-1} were made using the software OPUS at 4 cm^{-1} resolution with an average of 64 scans. Raman spectra of PLA samples were recorded using LabRAM HR spectrometer (Horiba Jobin Yvon). Laser light of 633 nm wavelength from He/Ne laser was used for excitation. A long working distance objective with magnification of 50x was used both to collect the scattered light and to focus the laser beam on the sample surface. The charge-coupled device (CCD) exposure time was 10 s, and an average of 5 cycles was used to increase S/N ratio.

Wide angle XRD analyses were conducted on a Panalytical Powder Diffractometer (X'Pert PRO) using Cu-Kα radiation (λ 1.5406 A°, 45 kV, 40 mA). Diffraction patterns were collected in the range of 5-70° 2θ scale, with a step size of 0.02°/s at ambient conditions. 1H NMR spectra were recorded using a 400 MHz Bruker spectrometer. $CDCl_3$ was used as solvent. Tetramethylsilane resonance at 0 ppm was used as the chemical shift reference. All experiments were carried out at 25 °C.

Thermal properties of the polymers were analyzed using TA Discovery Thermogravimetric Analyzer (TGA). Nitrogen was used as a carrier gas and the scans were obtained from 50 to 700 °C at a heating rate of 10 °C/min. Differential scanning calorimetric (DSC) measurements on the polymers were performed using a TA Discovery DSC under nitrogen atmosphere. Compression molded samples were used for the thermal analysis. The scans were obtained from 20 to 250 °C (heating rate of 20 °C.min^{-1}), 250 to -80 °C (cooling rate of 20 °C.min^{-1}) and -80 to 250 °C (heating rate of 10 °C.min^{-1}). 5-7 mg of the sample weight was used for the thermal analysis.

Tensile analysis of the polymers was carried out using universal testing machine (Instron, USA) (ASTM D638). The dumbbell-shaped samples were used. A loading rate of 10 mm/min was used and the tests were carried out at room temperature. The tensile strength and modulus of the composites were calculated using Win Test Analysis software. An average of five sample values was calculated. AR 2000 rheometer from TA Instruments was used to characterize the rheological properties such as storage modulus (G′), loss modulus (G″), as well as complex viscosity (η*). Disc-shaped samples of 8 mm diameter and 2 mm thickness were measured at 185 ° C using a gap opening of 1.2 mm. Strain sweep scans were recorded at ω = 1 rad.s^{-1} from 0.1% to 50% strain. For the comparison, frequency sweep scans (dynamic testing) of all the samples were recorded at 1% strain from ω = 0.1 to 100 rad.s^{-1}.

6.3 Results and Discussion

Structural modification of PLA presents many advantages over the other strategies to reinforce its properties, as it avoids the addition of significant extent of fillers, compatibilizers, processing aids, etc. However, it is also important to control the chain extension or hyperbranching process so as to tune the resulting properties of the polymer. Too high extent of these structural modifications may also lead to unwanted reduction in the modulus or strength of the polymer. To gain further insights into the impact of such structural modifications on the properties of PLA, two grades of PLA were modified in the current study using reactive extrusion and the resultant properties are compared with the source polymers.

The structure of PLAs was characterized by spectroscopic and diffraction techniques. Figure 6.4 shows the IR spectra of the pure and chain extended PLAs. The wavenumbers and vibrational assignments of each band in the PLA1 and PLA2 spectra are shown in Table 6.1. In addition to the normal vibrational features, both PLA1 and PLA2 exhibited additional bands at 3297 cm^{-1}, 1636 cm^{-1} and 1564 cm^{-1}. The

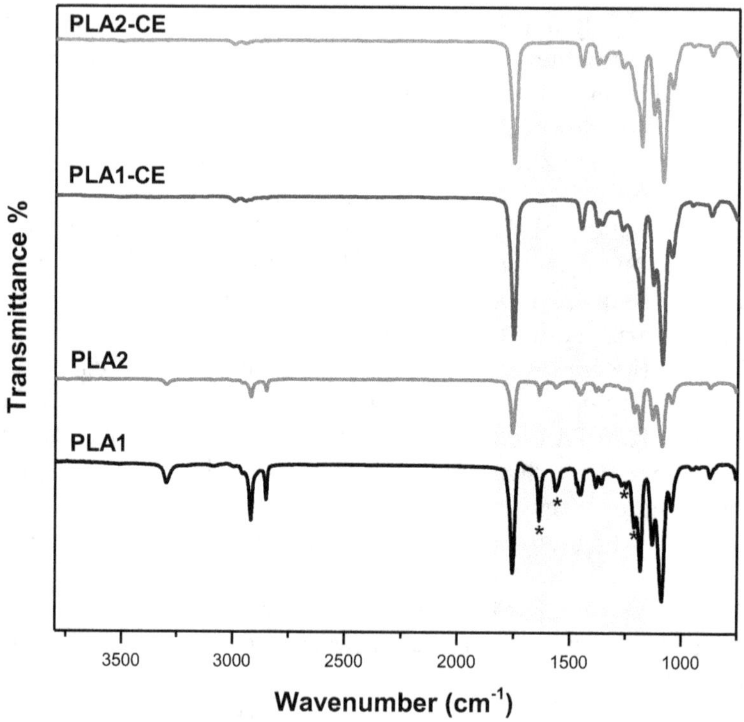

Figure 6.4 IR spectra of PLA1, PLA2, PLA1-CE, and PLA2-CE.

former one is attributed to the O-H stretching and the latter two bands are formed from –COOH vibrations, indicating the presence of end carboxylic acid and hydroxyl functional groups in the structure of pure PLAs, which provided reaction sites for the chain extension process [27,28]. These two vibrational bands corresponding to O-H and –COOH vibrations were completely absent in the chain extended PLAs, thus, confirming successful PLA chain extension using the end functional groups via reactive extrusion with epoxide. A sharp peak at 1212 cm^{-1} (due to the alkyl-ketone chain vibration), representative of the crystalline structure of pure PLA, changed to shoulder band after modification [29]. The doublet at 1269 cm^{-1} and 1250 cm^{-1}, corresponding to δCH and νCOC vibrations, was also observed to change to a single broad band at 1265 cm^{-1}. The stretching as well as bending vibrations of CH_3 and CH bonds in the region 3000 to 2850 cm^{-1} and at 1472 cm^{-1} became less IR active vibrations in the modified PLAs due to the changes in the chain conformations. A similar behavior of methyl and methane groups has also been reported in literature for the PLA polymers [30,31]. However, these vibrations were also observed to form sharp intense bands in the Raman spectra of PLAs at

Table 6.1 Wavenumber and vibrational band assignment of PLA1 and PLA2 polymers. Both IR and Raman bands are presented.

IR bands (cm^{-1})	Raman bands (cm^{-1})	Band assignment
3297		νOH
2919	2999, 2948	νCH_3
2850	2883	νCH
1755	1767	νC=O of ester
1637-1564		νC=O of COOH & COO^-
1458	1454	$\delta_{as} CH_3$
1385-1340	1387-1300	$\delta_s CH_3 + \delta CH$
1269-1250	1290	δCH+νCOC
1212-1184	1182	$\nu_{as} COC + r_{as} CH_3$
1131	1129	$r_{as} CH_3$
1088	1093	$\nu_s COC$
1046	1042	$\nu C\text{-}CH_3$
869-700	878	νC-COO + δC=O
	700-760	δC=O +rC=O
	600-500	$\delta C\text{-}CH_3 + \delta CCO$
	395- 411	δCCO
	304	$\delta COC + \delta C\text{-}CH_3$
	240-157	skeletal torsion

2883 cm^{-1}, 2948 cm^{-1}, & 2999 cm^{-1}, as demonstrated in Figure 6.5. The Raman band assignments are also presented in Table 6.1. The region below 1000 cm^{-1}, corresponding to the backbone vibrations of PLA chain, exhibited sharp and well defined peaks in the Raman spectra of PLA1 and PLA2. These bands were changed to broad vibrations in the chain extended PLAs, especially at 304 cm^{-1}, 395 cm^{-1}, 578 cm^{-1}, 706 cm^{-1} and 922 cm^{-1}, which corresponded to δCCO and δCOC, combined with $\delta C\text{-}CH_3$ vibrations (Table 6.1) [30]. The broadening of the Raman bands indicated that during the structural modification reaction, PLA chains underwent rearrangement to slightly disordered state in the PLA1-CE and PLA2-CE samples. Thus, the IR and Raman analyses confirmed the successful chain extension of PLA in a single and simple reactive extrusion step. Absence of any hydroxyl and carboxyl groups in the modified PLA chains also indicated that the amount of the chain extender was sufficient to carry out a significant extent of chain extension. Also, even though the PLA1 and PLA2 had differences in molecular weight and viscosity, however, these differences did not hinder the chain extension reaction. Complete reaction of the PLA functional groups also indicates that performing the chain extension reactions at lower extent of chain extender content may be considered to identify its optimal amount for the reaction. However, in an earlier study, the viscosity and storage modulus of the chain extended PLA was observed to be higher than the unmodified polymer

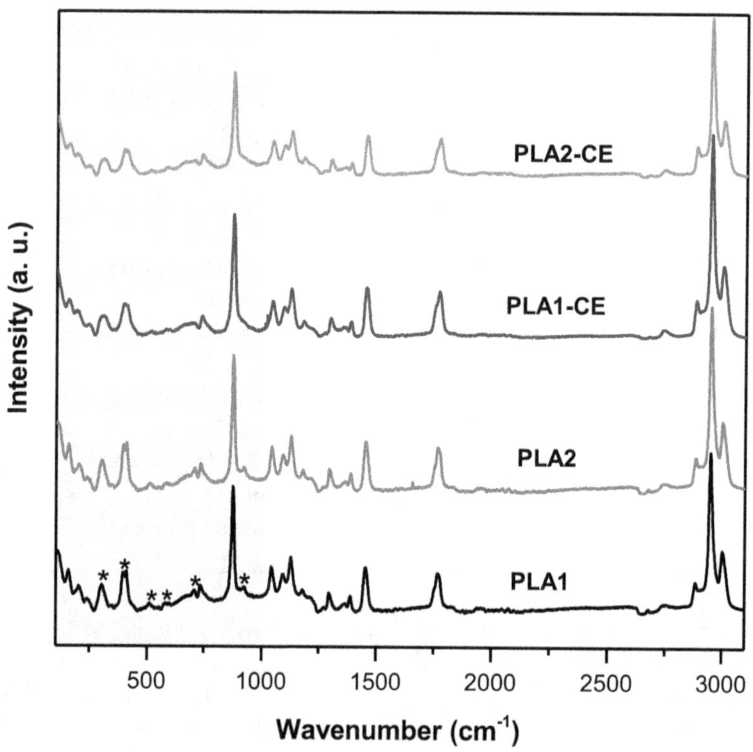

Figure 6.5 Raman spectra of PLA1, PLA2, PLA1-CE and PLA2-CE.

and the extent of increase was observed to enhance with the amount of the chain extender [25]. Thus, it is probably better to use a slightly higher extent of chain extender to gain increment in the physical properties, however, an excess has also to be avoided. For this purpose, a chain extension trial of PLA1 was also carried out with chain extender amount corresponding to 0.75 wt% of the PLA content. As expected, no change in the IR and Raman bands were observed for this material. However, changes in the mechanical, thermal and rheological properties can still be expected due to the increased extent of chain extension.

As observed in the IR/Raman analysis earlier, PLA chains underwent some rearrangement to slightly disordered state on performing the chain extension. However, there was no apparent change in the crystalline patterns of PLAs in the wide-angle X-ray diffractograms, as shown in Figure 6.6. Both PLA1 and PLA2 exhibited X-ray diffraction peaks at 16.8° and 19.1° corresponding respectively to (200)/(110) and (203) planes of the α-form of PLA crystals [32,33]. These diffraction peaks were also observed in the modified PLAs at slightly shifted peak maximum locations of 17.1° and 19.6° respectively, which indicated that the chain extension processes did not modify the PLA crystal structure significantly, though reduction in

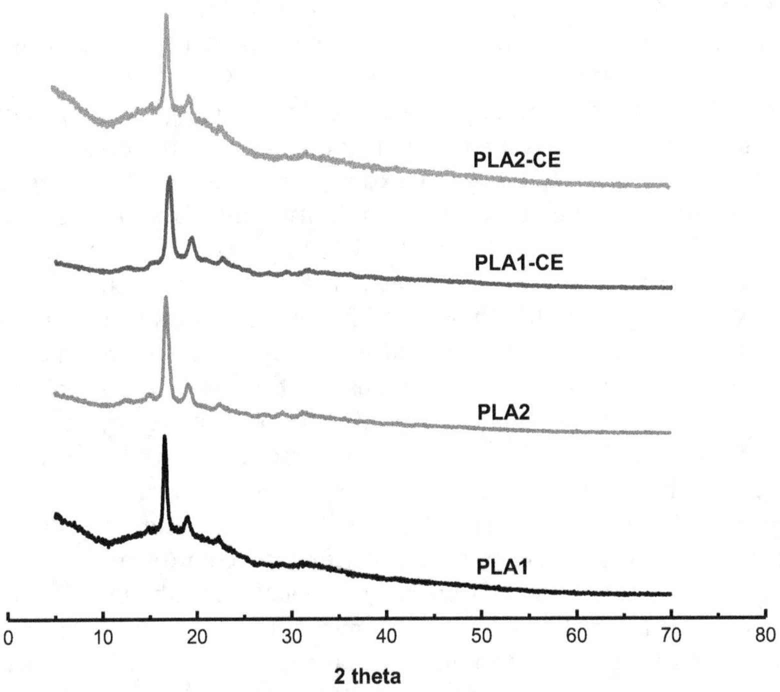

Figure 6.6 X-ray diffractograms of PLA1, PLA2, PLA1-CE and PLA2-CE.

the degree of crystallinity could still be expected due to the chain extension, which would hinder the PLA chains from forming crystals. Similar diffractogram was also observed for the chain extended PLA1 sample with 0.75% chain extender, which indicated that the PLA chains still retained the ability to crystallize even with larger extent of chain extension. Overall, a small degree of broadening of the diffraction peaks was observed in the chain extended samples probably due to some change in lamellar thickness or crystal size due to the rearrangement of chains, as observed earlier.

Further insights into the structure of the PLAs were also gained through NMR analysis. Figure 6.7 presents the 1H NMR spectra of the unmodified and chain extended PLAs. All spectra exhibited the resonances of methane (CH) as a quadruplet at 5.2 ppm and methyl (CH_3) as a doublet at 1.9 ppm [29,34,35]. In addition, both PLA1 and PLA2 also exhibited end hydroxyl vibrations at 2.1 ppm, which was observed to disappear in the spectra of modified PLAs due to the chemical reaction with the multi-functional epoxide during the chain extension process, thus, supporting further the aforementioned IR results. The chain extended PLA1 with higher amount of chain extender also exhibited the similar pattern as PLA1-CE. Thus, overall, spectroscopic, diffraction and nuclear magnetic resonance characterization of the samples confirmed the chain extension process. Though some changes in the chain arrangement or diffraction angle occurred owing to the chain extension, no significant change in the crystal structure was observed. The chain extension process is, however, expected to reduce the degree of crystallinity, which is also expected to impact the properties of the modified PLAs, especially the mechanical performance. The thermal and mechanical properties of the modified PLAs are compared with the unmodified PLAs in the subsequent sections in order to relate the structure development after chain extension with the subsequent properties. Furthermore, the rheological performance of PLA samples has also been characterized so as to quantify the buildup of storage and loss moduli, as well as viscosity after the chain extension.

PLA exhibits lower thermal stability than many other biopolymers such as polycaprolactone, poly(butylene succinate), etc. Thermal degradation leads to reduction in the molecular weight and viscosity, thus, resulting in the reduction of mechanical performance. Therefore, it is of importance to ensure that the chain extension process does not lead to further reduction in the thermal stability of the polymer. Al-Itry *et al.* [25] have also reported an increase in the thermal

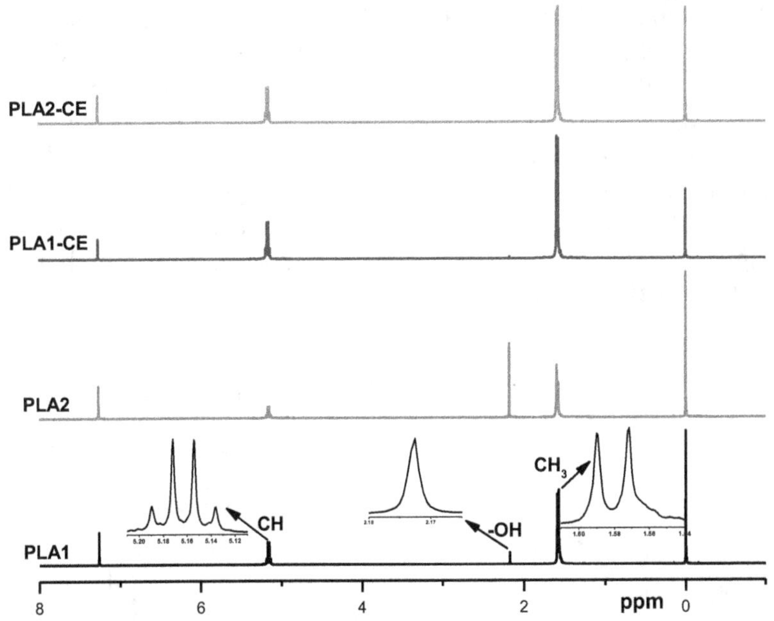

Figure 6.7 1H NMR spectra of unmodified and chain extended PLA samples.

stability of PLA after chain extension with multi-functional epoxide due to the branched chains leading to the enhancement in the molar mass, intrinsic viscosity, shear thinning and elasticity. Figure 6.8 demonstrates the TGA thermograms of the unmodified and chain extended PLA samples. Both PLA1 and PLA2 exhibited a single degradation signal, however PLA1 was observed to be more stable than PLA2. The onset and peak degradation values for PLA1 were observed to be 290 °C and 350 °C respectively, whereas these were 260 °C and 325 °C respectively in the case of PLA2. In comparison with the PLA1, the PLA1-CE was observed to have slight enhancement in the thermal degradation temperatures, thus, indicating the enhancement of thermal stability after chain extension. PLA2-CE exhibited somewhat lower onset degradation temperature than PLA2, however, its peak degradation temperature was higher than PLA2. It indicated firstly that there were differences in the thermal degradation of the unmodified PLAs, which also resulted in the corresponding degradation patterns of the modified polymers. Secondly, the overall thermal stability of the PLAs was improved after chain extension due to enhanced degree of branching and inter-connection between the chains. Probably the epoxide groups may have also led to a small degree of crosslinking of the PLA chains, thus, resulting in enhanced thermal stability. Also, no significant change in the shape of the degradation pattern after chain extension confirmed the absence of any structural change in PLAs, as also confirmed earlier through diffraction and spectroscopy. The PLA1 chain extended sample with 0.75% chain extender also revealed similar pattern as PLA1-CE, thus, indicating that the thermal stability was not impacted even at higher fraction of chain extension.

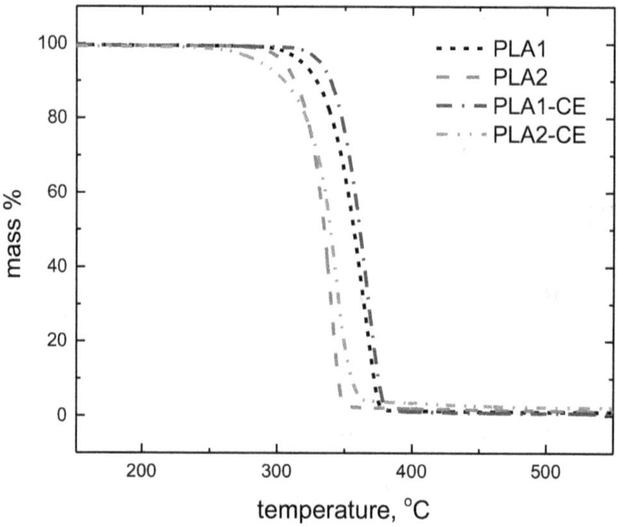

Figure 6.8 TGA thermograms of unmodified and chain extended PLAs.

The calorimetric behavior of the PLA samples is also demonstrated in Figure 6.9. As evident in Figure 6.9a, the PLA1 exhibited the peak melting point of 156 °C, along with the melt enthalpy of 7.78 cal/g. On the other hand, both peak melting point and melt enthalpy were observed to decrease after chain extension to 149 °C and 2.49 cal/g respectively. It indicated that there was an impact on the degree of crystallinity of PLA after chain extension, also opined earlier. In general, a decrease in the melting point in a polymer can be due to both morphological effects (decrease in lamellar thickness) and to thermodynamic factors (polymer-polymer interactions). From thermodynamics perspective, it has been reported that such decrease in chemical potential of the crystallizable polymer may be due to the enhancement of amorphous content, resulting in a decrease in the melting point [36]. These hypotheses can be inferred to explain the depression in endothermic melting point of the chain extended polymer in the current study owing to the presence of the amorphous content generated due to the epoxide groups, which also subsequently hindered the PLA chains to crystallize. The earlier suggested crosslinking of the PLA chains due to the reaction with the multi-functional epoxide would also have resulted in the reduced extent of crystallinity. As expected, the PLA1 sample modified with 0.75% chain

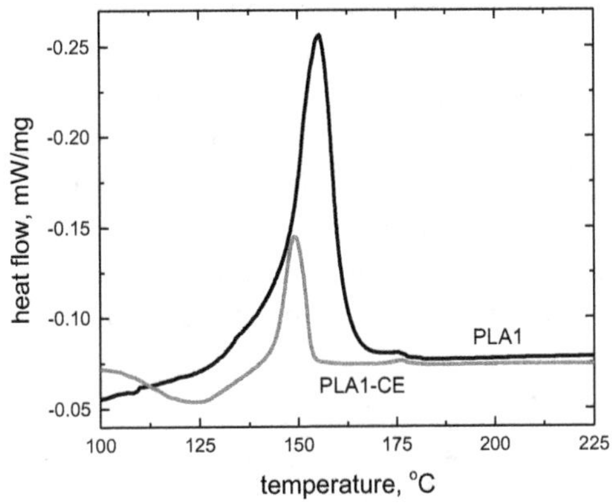

(a)

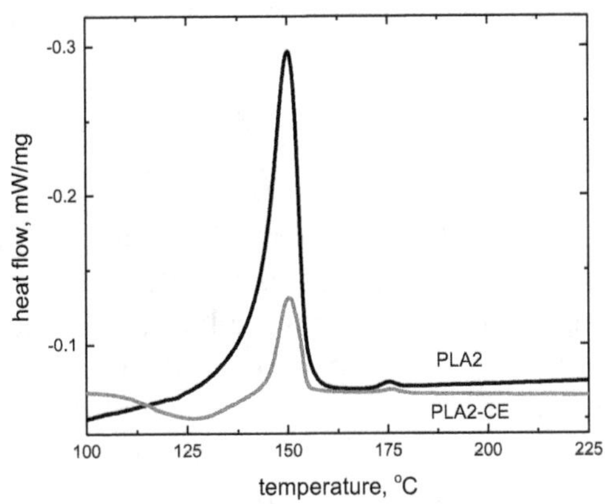

(b)

Figure 6.9 DSC thermograms of (a) PLA1 and PLA1-CE; (b) PLA2 and PLA2-CE.

extender amount was observed to reduce the melt enthalpy of PLA1 further to 2.18 cal/g, though the peak melting point remained unchanged. In the case of PLA2, as shown in Figure 6.9b, the polymer exhibited a peak melting point of 150 °C, which was lower than PLA1. After chain extension, the melting point of the modified polymer was observed to remain unchanged. On the other hand, the melt enthalpy of the polymer was measured to be 7.57 cal/g, which was reduced to 2.32 cal/g in the chain extended sample, thus, indicating similar reduction in the degree of crystallinity of the polymer as PLA1-CE. Overall, the DSC results supported the earlier findings from the structural analysis about some change in the arrangement of chains and extent of crystallinity, but no change in the crystal structure.

Table 6.2 shows the tensile properties of the modified polymers in comparison with the pristine polymers. Figure 6.10 also shows the corresponding stress-strain plots to demonstrate the effect of chain extension. PLA1 exhibited a modulus of 777 MPa, in comparison to 948 MPa for PLA2. Both polymers had lower tensile modulus than typically observed for PLA polymers. However, the peak melting point of these polymers was also observed to be lower than usually

Table 6.2 Tensile properties of the PLAs

Sample	Tensile modulus, MPa	Elongation, %	Ultimate tensile strength, MPa
PLA1	777±20	6.8±2.4	49.6±12.1
PLA2	948±6	5.5±0.7	44.5±5.8
PLA1-CE	874±53	8.0±0.5	64.7±5.3
PLA2-CE	961±35	6.1±1.2	52.0±13.2

observed value around 170 °C. Even the PLA samples unprocessed in the compounder exhibited similar peak melting temperatures. Thus, the observed tensile properties may be largely due to the nature of the samples, however, some structural degradation during processing may have also occurred. The tensile modulus of the PLA samples was observed to increase after chain extension. Similar observation has also been reported earlier by Al-Itry *et al.* [25], which was attributed to extended and branched chains able to withstand more stress. Some extent of crosslinking, as opined earlier, would have also resulted in the enhancement of the modulus. However, it should also be noted that degree of crystallinity in the chain extended polymers decreased, which generally is expected to decrease the modulus. Thus, the observed tensile modulus resulted due to the interplay of different factors, such as increased modulus due to extended (and partially crosslinked) chains and reduced modulus due to reduction in crystallinity. Interestingly, the enhancement of modulus did not lead to reduction in the elongation of chain extended polymers, as observed in Figure 6.10. In fact, the elongation in the modified polymers was observed to increase along with the modulus, which is generally very challenging to achieve. The ultimate strength of the modified polymers was also higher as compared to the unmodified PLAs. For

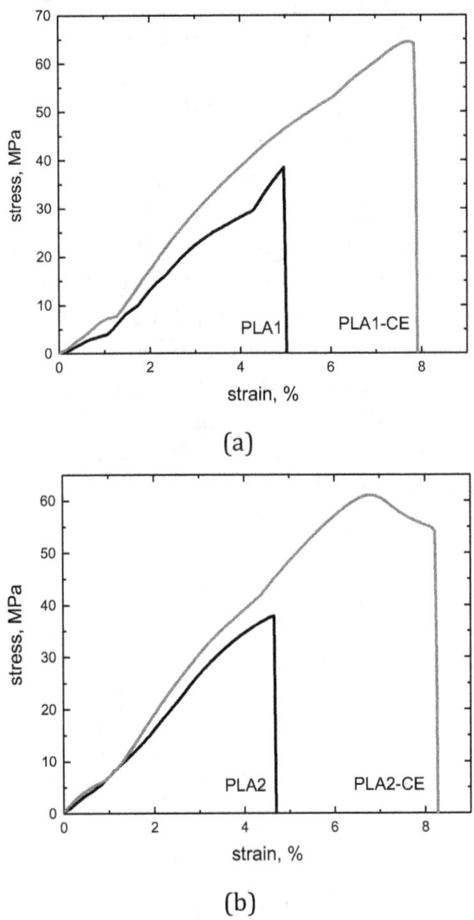

Figure 6.10 Stress-strain plots of (a) PLA1 and PLA1-CE; (b) PLA2 and PLA2-CE.

instance, the ultimate strength in PLA1 was measured to be 49.6 MPa, which was enhanced to 64.7 MPa in PLA1-CE sample. Similarly, the mechanical response of the chain extended PLA1 sample with 0.75% chain extender was further improved with a modulus of 980 MPa. The elongation was also observed to increase to 9.1%, along with enhanced value of 71.2 MPa for ultimate tensile strength, indicating good degree of property enhancement.

Figure 6.11 shows the strain sweep performance (at angular frequency of 1 rad.s^{-1}) of the PLA polymers, along with the buildup of the oscillatory torque as a function of angular frequency. The polymers were largely stable till 50% strain, though the storage modulus of PLA1 and PLA2 exhibited disturbance especially at higher strain. The storage modulus of the chain extended samples as well as loss modulus and complex viscosity of all the samples did not exhibit any strain-dependent behavior. Thus, for the comparison of the samples for the frequency sweep analysis, a strain value of 1% was chosen. In addition, the chain extended samples were also observed to exhibit higher modulus and viscosity at all strain values. Figure 6.11d also shows that the oscillatory torque increased for all the samples as a function of angular frequency during the rheological testing due to enhanced resistance of the melt at higher rates of deformation. The buildup of torque was observed to be higher for chain extended samples, especially at lower angular frequency values, which also supported the observed enhancement in the modulus and viscosity. At higher angular frequency values, the differences between the samples were not retained as the oscillatory torque response of all the samples merged into a single curve.

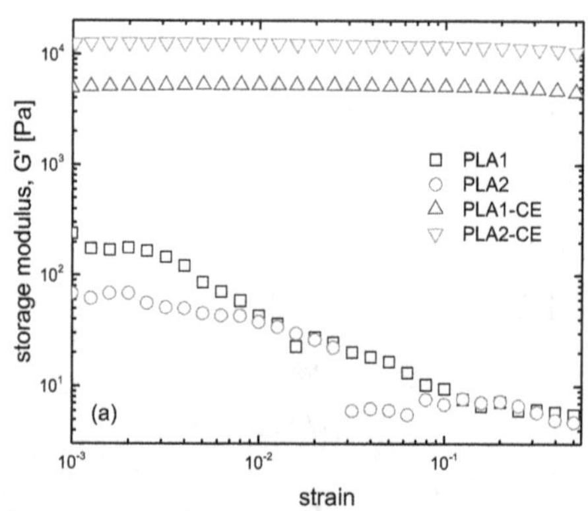

Figure 6.11 (continued on next page)

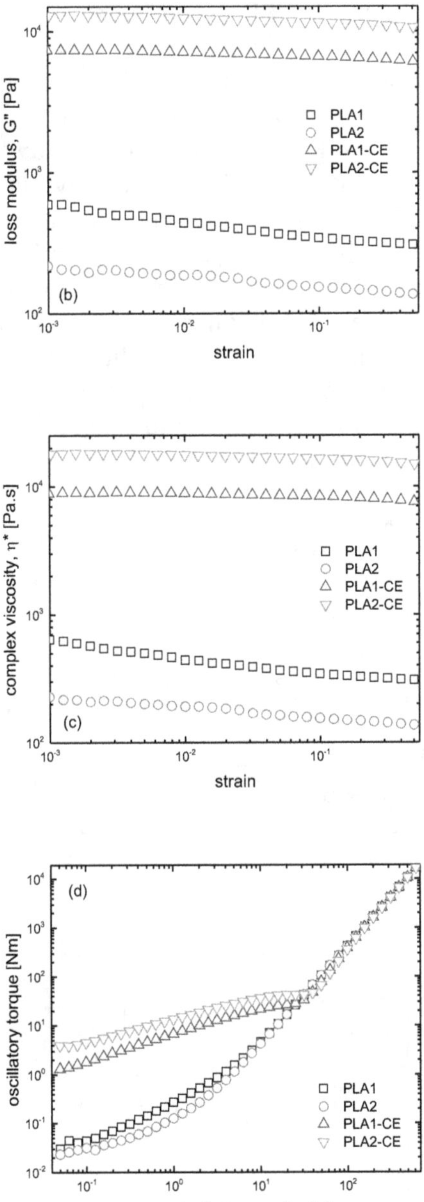

Figure 6.11 Strain sweep analysis exhibiting the (a) storage modulus, (b) loss modulus and (c) complex viscosity of the PLA samples; (b) buildup of the oscillatory torque during the rheological analysis as a function of angular frequency.

Figure 6.12 demonstrates the rheological performance of the unmodified and chain extended PLA samples, as a function of angular frequency. In case of PLA1, G' was always lower than G" over the whole frequency range, indicating the dominance of the viscous character in the sample. The chain extended sample exhibited significantly higher storage and loss modulus values than the PLA1 due to extended chains. In addition, the G" was slightly higher than G' only at lower angular frequency. A cross-over (where the G' became higher than G") was observed around 20 rad.s^{-1} indicating dominance of elastic character in the polymers. PLA2 also exhibited similar behavior as the PLA1 sample. The chain extended PLA2 had even higher values of storage and loss modulus than the corresponding PLA-CE sample. In addition, the cross-over point was observed at much lower angular frequency of ~ 1 rad.s^{-1}. Thus, the chain extension of PLA polymers not only increased the modulus, but also enhanced the elastic

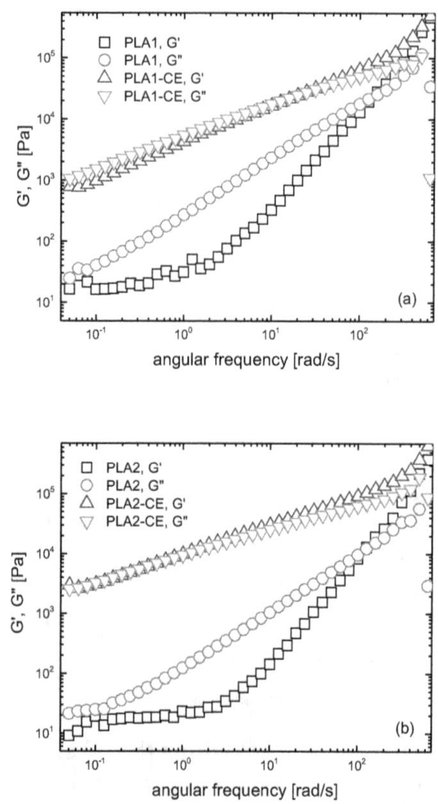

Figure 6.12 (continued on next page)

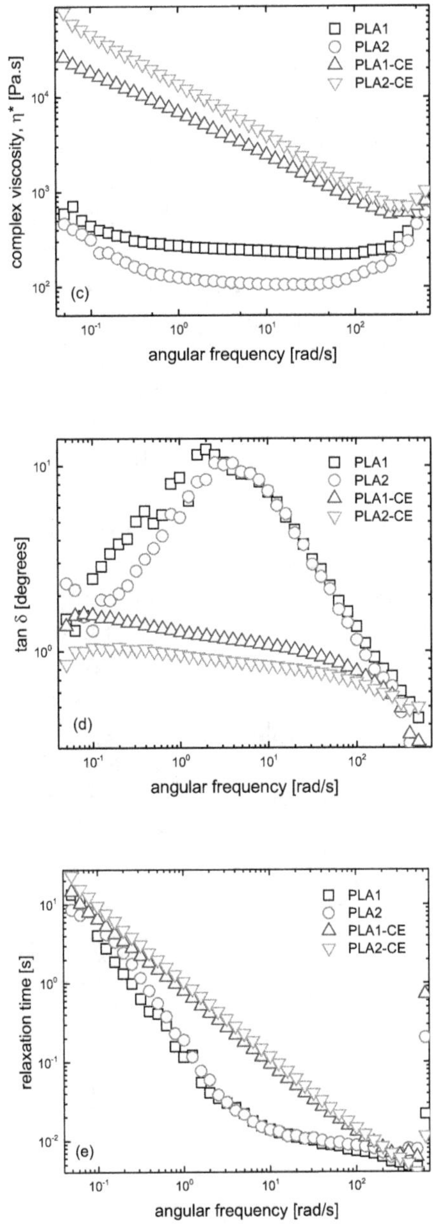

Figure 6.12 Storage and loss modulus as a function of angular frequency for (a) PLA1 and PLA1-CE, (b) PLA2 and PLA2-CE; (c) complex viscosity of the PLA samples as a function of angular frequency; variation in the (d) tan δ values and (e) relaxation time of the samples with angular frequency.

nature. Similar to the modulus, the viscosity of the chain extended samples was also higher than unmodified PLAs, with PLA2-CE exhibiting higher viscosity. The observed changes in the modulus after chain extension also impacted the tan δ transition in the polymers. As observed in Figure 6.12d, the transition exhibited by the unmodified polymers was not present in the chain extended samples. Similarly, due to the extended chains, the relaxation time of the chain extended PLAs was observed to be longer than the unmodified PLAs, though PLA1-CE and PLA2-CE did not exhibit any significant difference in the relaxation time. These findings also reinforced the earlier observations about the extended chains improving the polymer performance. The chain extended PLA1 sample with 0.75% chain extender had ~ 40% higher storage and loss modulus than PLA1-CE sample. In addition, the cross-over between G' and G" was also observed at lower angular frequency of 10 rad.s^{-1}. Similarly, the viscosity of the sample was also enhanced.

As mentioned earlier, the comparison of the chain extension and hyperbranching of PLA1 was also carried out. The success of the hyperbranching trial was also confirmed through structural analysis. Specifically, the performance of the hyperbranched PLA (PLA1-HB) was compared with PLA1 as well as PLA1-CE. For instance, Figure 6.13 shows the TGA thermograms of these samples. The hyperbranched sample exhibited similar onset of degradation temperature as PLA1, however, a slightly lower peak degradation temperature was observed. Thus, in comparison with PLA1-CE, the hyperbranching led to a slight reduction in the thermal stability of PLA, however,

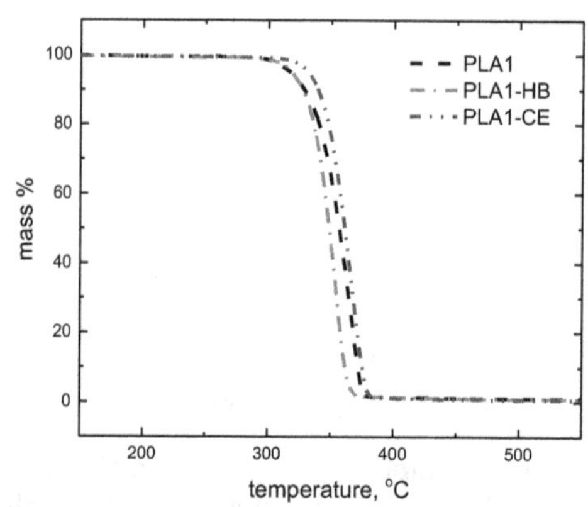

Figure 6.13 Comparison of the TGA thermograms of PLA1, PLA1-HB and PLA1-CE.

the reduction was not significant and the hyperbranched polymer was suitable for high temperature processing. Also, as no lower temperature degradation peak was observed in the TGA thermogram of PLA1-HB, it was evident that the hyperbranching agent was bound to the PLA chains, which in a free state would otherwise lead to the degradation at a lower temperature. The calorimetric behavior of the samples is also compared in Figure 6.14a. The hyperbranching also did not lead to significant reduction in the degree of crystallinity as the melt enthalpy of the hyperbranched polymer was measured to be 4.1 cal/g, in comparison with 7.78 cal/g for PLA1. The enthalpy was

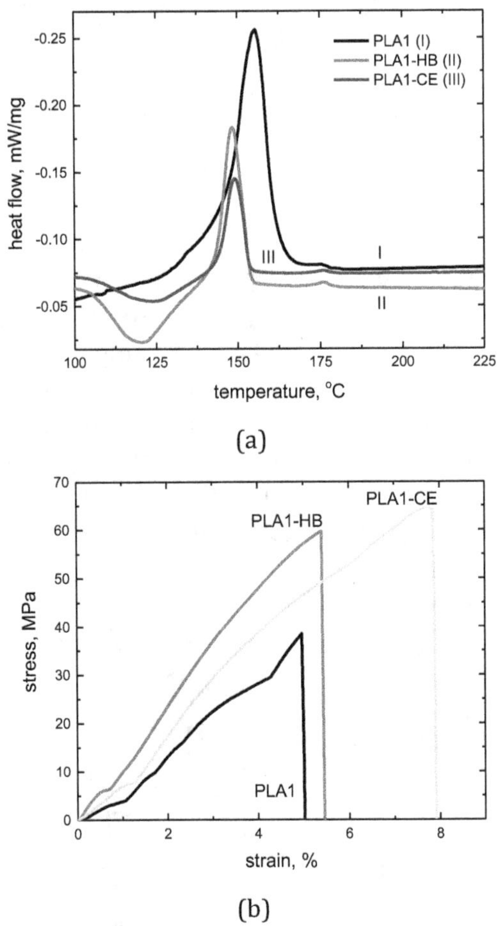

Figure 6.14 (a) DSC thermograms and (b) stress-strain behavior of PLA1, PLA1-HB and PLA1-CE.

also higher than the PLA1-CE sample, which exhibited the value of 2.49 cal/g. The peak melting point of PLA1-HB sample was 149 °C, similar to PLA1-CE. Thus, it was evident that similar to PLA1-CE, PLA1-HB also had some changes in the chain arrangement, however, no significant change in the crystal structure. The hyperbranched chains did hinder the crystallization to some extent, however, the extent of reduction was lower than the chain extended sample. The observed higher extent of crystallinity is also expected to result in higher mechanical response of the hyperbranched sample, which is compared with PLA1 and PLA1-CE in Figure 6.14b. As expected, owing to the higher extent of crystallinity than PLA1-CE, the PLA1-HB sample had higher tensile modulus and correspondingly lower elongation. In comparison with 874 MPa modulus for PLA1-CE, the PLA1-HB sample had a modulus of 1109 MPa. Similarly, the elongation for PLA1-CE and PLA1-HB samples was 8.0% and 5.8% respectively. Thus, hyperbranching had different impact on the polymer properties than chain extension, and a choice between the two strategies can be made depending on the requirement.

Figure 6.15 also compares the rheological behavior of PLA1-HB sample with PLA1 and PLA1-CE. The hyperbranched sample exhibited lower storage modulus than the chain extended PLA. In addition, the G" was higher than G' in the PLA1-HB sample for most of the angular frequency range. Similarly, the complex viscosity of the PLA1-HB samples was significantly lower than PLA1-CE. The tan δ transition was also observed to occur differently in PLA1-HB, as compared to PLA1 and PLA1-CE. The chain in PLA1-HB had faster relaxation than the corresponding PLA1-CE chains. Overall, comparisons of the

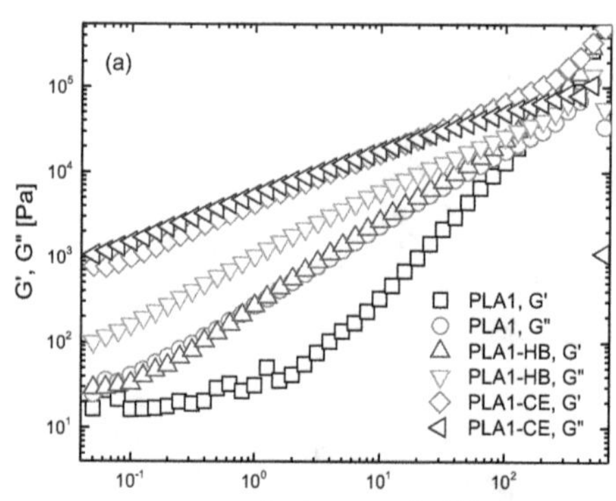

Figure 6.15 (continued on next page)

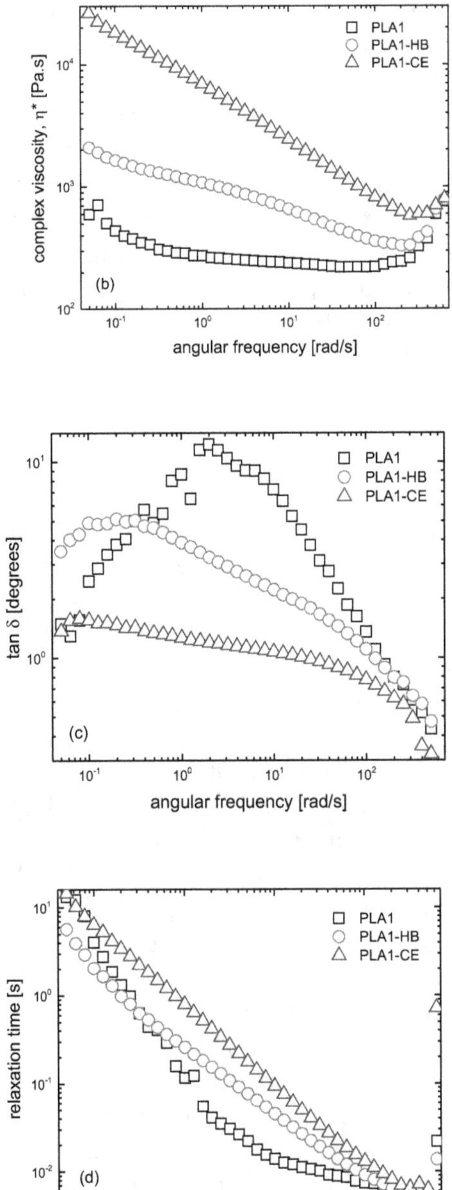

Figure 6.15 Comparison of the performance of PLA1, PLA1-HB and PLA1-CE: (a) storage and loss modulus, (b) complex viscosity, (c) tan δ, and (d) relaxation time.

properties of PLA1-CE and PLA1-HB samples revealed that the chain extension had more optimal impact in modifying the properties of PLA. Though the hyperbranching led to increased extent of tensile modulus, however, from the overall perspective of simultaneous enhancement in the toughness and stiffness, chain extension had more balanced effect. However, it should be noted that the property enhancement achieved by hyperbranching, especially in comparison with unmodified PLA, still presented an effective strategy for structural modification of PLA. Probably more trials of structural modification are needed with a lower amount of hyperbranching agent. Altering the functionality of the hyperbranching agent is another route to achieve further tuning in the polymer behavior.

6.4 Conclusions

In the current study, structural modification of poly-L-lactide has been successfully demonstrated by employing chain extension and hyperbranching strategies. Chain extension of two different PLA grades with a multifunctional epoxide resulted in the loss of IR vibrational bands corresponding to O-H and –COOH vibrations in the chain extended samples due to the chemical reaction. Raman spectroscopy indicated some degree of re-arrangement or disordering of chains in the chain extended samples. NMR analysis also demonstrated the disappearance of hydroxyl vibrations at 2.1 ppm in the modified PLAs. The X-ray diffraction studies however revealed no impact on the crystal structure of the PLA samples after chain extension. Chain extension resulted in a decrease in the degree of crystallinity, thus, also confirming the findings from the structural characterization. The thermal stability of the PLAs was either enhanced or remained largely unaffected by the chain extension process. The chain extension resulted in a simultaneous increment in the stiffness and flexibility as compared to unmodified PLAs due to extended chains as well as possibility of a small extent of crosslinking in the polymer chains due to the reaction with the multifunctional epoxide. The rheological performance of the chain extended samples was also superior to the unmodified PLAs. The chain extended PLAs exhibited higher modulus and viscosity as a function of angular frequency, along with dominance of elastic nature. Hyperbranching with a hyperbranched polyester polyol also impacted the degree of crystallinity of PLA, though the effect was less prominent than the chain extended sample. In further comparison with the chain extended sample, the hyperbranched sample exhibited higher tensile modulus, but reduced elongation, storage modulus and viscosity. Though both techniques presented fundamental routes for the structural modification of PLA to enhance its property profile, chain extension was observed to be more optimal in property enhancement. Further insights need to be gained for the hyperbranching strategy through the use of different hyperbranching agents in varying amounts.

Acknowledgement

The authors are indebted to Dr. Gigi George and Mr. Fakhruddin Patwary for experimental support.

References

1. Averous, L., and Boquillon, N. (2004) Biocomposites based on plasticized starch: thermal and mechanical behaviours. *Carbohydrate Polymers*, **56**, 111-122.
2. Klun, U., Friedrich, J., and Krzan, A. (2003) Polyamide 6 fibre degradation by a lignolytic fungus. *Polymer Degradation and Stability*, **79**, 99-104.
3. Esquivel, D., Bou, J. J., and Munoz-Guerra, S. (2003) Synthesis, characterization and degradability of polyamides derived from tartaric acid and diaminoethers. *Polymer*, **44**, 6169-6177.
4. Kricheldorf, H. R., and Jonte, J. M. (1983) New polymer syntheses. 8. Synthesis and polymerization of L-lactic acid O-carboxyanhydride (5-methyl-dioxolan-2,4-dione). *Polymer Bulletin*, **9**(6), 276-283.
5. Anders, S., and Stolt, M. (2002) Properties of lactic acid based polymers and their correlation with composition. *Progress in Polymer Science*, **27**, 1123-1163.
6. Mittal, V., Chaudhry, A. U., and Luckachan, G. E. (2014) Biopolymer-thermally reduced graphene nanocomposites: Structural characterization and properties. *Materials Chemistry and Physics*, **147**(1-2), 319-332.
7. Wu, T.-M., and Wu, C.-Y. (2006) Biodegradable poly(lactic acid)/chitosan-modified montmorillonite nanocomposites: Preparation and characterization. *Polymer Degradation and Stability*, **91**, 2198-2204.
8. Chun, K. S., Husseinsyah, S., and Osman, H. (2013) Properties of coconut shell powder-filled polylactic acid ecocomposites: Effect of maleic acid. *Polymer Engineering and Science*, **53**, 1109-1116.
9. Chun, K. S., and Husseinsyah, S. (2014) Polylactic acid/corn cob ecocomposites: Effect of new organic coupling agent. *Journal of Thermoplastic Composites Materials*, **27**(12), 1667-1678.

10. Wang, Y., Qi, R., Xiong, C., and Huang, M. (2011) Effects of coupling agent and interfacial modifiers on mechanical properties of poly(lactic acid) and wood flour biocomposites. *Iranian Polymer Journal*, **20**, 281-294.
11. Mittal, V. Luckachan, G. E., Chernev, B., and Matsko, N. B. (2015) Biopolyester–date seed powder composites: Morphology and component migration. *Polymer Engineering and Science*, **55**(4), 877-888.
12. Mittal, V., Chaudhry, A. U., and Matsko, N. B. (2014) "True" biocomposites with biopolyesters and date seed powder: Mechanical, thermal, and degradation properties. *Journal of Applied Polymer Science*, **131**(19), doi:10.1002/app.40816
13. Xu, Y. X., and Hanna, M. A. (2005) Preparation and properties of biodegradable foams from starch acetate and poly (tetramethylene adipate-co-terephthalate). *Carbohydrate Polymers*, **59**, 521-529.
14. Quero, E., Muller, A. J., Signori, F., Coltelli, M. B., and Bronco, S. (2012) Isothermal cold-crystallization of PLA/PBAT blends with and without the addition of acetyl tributyl citrate. *Macromolecular Chemistry and Physics*, **213**, 36-48.
15. Bhatia, A., Gupta, R. K., Bhattacharya, S. N., and Choi, H. J. (2007) An investigation of melt rheology and thermal stability of poly(lactic acid)/ poly(butylene succinate) nanocomposites. *Journal of Applied Polymer Science*, **114**(5), 2837-2847.
16. Dell'Erba, R., Groeninckx, G., Maglio, G., Malinconico, M., and Migliozzi, A. (2001) Immiscible polymer blends of semicrystalline biocompatible components: thermal properties and phase morphology analysis of PLLA/PCL blends. *Polymer*, **42**, 7831-7840.
17. Maglio, G., Malinconico, M., Migliozzi, A., and Groeninckx, G. (2004) Immiscible poly(L-lactide)/poly(ε-caprolactone) blends: Influence of the addition of a poly(L-lactide)-poly(oxyethylene) block copolymer on thermal behavior and morphology. *Macromolecular Chemistry and Physics*, **205**(7), 946-950.
18. Sarazin, P., Li, G., Orts, W. J., and Favis, B. D. (2008) Binary and ternary blends of polylactide, polycaprolactone and thermoplastic starch. *Polymer*, **49**, 599-609.
19. Liao, H.-T., and Wu, C.-S. (2009) Preparation and characterization of ternary blends composed of polylactide, poly(ε-caprolactone) and starch. *Materials Science and Engineering A*, **515**, 207-214.
20. Mittal, V., Akhtar, T., Luckachan, G., and Matsko, N. (2015) PLA, TPS and PCL binary and ternary blends: structural characterization and time-dependent morphological changes. *Colloid and Polymer Science*, **293**(2), 573-585.
21. Mittal, V., Akhtar, T., and Matsko, N. (2015) Mechanical, thermal, rheological and morphological properties of binary and ternary blends of PLA, TPS and PCL. *Macromolecular Materials and Engineering*, **300**, 423-435.

22. Mittal, V., Manna, Y., and Matsko, N. (2017) Morphology and properties of PCL, PLA and PBS ternary blends. In: *Biopolymers Based Advanced Materials*, Mittal, V. (ed.), Central West Publishing, Australia, pp. 115-144.
23. Zhou, Z. F., Huang, G. Q., Xu, W. B., and Ren, F. M. (2007) Chain extension and branching of poly(L-lactic acid) produced by reaction with a DGEBA-based epoxy resin. *Express Polymer Letters*, **1**(11), 734-739.
24. Elhassan, A. S. M., Saeed, H. A. M., Eltahir, Y. A., Xia, Y. M., and Wang, Y. P. (2015) Modification of PLA with chain extender. *Applied Mechanics and Materials*, **716-717**, 44-47.
25. Al-Itry, R., Lamnawar, K., and Maazouz, A. (2012) Improvement of thermal stability, rheological and mechanical properties of PLA, PBAT and their blends by reactive extrusion with functionalized epoxy. *Polymer Degradation and Stability*, **97**(10), 1898-1914.
26. Mesias, R., and Murillo, E. A. (2015) Hyperbranched polyester polyol modified with polylactic acid. *Journal of Applied Polymer Science*, **132**(10), doi:10.1002/app.41589.
27. Chen, F., Gao, Q., Hong, G., and Ni, J. (2008) Synthesis of magnetite core–shell nanoparticles by surface-initiated ring-opening polymerization of l-lactide. *Journal of Magnetism and Magnetic Materials*, **320**, 1921-1927.
28. Socrates, G. (2004) *Infrared and Raman Characteristic Group Frequencies: Tables and Charts*, 3rd edition, John Wiley & Sons Ltd., USA.
29. Pivsa-Art, S., Phansroy, N., Thodsaratpiyakul, W., Sukkaew, C., Pivsa-Art, W., Lintong, S., and Dedgheng, T. (2014) Preparation of biodegradable polymer copolyesteramides from L-lactic acid oligomers and polyamide monomers. *Energy Procedia*, **56**, 648-658.
30. Kister, G., Cassanas, G., and Vert, M. (1998) Effects of morphology, conformation and configuration on the IR and Raman spectra of various poly(lactic acid)s. *Polymer*, **39**(2), 267-273.
31. Kaynak, C., and Sari, B. (2016) Accelerated weathering performance of polylactide and its montmorillonite nanocomposite. *Applied Clay Science*, **121-122**, 86-94.
32. Luiz de Paula, E., Roig, F., Mas, A., Habas, J.-P., Mano, V., Pereira, F. V., and Robin, J. J. (2016) Effect of surface-grafted cellulose nanocrystals on the thermal and mechanical properties of PLLA based nanocomposites. *European Polymer Journal*, **84**, 173-187.
33. Wu, H., Nagarajan, S., Zhou, L., Duan, Y., and Zhang, J. (2016) Synthesis and characterization of cellulose nanocrystal-graft-poly(d-lactide) and its nanocomposite with poly(l-lactide). *Polymer*, **103**, 365-375.
34. Olewnik-Kruszkowska, E., Nowaczyk, J., and Kadac, K. (2016) Effect of ozone exposure on thermal and structural properties of polylactide based composites. *Polymer Testing*, **56**, 299-307.

35. Olewnik-Kruszkowska, E., Koter, I., Wisniewska, J. S., and Richert, J. (2015) Degradation of polylactide composites under UV irradiation at 254 nm. *Journal of Photochemistry and Photobiology A: Chemistry*, **311**, 144-153.
36. Broz, M. E., van der Hart, D. L., and Washburn, N. R. (2003) Structure and mechanical properties of poly(D,L-lactic acid)/poly(epsilon - caprolactone) blends. *Biomaterials*, **24**(23), 4181-4190.

7

PLA/PBAT Blend Nanocomposites with Graphene: Structure and Properties

7.1 Introduction

Biopolymers (bio-based and biodegradable) have become polymer matrices of interest in recent years. The primary reason for their increased use is due to the fear of depleting fossil resources, which are by far the largest source of monomers from which the majority pf polymers are generated. Also, increasing consciousness towards the use of more environmentally friendly materials and allied legislations have shifted the focus from conventional non-biodegradable polymers to 'green' or biopolymers. Conventionally used biopolymers like polypropylene, polystyrene, polyethylene and poly(methyl methacrylate) are non-biodegradable and recycling or reusing them is very challenging, thus, contributing to piles of non-biodegradable wastes all over the world. As most biodegradable polymers find their origin in other alternative renewable sources, their use presents an attractive option, however, their use as an alternative to the commercial non-biodegradable polymers has emerged as both a dream and a challenge. There is need to develop more functional biopolymers as well as to enhance the property profiles of the existing ones so as to make them competitive to the conventional polymers.

Poly-l-lactide (PLA) is one of the most abundantly used biopolymer and has received much attention due to its superior property profiles [1,2]. PLA is linear aliphatic thermoplastic polyester, produced from renewable resources such as corn, sugar beets, etc. It is produced by ring-opening polymerization of lactides and the lactic acid monomers, obtained from the fermentation of sugar feed stocks. Due to mechanical properties comparable to many traditional polymers, PLA has the potential to reasonably substitute these conven-

V. Mittal[a,], M. Pannirselvam[b] and Gregory Griffin[b]*
[a]The Petroleum Institute (part of Khalifa University of Science and Technology), Abu Dhabi, UAE; [b]Royal Melbourne Institute of Technology, Melbourne, Australia
**Current address: Bletchington, Wellington County, Australia*

tional polymers in many application domains, e.g. packaging. However, it suffers from brittle behavior and has low resistance to heat and humidity. The flexibility of PLA is improved by the addition of plasticizers or by blending it with various non-biodegradable and biodegradable polymers [3-8]. Poly(butylene adipate-*co*-terephthalate) (PBAT) is one of the most common biopolymers used to blend with PLA. It is an excellent biodegradable aliphatic-aromatic copolyester, which rapidly degrades in the presence of naturally occurring enzymes and has excellent flexibility and impact properties. Owing to their comparable structural properties, PBAT is an ideal candidate as blending component to enhance flexibility of PLA. Many studies have reported the blends of PLA with PBAT, which combined the toughness and flexibility characteristics of PBAT with the mechanical properties of PLA [9-16]. The resulting blends were observed to have ductile failure and reduced brittleness depending on the amount of PBAT. In addition, it is well studied that addition of fillers can lead to enhancement in structural properties of polymer matrices [17-23]. Various types of fillers, mainly categorized as natural [18,23-25] and synthetic [22,26-27] fillers, have been developed as reinforcement for polymer matrices. Thus, incorporation of various fillers in PLA has also been reported so as to enhance its thermal and mechanical properties. For instance, Mittal *et al.* [28] reported the bio-nanocomposites of biopolymers PLA, polyamide (PA) and PBAT with varying amounts of graphene, generated by melt mixing. Moderate enhancement in the thermal and mechanical properties of PLA was observed, though the flexibility of the polymer was negatively impacted. Figure 7.1 also demonstrates the relative storage modulus of these composites as a function of filler fraction.

In a recent study, Pawar *et al.* [29] reported simultaneous enhancement of flexibility and mechanical properties of poly(3-hydroxybutyrate-*co*-3-hydroxyvalerate) (PHBV) by blending with PBAT and reinforcing with fillers like graphene nano-sheets and vermiculite platelets. In another study, Kumar *et al.* [30] also reported an increase in the tensile modulus from 1841 MPa to 2107 MPa in PLA/PBAT blends on incorporation of nanoclay. Jiang *et al.* [31] also reported that the use of maleic anhydride (MA) grafted PLA as a compatibilizer in the blend composites of PLA/PBAT with montmorillonite led to enhanced elongation of the ternary composites due to improved dispersion of the nanoparticles. Thus, blend composites open the possibilities to combine the positive effects of polymer blending and filler incorporation. The polymer phase leads to enhancement of

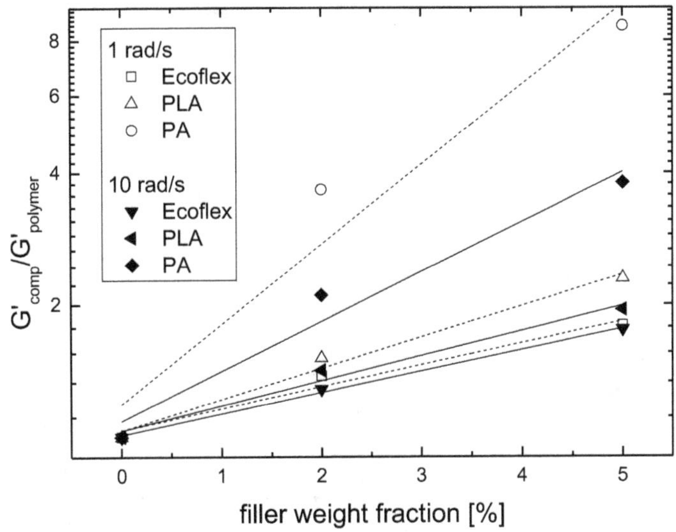

Figure 7.1 Relative storage modulus of PLA, PBAT (Ecoflex) and polyamide (PA) as a function of filler weight fraction at 1 rad/s and 10 rad/s. Reproduced from Reference 28 with permission from Elsevier.

flexibility, whereas the filler phase retains or enhances the stiffness, thus, providing increment in both stiffness and toughness, which is generally challenging to achieve.

It should also be noted that most of the reported studies have focused on the enhancement of mechanical properties of biopolymers by the incorporation of fillers, however, the effect of filler on the biodegradability of the polymer has largely been ignored. A decrease in the biodegradation rate of the polymer in the presence of conventional fillers has generally been reported [32], thus, negating the positive effect of filler on other properties. Pawar *et al.* [29] also reported a slow-down in the rate of biodegradation of PHBV/PBAT blend in the presence of graphene nano-sheets, whereas the incorporation of vermiculite was observed to enhance the rate of biodegradation due to its polar nature.

In another study, Mittal *et al.* [33] also reported the enhancement of biodegradation of PLA and PBAT in the presence of bio-filler based on date seed powder (DSP), as shown in Figure 7.2. Thus, it is evident that the nature of the fillers has a significant effect on biodegradation

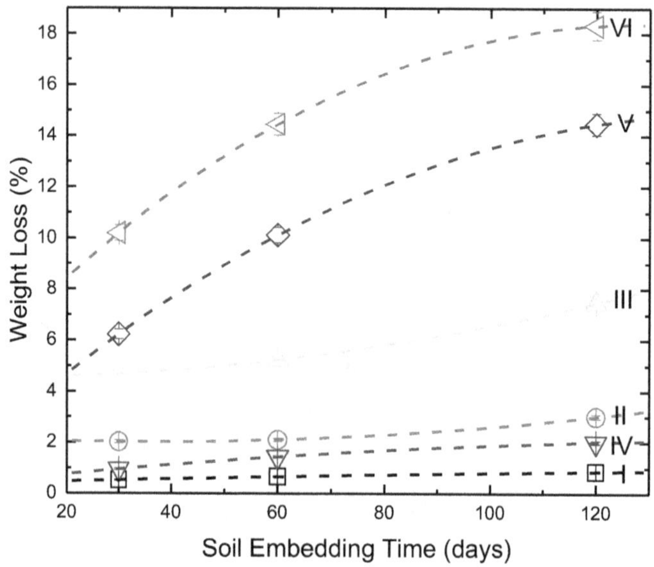

Figure 7.2 Weight loss versus soil embedding time for PLA, PBAT and date seed powder (DSP) composites. I: PLA; II: PLA + 20% DSP; III: PLA + 40% DSP; IV: PBAT; V: PBAT + 20% DSP; VI: PBAT + 40% DSP. Reproduced from Reference 33 with permission from Wiley.

of the polymers and it is of importance to at least retain the biodegradability of the polymers in the composites.

In the current study, successful attempt to enhance the mechanical properties of PLA has been made by combining the polymer blending and filler incorporation approaches. PLA was blended with highly flexible PBAT polymer to enhance ductility. In addition, in order to enhance modulus, graphene nano-sheets based reinforcement was introduced in the blend in different weight fractions by melt mixing. Mechanical properties like tensile modulus and elongation at break of PLA/PBAT (60/40 wt%/wt%) blend and nanocomposites were compared with neat polymers. In addition, effect of filler fraction in the nanocomposites on the rheological properties was studied systematically. Thermal properties were assessed through differential scanning calorimetry (DSC) and thermogravimetric analysis (TGA) techniques. Effect of filler loading on the biodegradation rate of blend composites, in comparison with the pure polymers, was also studied through soil embedding method.

7.2 Experimental

7.2.1 Materials

Biopolyester PBAT with trade name Ecoflex® F Blend C1200 was supplied by BASF SE, Germany. As mentioned earlier, it is a biodegradable aliphatic-aromatic copolyester which biodegrades to the basic monomers and eventually to carbon dioxide, water and biomass when metabolized in the soil or compost under standard conditions. PLA was procured from Biomer, Germany and had a density of 1.25 g/cm^3, glass transition temperature (T_g) of 55 °C and peak melting temperature (T_m) of 168 °C. Graphene sample (exfoliated graphite in the form of a fluffy powder) with trade name Grafen-iGP was kindly provided by Grafen Chemical Industries, Turkey. The polymers and filler materials were used as commercially received, without further modification.

7.2.2 Generation of PLA/PBAT Blend and Nanocomposites

Mini twin conical screw extruder (MiniLab HAAKE Rheomex CTW5, Germany) was used for the melt mixing of PLA, PBAT and graphene to generate PLA/PBAT (60:40) blend and its composites with 1, 2, 3 and 5 wt% graphene. The screw length and screw diameter were 109.5 mm and 5/14 mm conical respectively. A mixing temperature of 180 °C, rotational speed of 70 rpm and mixing time of 8 min were used. Approximately 20 g of each sample was prepared and dry mixed prior to extrusion, in order to ensure proper ratios of the components. An initial mass of 6 g was added to the compounder and blended. After 5 min, 3 g was flushed from the machine so that another 3 g could be added. This process was repeated until 20 g material was blended. This method ensured that the compounder would not be overloaded and also provided small batches to ensure more intense mixing. Pure polymers were also extruded similarly for effective comparison of the properties with the blends, though different compounding temperatures were used (180 °C for PLA and 145 °C for PBAT).

For the subsequent characterization, the disc, dumbbell and bar-shaped test samples were prepared by hot pressing the extruded materials at 180 °C for 5 min. The hot-pressed samples were cooled down to 50 °C by using flowing water through the plates following a cooling rate of 10 °C.min^{-1}.

7.2.3 Material Characterization

For the bright field transmission electron microscopy (TEM) analysis of the composite samples, Philips CM 20 (Philips/FEI, Eindhoven) electron microscope at 120 kV and 200 kV accelerating voltages was used. Thin sections of 70-90 nm thickness were microtomed from the sample block and were supported on 100 mesh grids sputter coated with a 3 nm thick carbon layer.

X-ray diffraction analysis was conducted using a Bruker AXS D4 Endeavour system using Cu-Kα radiation operated at 40 kV and 40 mA and a Lynxeye linear strip detector. All polymer samples (hot pressed) were loaded onto the sample holder. Microscope slides were used to flatten the surface of all samples before loading. All samples were tested between 10 and 90° 2θ with variable slit v20 with step size 0.02 s. The instrument was calibrated using a NIST standard of Al_2O_3 (corundum ceramic block). Endeavour is equipped with a large sample magazine that allows a wide range of samples to be docked in a single stream (manual). In the current study, a search/match technique was used to compare the constituents present in the samples (ICDD Database, Bruker DIFFRAC.EVA software version 3.0). It should be noted that background was removed from all spectra before further processing. Diffraction peaks were not smoothened for any spectra as smoothing might vary area of the peaks of interest.

Thermal properties of the samples were analyzed using TA Discovery thermogravimetric analyzer (TGA). Nitrogen was used as a carrier gas and the scans were obtained from 50 to 750 °C at a heating rate of 10 °C.min^{-1}. Differential scanning calorimetric (DSC) measurements on the polymer blend and bio-nanocomposites were performed using TA Discovery DSC under nitrogen atmosphere. The second heating scans were obtained by heating the samples from -20°C to 200°C at a heating rate of 10°C.min^{-1}, followed by cooling from 200°C to -20°C at a cooling rate of 10°C.min^{-1}.

Tensile analysis of the samples was carried out using universal testing machine (Instron, USA) (ASTM D638). The dumbbell-shaped samples were used. A loading rate of 10 mm/min was used and the tests were carried out at room temperature. The tensile strength and modulus of the composites were calculated using Win Test Analysis software. An average of five sample values was reported. ARES rheometer was used to characterize the rheological properties such as storage modulus (G′), loss modulus (G″), as well as complex viscosity

(η^*). Disc-shaped samples of 8 mm diameter and 2 mm thickness were measured using a gap opening of 1.2 mm. Strain sweep scans were recorded at 180 °C with ω = 1 rad.s^{-1} and 0.1% to 100% strain. For the comparison, temperature and frequency sweep scans (dynamic testing) of all the samples were recorded at 1% strain from ω = 0.1 to 100 rad.s^{-1} in the temperature range of 180-210 °C. DMA (dynamic mechanical analyzer) RSA3 of TA Instruments was used to study the change in modulus with temperature (-100 to 125 °C). Dynamic temperature ramp test was performed at a frequency of 5 rad.s^{-1}, ramp rate 3 °C.min^{-1} and 1% strain for all samples.

Biodegradability of the blend and nanocomposites was analyzed with soil burial tests, as reported earlier [34]. The tests were carried out under natural environmental conditions and no additional effort was made to modify the conditions. The samples were buried in commercially available compost soil for gardening (Bestgreen Universal potting soil). Two pots of 52*14*10 cm dimension were used for the soil burial test and placed in an open place usually used for gardening. Soil was filled to a height of 10 cm in the pots. Samples with dimensions of 17*9.5*1.5 mm were buried 5 cm deep, 4 cm apart in longitudinal direction and 5 cm apart in transverse direction. The soil moisture content was 60 to 70% on a normal day. On a sunny day, the soil temperature was usually between 33 to 37 °C at mid-day and at night between 20 to 25 °C. The samples were dug out every 15 d, cleaned from soil particles and dried in a vacuum oven at 70 °C, followed by measurement of their weight.

7.3 Results and Discussion

In the current study, efforts have been made to obtain optimal enhancement in the PLA toughness and stiffness by using the blend composites concept. PBAT phase was melt mixed with PLA to enhance the flexibility and owing to its 40% weight fraction in the blend as well as immiscibility with PLA, it was expected to form dispersed domains inside the PLA phase, thus, providing toughness. On the other hand, varying fractions of graphene were added to enhance the modulus and high temperature performance. However, as mentioned earlier, it was also important to either enhance or retain the biodegradability of PLA in the blend composites. The localization of filler in the blend components, in addition, is also an important factor affecting the overall evolution of the composite performance. Also, owing to lower thermal stability of PLA polymer at high temperatures, the compounding was carried out at 180 °C for a short period of time so as to provide good mixing but also to avoid the polymer degradation. The extruder was also used in the continuous mode for avoiding longer residence time for the polymer in the compounder. The materials were dried at 80 °C under vacuum overnight to remove any moisture bound to the surface, which can otherwise lead to unwanted degradation reactions during high temperature compounding. The blend composites were extensively characterized for the morphology, thermal properties, mechanical properties as well as biodegradation and the performance was compared with the pure polymers as well as polymer blend as a function of filler fraction.

Figure 7.3 shows the morphology of the blend and composite with

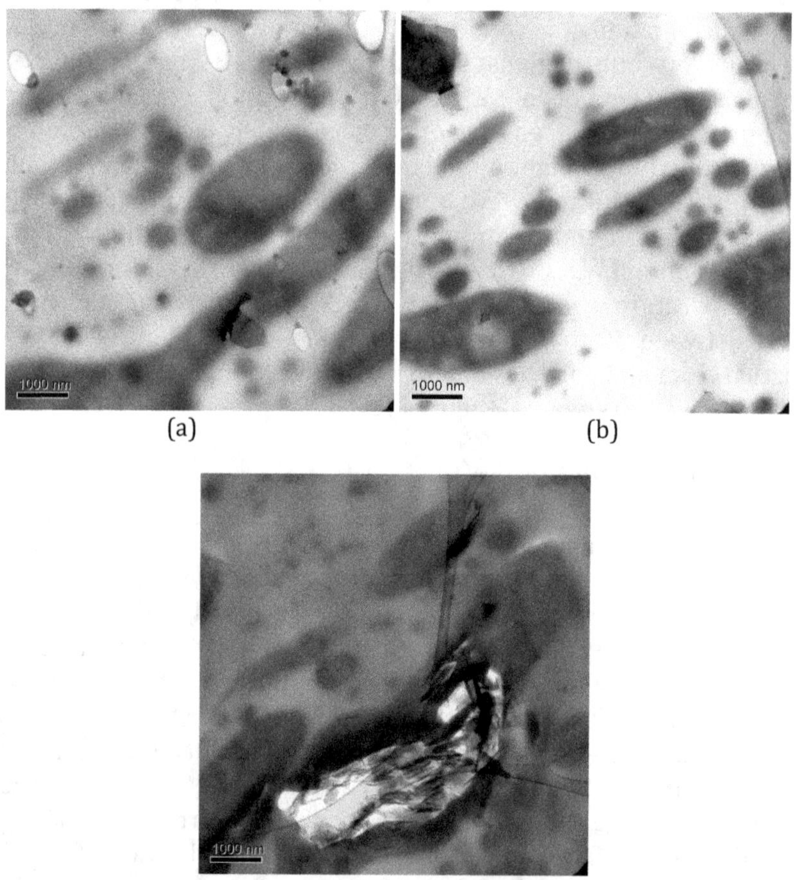

Figure 7.3 (a-c) TEM micrographs of PLA/PBAT/5%G composite.

5% filler content. PBAT (dark phase in the images) was observed as dispersed domains in PLA (light phase in the images), indicating phase immiscibility. Also, the PBAT domains were uniformly dispersed in PLA without any large phase separation. The phase was also observed to be retained after multiple processing steps, indicating the phase stability, even though the polymers were only kinetically mixed in the compounder. As evident from the micrographs, the filler was observed to be localized specifically in the PBAT phase. Graphene is not expected to have any interaction with the PBAT phase, thus, the observed filler dispersion primarily in the PBAT phase is due to the kinetic mixing in the compounder. As PBAT has lower melting point than PLA, thus, its melt would mix with the filler particles much before the PLA pellets formed the melt. An earlier study has also reported the preferential dispersion of graphene and vermiculite in the PBAT phase in PHBV/PBAT blends [29]. However, the dispersion of filler in relatively amorphous PBAT helped to have optimal mechanical performance as the dispersion of filler in the lower elongation polymer would have resulted in even further reduction in flexibility, and probably biodegradation. Thus, in the current study, PBAT can be visualized to act as a rubbery phase holding graphene, which is expected to retain flexibility of the system, along with providing better stress transfer to the polymer chains to simultaneously enhance the stiffness. The filler was also observed to have varying degree of dispersion in the PBAT phase. Occasionally, large stacks of graphene nano-sheets were observed, however, single or few layered stacks were also simultaneously noticed. As shown in Figure 7.4 demonstrating the graphene dispersion in the PBAT phase of the nanocomposite with 5% graphene content, a good degree of filler dispersion could be achieved by one step melt mixing process.

Further insights into the morphology of the blend and composites was also gained through X-ray diffraction analysis, as presented in Figure 7.5. The diffractogram of pure PLA had two diffraction peaks at 2θ values of 16.8° and 19°. The peak recorded at 2θ 16.8° was recognized as the (200, 110) reflection of the α-form of crystals, and the other peak appearing at 19° corresponded to the (010) reflection [35]. On the other hand, PBAT did not exhibit any defined diffraction peak owing to its largely amorphous nature, with a less intense peak at 23.5°. The diffraction pattern of the PLA/PBAT blend was observed to contain the peaks of both polymers, indicating no change in their crystalline nature on blending. In the blend composites, some change in diffraction peaks was observed. The PLA characteristic peak at 19°

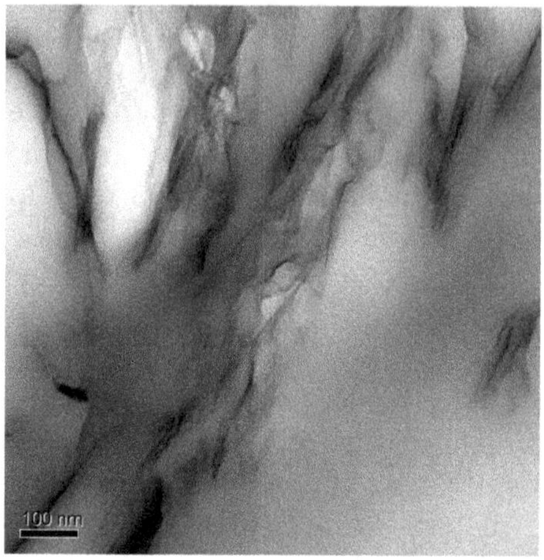

Figure 7.4 TEM micrograph of the PBAT phase of PLA/PBAT/5%G composite depicting the distribution of graphene.

was observed as a weak signal in the composite with 1% filler content. Further increasing the amount of graphene in the composites led to complete disappearance of this peak, which indicated a change in the crystal structure of PLA in the composites. On the other hand, the characteristic peak at 16.8° was not affected and was observed as a sharp signal in the diffractograms. As the filler is observed largely in the PBAT phase, the impact on PLA crystal structure in the composites is surprising. Furthermore, the weak signal corresponding to PBAT at 23.5° was also observed in the composites. In addition, the diffraction peak corresponding to graphene at 27° was observed as a sharp signal in the composites indicating the absence of complete exfoliation, which was also observed in the TEM analysis. Thus, the diffraction analysis confirmed that the crystal structure of PLA underwent some changes in the composites, though it was unaffected in the polymer blend. In addition, there was apparently no impact on the crystal structure of largely amorphous PBAT.

Figure 7.6 also shows the calorimetric behavior of the materials. As evident in Figure 7.6a, PLA exhibited semi-crystalline nature with peak melting point of 171.3 °C. On the other hand, a diffuse melt transition was observed for PBAT due to its largely amorphous nature, with a peak melting temperature of around 120 °C. The thermograms of the blend and composites exhibited a single melt transition corresponding to PLA. The melt transition of PBAT was almost absent from the thermograms, which indicated that the filler present in the PBAT

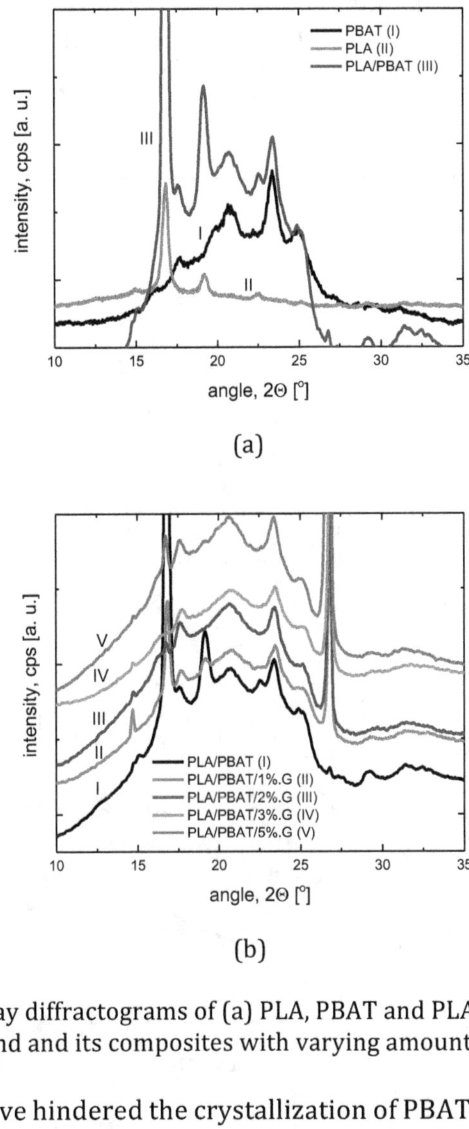

(a)

(b)

Figure 7.5 X-ray diffractograms of (a) PLA, PBAT and PLA/PBAT blend; (b) PLA/PBAT blend and its composites with varying amounts of graphene.

phase may have hindered the crystallization of PBAT further. Though the X-ray diffractograms exhibited the presence of crystalline diffraction peaks corresponding to PBAT, however, its degree of crystallinity may have further reduced in the presence of filler, as depicted by DSC. Furthermore, the melt transition corresponding to PLA was also observed at a lower temperature around 169 °C, which also supported the findings from the diffraction analysis. The degree of crystallinity of the PLA phase in the composites was also observed to gradually decrease with increasing filler content. Thus, overall, the diffraction and DSC studies confirmed a change in the crystal structure and extent of crystallinity of PLA in the bio-nanocomposites, thus, also indicating corresponding impact on the mechanical performance.

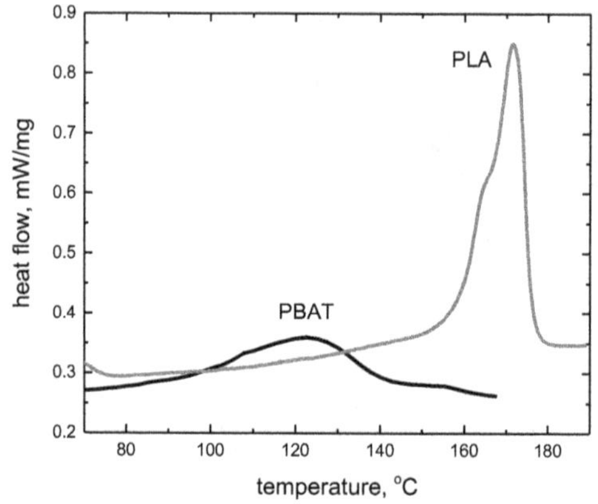

(a)

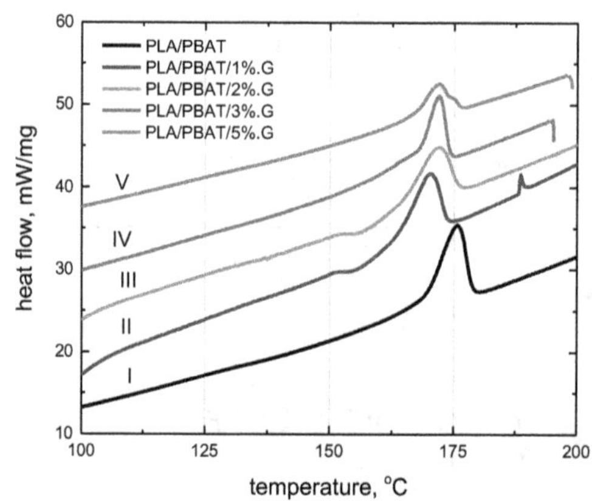

(b)

Figure 7.6 DSC thermograms of (a) PLA and PBAT; (b) PLA/PBAT blend and composites.

Figure 7.7 demonstrates the thermal degradation behavior of the pure polymers, blend and nanocomposites. PBAT was observed to be more thermally stable than PLA. The peak degradation temperatures for PLA and PBAT were observed as 378 and 425 °C respectively. The thermal stability of the blend was observed to be lower than both PBA and PLA, with a peak degradation around 325 °C. Probably, the mechanism of polymer degradation underwent some changes in the blend which resulted in a lower extent of thermal stability. In addition, the thermogram also exhibited two degradation steps corresponding to PLA (lower temperature degradation step) and PBAT (higher temperature degradation step). In the nanocomposites, the thermal stability enhanced significantly, though the impact of filler content was less significant. The thermograms of composites also exhibited a two

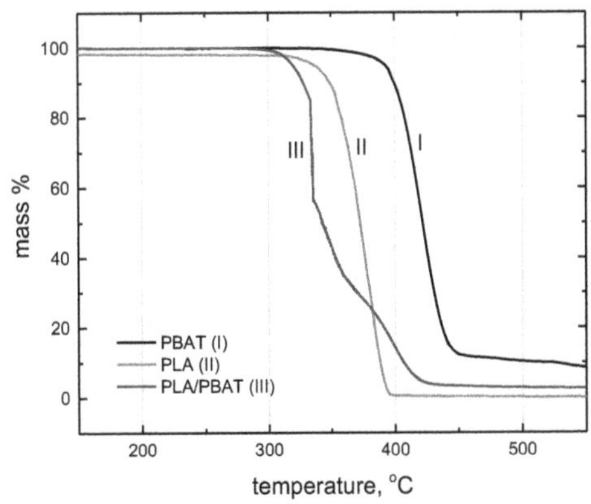

(a)

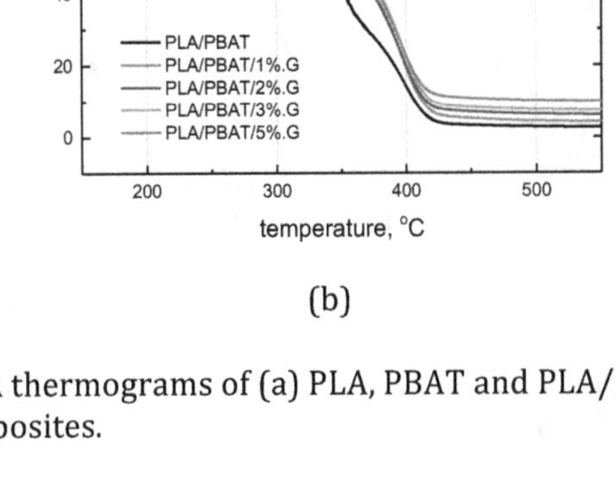

(b)

Figure 7.7 TGA thermograms of (a) PLA, PBAT and PLA/PBAT blend; (b) blend and composites.

step degradation pattern. Overall, though the thermal stability was improved in the composites, however, these were still less stable than the pure PBAT.

Figure 7.8 and Table 7.1 demonstrate the tensile properties of the materials. PLA exhibited the tensile modulus of 3306 MPa, in comparison to 78 MPa for PBAT. On the other hand, the yield strain of 3% was observed for PLA, in comparison with 13.1% for PBAT. In the blend, the tensile properties were intermediate between the two polymers, as the blend exhibited the tensile modulus and yield strain of

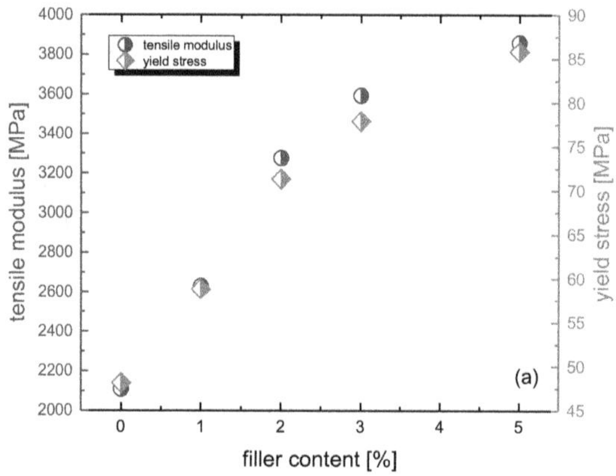

(a)

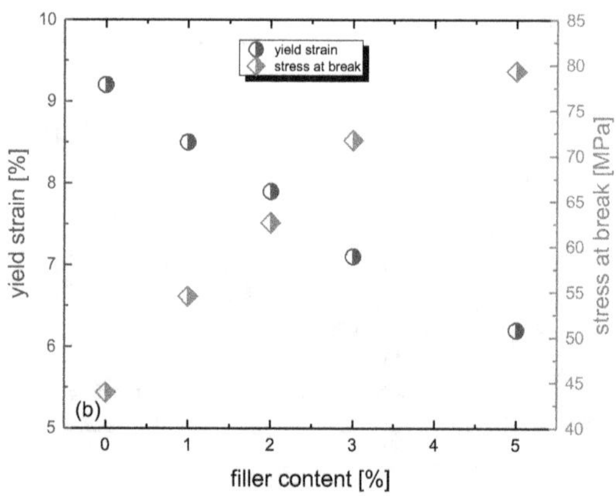

(b)

Figure 7.8 (a) Tensile modulus and yield stress; (b) yield strain and stress at break of the nanocomposites as a function of filler content.

Table 7.1 Tensile properties of pure polymers, blend and composites

Material	Tensile modulus[a], MPa	Yield stress[b], MPa	Yield strain[a], %	Stress at break[c], MPa
PLA	3306	70.3	3.0	63.7
PBAT	78	9.1	13.1	20.9
PLA/PBAT	2108	48.1	9.2	43.9
PLA/PBAT/1%G	2629	58.8	8.5	54.5
PLA/PBAT/2%G	3276	71.3	7.9	62.6
PLA/PBAT/3%G	3593	77.9	7.1	71.7
PLA/PBAT/5%G	3859	85.8	6.2	79.3

[a]Relative probable error 5%
[b]Relative probable error 2%
[c]Relative probable error 15%

2108 MPa and 9.2% respectively. The incorporation of graphene in the blend resulted in gradual increase in the tensile modulus. For instance, the modulus of the blend composite with 5% graphene content was measured to be 3859 MPa, which was >15% higher than pure PLA. Similarly, the yield stress and stress at break of the composites increased as a function of graphene content. The yield strain of the composites was observed to decrease with filler content, however, it was still higher than the pure PLA. For comparison, the composite with 5% filler content had a yield strain of 6.2%, which was >100% higher than the yield strain of PLA. This indicated simultaneous increment of modulus and flexibility in the composites, which was not hindered even at higher extent of filler. Thus, the particular morphology of filler localization in PBAT phase helped to achieve optimal enhancement in the mechanical performance. The observed increment was also much superior than earlier reported graphene and date seed powder based composites with PLA [28,33], thus, confirming the benefit of blending with PBAT and specific morphology of the composite. The filler dispersion in these studies was reported to be less optimal, thus, the observed performance in the current study indicates effective filler dispersion. In an earlier study on the PHBV/PBAT blend based composites with graphene, increment of 30% and 225% in the tensile modulus and flexibility in the composite with 5% graphene content was observed, in comparison with pure PHBV [29]. Though, similar enhancements in the mechanical performance of the composites were not observed in the current study, however, the observed improvements are still significant due to the

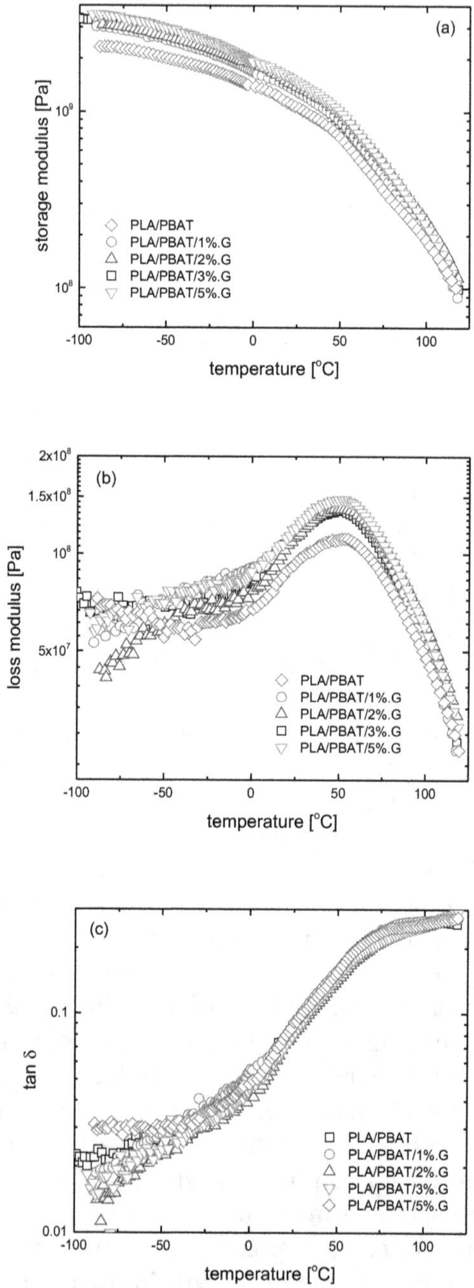

Figure 7.9 Dynamic mechanical analysis of blend and its composites: (a) storage modulus, (b) loss modulus and (c) tan δ.

challenge associated with the enhancement of much higher modulus of PLA, while also improving the flexibility.

Figure 7.9 also shows the dynamic mechanical performance of the composites. As evident from Figure 7.9a, the storage modulus of the composites increased as a function of filler content, similar to the findings from the tensile analysis. The modulus exhibited a gradual decrease with temperature, but still remaining high even at 125 °C. In addition, the storage modulus did not exhibit any increase in the value around the transition temperature of ~60 °C. An increase in the storage modulus around the transition temperature is attributed to the crystallization of the PLA phase occurring during the DMA measurements [36]. Absence of this phenomenon in the current study indicated that the crystallization of PLA at the measurement conditions did not take place. The storage modulus was higher in magnitude than the loss modulus over the complete temperature range. In addition, the loss modulus was observed to drop sharply beyond 50 °C. The tan δ curves also exhibited a weak transition for PLA around 70 °C, which was not impacted by the presence of filler.

Further insights into the viscoelastic behavior of the blend and nanocomposites were gained through rheological analysis, demonstrated in Figures 7.10-7.13. Figure 7.10 presents the strain sweep analysis (0.1-100%) at an angular frequency of 1 rad.s^{-1}. The performance of the blend and blend composites with 1% and 3% filler content was analyzed. The modulus and viscosity of the materials had a weak strain dependence as the values were observed to gradually decrease with strain. A value of 1% strain was observed to be optimal for both frequency and temperature sweep analysis. In addition, loss

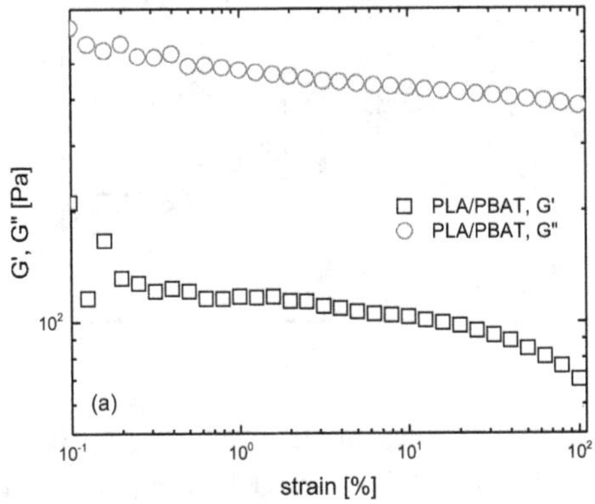

Figure 7.10 (continued on next page)

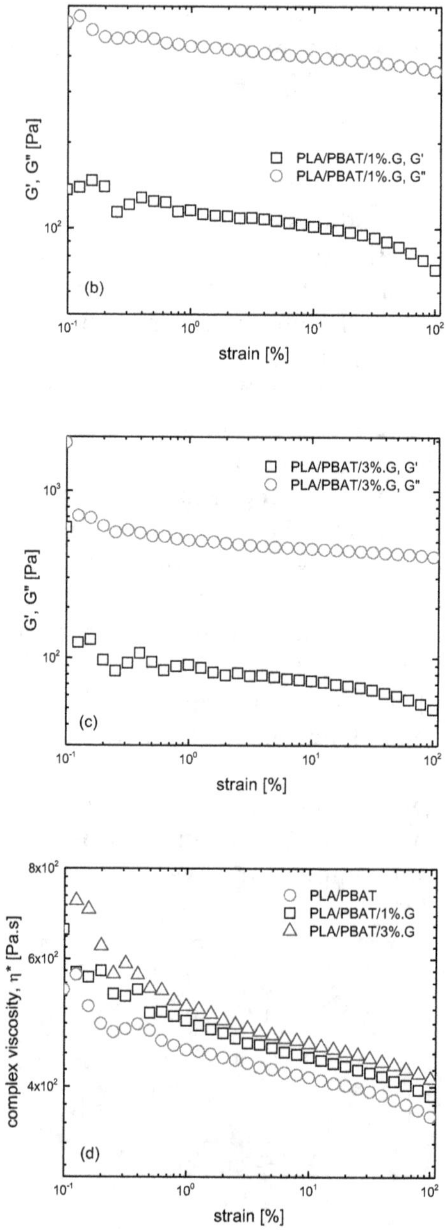

Figure 7.10 Strain sweep analysis at an angular frequency of 1 rad.s^{-1}. Storage and loss modulus of (a) PLA/PBAT blend, (b) PLA/PBAT blend composite with 1% filler content and (c) PLA/PBAT blend composite with 3% filler content; (d) complex viscosity of these materials.

modulus was higher than storage modulus over the whole angular frequency range, indicating the dominance of viscous nature. The modulus and viscosity were also observed to increase as a function of filler content in the composites.

Figure 7.11 shows the storage and loss modulus of the materials as a function of temperature (180-210 °C) and angular frequency (0.1-100 rad.s^{-1}). The first curve in the set of four represented the behavior at 180 °C, whereas second, third and fourth curves (moving downwards) exhibited the property values at 190, 200 and 210 °C respectively. As expected, the modulus increased with angular frequency, but decreased with temperature. Comparing among the different samples, the modulus was observed to increase with filler content, though the enhancement was not significant. In addition, the loss modulus was higher than the storage modulus at all frequencies

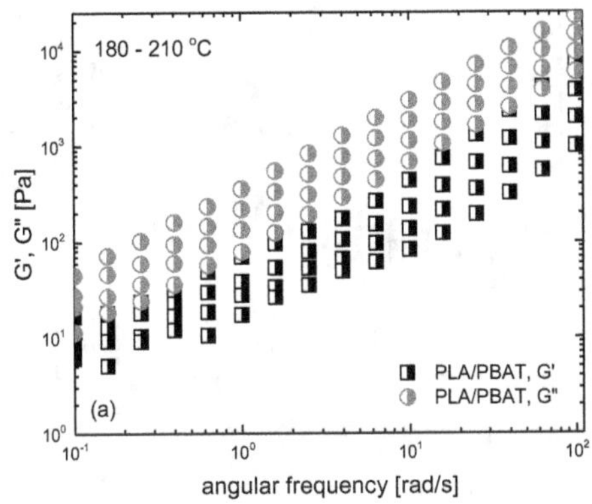

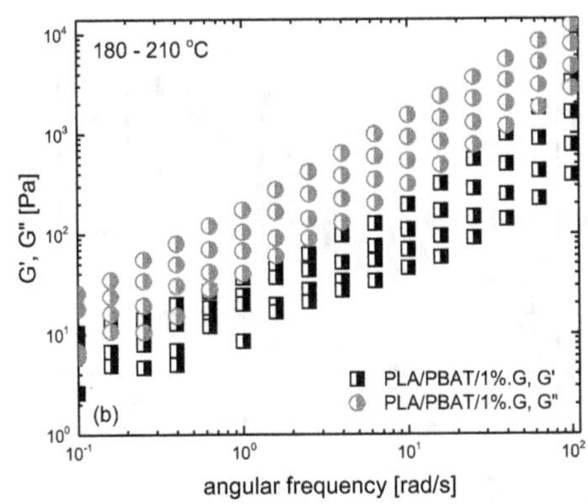

Figure 7.11 (continued on next page)

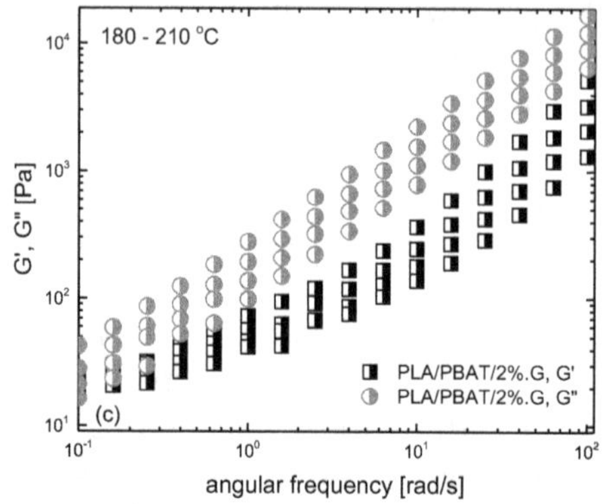

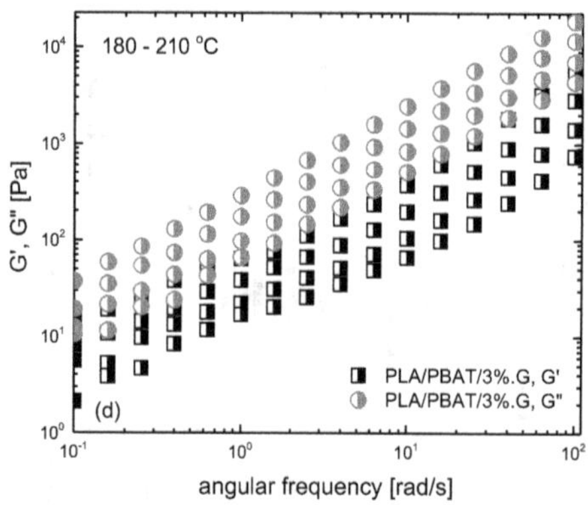

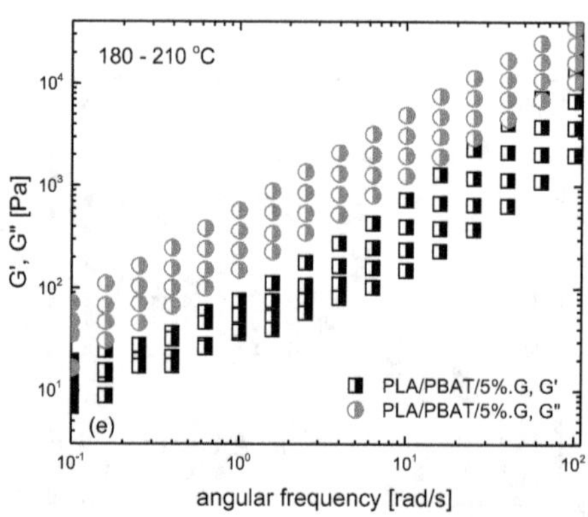

Figure 7.11 Storage and loss modulus as a function of temperature and angular frequency: (a) PLA/PBAT blend, (b) PLA/PBAT/1%G, (c) PLA/PBAT/2%G, (d) PLA/PBAT/3%G and (e) PLA/PBAT/5%G.

due to dominant viscous nature at the shear rheology conditions. As shown in Figure 7.12, the complex viscosity of the materials also exhibited an increase with increasing graphene content, whereas it reduced with temperature and angular frequency. The observed modulus and viscosity behavior also indicated that the materials did not suffer melt degradation even at 210 °C. Figure 7.13 also demonstrates the tan δ of the materials as a function of temperature and angular frequency. The materials did not exhibit any significant change in the transition point with temperature as well as filler fraction, as also observed earlier in the dynamic mechanical analysis.

Figure 7.14 demonstrates the biodegradation behavior of the pure polymers, PLA/PBAT blend and selected nanocomposites, analyzed using soil embedding test as a function of time. PBAT was observed to have the highest extent of biodegradation, as evident from the 2%

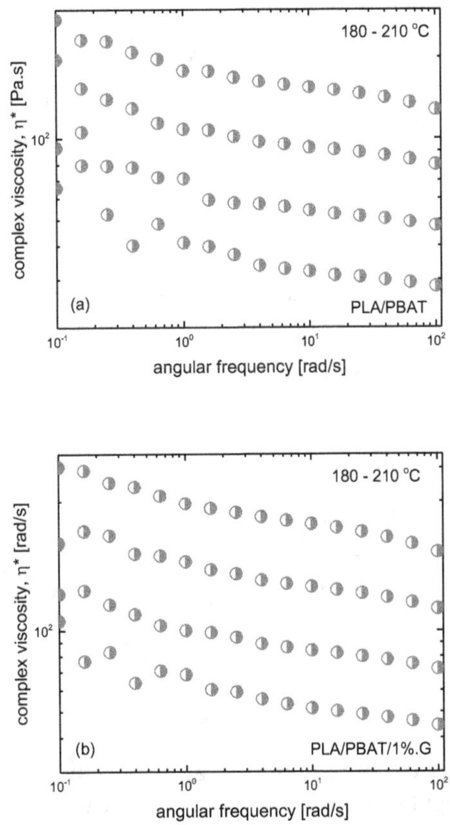

Figure 7.12 (continued on next page)

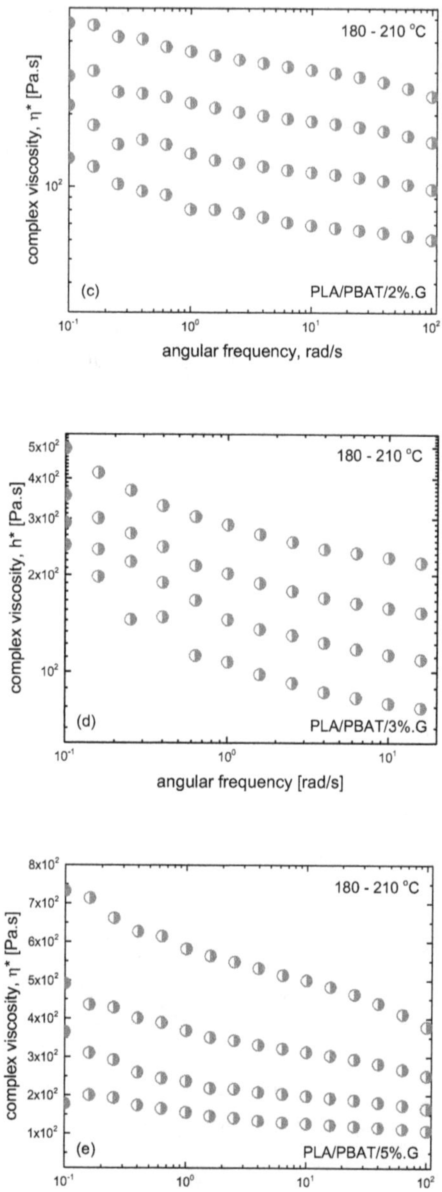

Figure 7.12 Complex viscosity of the materials as a function of temperature and angular frequency: (a) PLA/PBAT blend, (b) PLA/PBAT/1%G, (c) PLA/PBAT/2%G, (d) PLA/PBAT/3%G and (e) PLA/PBAT/5%G.

weight loss after 120 days of embedding. On the other hand, PLA exhibited <0.9% weight loss during this period. As expected, the incorporation of PBAT in PLA in the blend led to the enhancement in its biodegradation behavior. The incorporation of 2% graphene in the blend led to the reduction in the extent of weight loss with time. For instance, the composites exhibited a cumulative weight loss of ~1.2% after 120 days of embedding as compared to ~1.4% for the PLA/PBAT blend, which was still higher than pure PLA. Further increasing the filler content to 5% also led to corresponding decrease in the extent of biodegradation, with the composite behavior marginally better than pure PLA. Thus, it indicated that the blend composites till 3% filler content had higher biodegradation than PLA, and biodegradation of PLA was maintained to the similar value in the blend composite with 5% graphene. In an earlier study by Patwary *et al.* [34], a

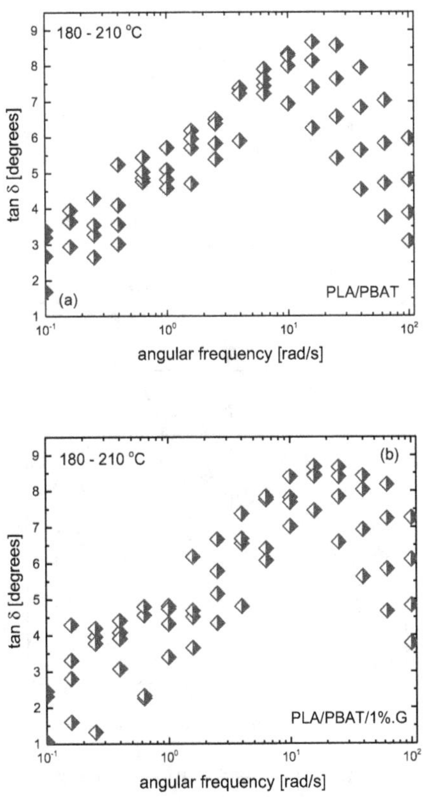

Figure 7.13 (continued on next page)

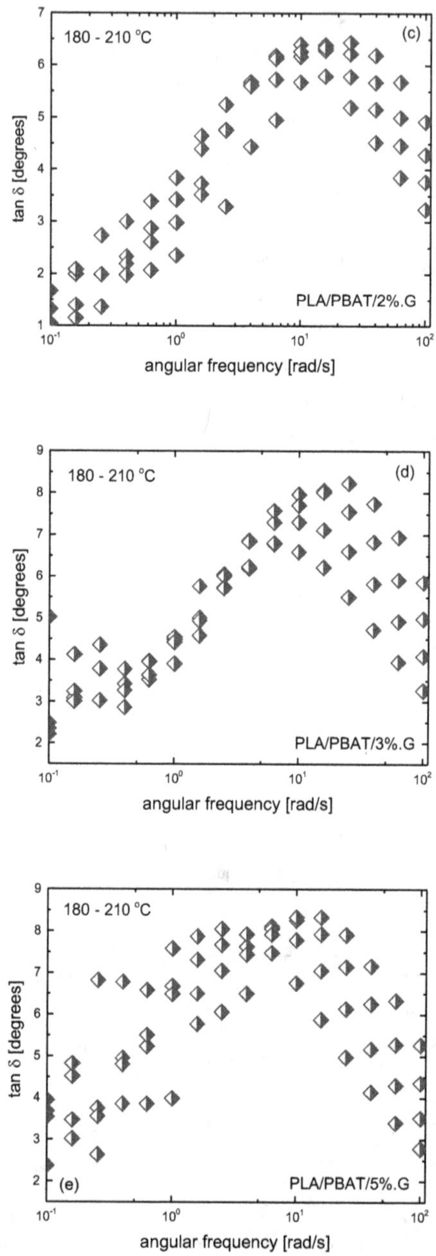

Figure 7.13 Tan δ of the materials as a function of temperature and angular frequency: (a) PLA/PBAT blend, (b) PLA/PBAT/1%G, (c) PLA/PBAT/2%G, (d) PLA/PBAT/3%G and (e) PLA/PBAT/5%G.

lower extent of weight loss than the pure blend in the melt processed bio-nanocomposites of poly(butylene succinate) (PBS)-chitosan blend with varying amounts of silica, alumino-silicate and thermally reduced graphene was also reported. However, extensive surface roughness and cracks were still observed for all systems indicating the initiation of biodegradation, which also holds true for the nanocomposites developed in the current study. It, thus, indicates that even though the weight loss % may be low, it is still enough to initiate structural degradation in the polymers. It should also be noted that normal environmental conditions without any additional effort were employed for the degradation study, thus, it is expected that the degradation would become much faster if the severity of the conditions is further increased. For instance, in an earlier study, the composites degraded at higher environmental temperatures exhibited much higher weight loss as a function of time [37]. In addition, though the extent of biodegradation was observed to be lesser than the reported studies on PLA composites with biofillers such as date seed powder

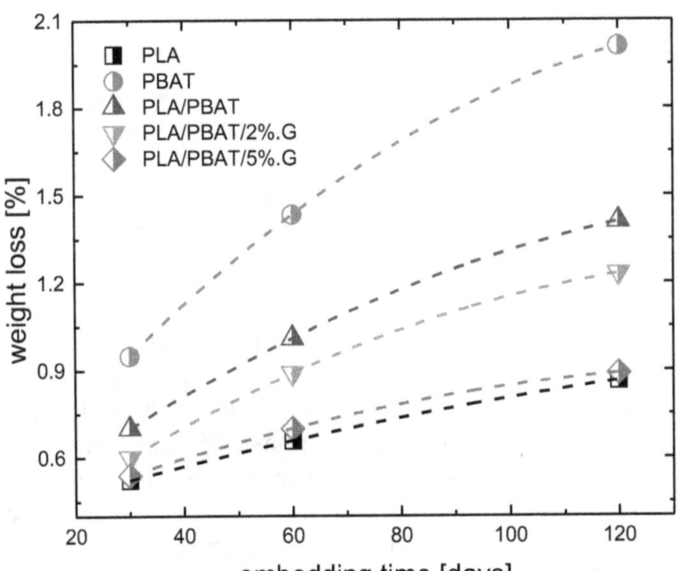

Figure 7.14 Biodegradation behavior of the pure polymers, PLA/PBAT blend and selected nanocomposites, analyzed using soil embedding test as a function of time.

[33], however, the optimal set of overall properties attained in the current study still indicate much better performance profile of these nanocomposites. Earlier studies also exhibited a slowdown in the rate of biodegradation with time due to faster degradation of the amorphous phase followed by the slower degradation of the crystalline phase [34]. However, no such slowdown was apparent in the current study, probably due to the morphology of the composites where the relatively amorphous polymer was embedded as dispersed phase in the highly crystalline polymer, thus, not readily available to the degradation agents during soil embedding analysis. Overall, the developed nanocomposites represent advanced materials with optimal set of stiffness, toughness and biodegradation performance, which opens many opportunities for the commercial application of these materials in not only commodity fields, but also in many engineering areas. Further tuning in the performance can be achieved by varying the blend composition as well as by studying the biodegradation in different environments.

7.4 Conclusions

In the current study, the usefulness of blend composites approach has been demonstrated to achieve simultaneous improvement in toughness and stiffness of the biopolymers, along with retaining the biodegradability. Specifically, a 60:40 blend of PLA with PBAT was generated to achieve good degree of flexibility, which was subsequently reinforced with varying amounts of graphene in order to achieve high stiffness. Conventional melt mixing process was used to generate the materials and one-step mixing was used. As a result, the morphology of the composites exhibited localization of the filler in the PBAT phase due to earlier melting of PBAT in the compounder during the mixing process. PBAT was observed as dispersed phase in PLA, and the phases were immiscible, though the phase domains were smaller in size and were stable even after few processing cycles. Varying degree of filler dispersion was achieved in the composites, ranging from occasional thick stacks to single or few-layered stacks of graphene nano-sheets. The diffraction analysis indicated alteration in the crystal structure of PLA as the characteristic diffraction peak at 19° was observed to disappear as the amount of graphene was increased in the composites. There was no change in the other lower angle diffraction peak of PLA at 16.8° as well as the diffuse peak at 23.5° for PBAT. Also, the characteristic diffraction peak corresponding to graphene was observed in the composite patterns, indicating that complete graphene delamination could not be achieved in the melt mixing process. In addition, the DSC thermograms of the composites did not exhibit any melt transition corresponding to PBAT indicating its reduced degree of crystallinity in the presence of filler. Also, the peak melting point of PLA was observed at lower temperatures in the composites, along with reduction in its melt enthalpy as the filler fraction was increased. The blend was also observed to exhibit lower thermal stability than the pure polymers, probably owing to different degradation mechanism. In the composites, the thermal stability was observed to be enhanced, however, it was still lower than pure PBAT. In addition, there was no significant effect of filler fraction on the thermal stability. Also, the thermograms exhibited two step degradation pattern owing to the presence of PBAT and PLA polymers as immiscible phases.

The tensile properties of the composites were observed to significantly enhance with filler amount. The composite with 5% graphene content exhibited >15% higher modulus and >100% higher yield strain as compared to pure PLA. Thus, it indicated simultaneous increment in the stiffness and flexibility in the blend composites. Specific localization of graphene in PBAT phase also contributed to the observed mechanical behavior as PBAT acted as a rubbery phase holding graphene particles, thereby effectively reinforcing the PLA phase. Dispersion of graphene in PLA would have otherwise resulted in further reduction in flexibility. It should also be noted that the observed increment in mechanical properties was observed even after reduction in the polymer crystallinity. Dynamic mechanical analysis also confirmed enhanced modulus as a function of filler content. The composites exhibited high modulus even at 125 °C. The rheological analysis also confirmed slight increment in the storage and loss modulus as well as viscosity as a function of filler content. The blend composites had higher extent of biodegradation as compared to pure PLA. Even the composite with 5% filler was observed to have slightly higher extent of biodegradation than PLA. This indicated that the generated blend nanocomposites had not only good balance of stiffness and toughness, but also had same or enhanced extent of biodegradation as PLA.

These bio-nanocomposites, thus, exhibit synergistic combination of the physical properties of their component polymers and represent high potential of use in a wide variety of advanced applications in diverse areas.

Acknowledgement

The authors are indebted to Bradley Burren for support in material preparation.

References

1. Kricheldorf, H. R., and Jonte, J. M. (1983) New polymer syntheses. 8. Synthesis and polymeri- zation of L-lactic acid O-carboxyanhydride (5-methyl-dioxolan-2,4-dione). *Polymer Bulletin*, **9**(6), 276-283.
2. Anders, S., and Stolt, M. (2002) Properties of lactic acid based polymers and their correlation with composition. *Progress in Polymer Science*, **27**, 1123-1163.
3. Labrecque, L. V., Kumar, R. A., Dave, V., Gross, R. A., and McCarthy, S. P. (1997) Citrate esters as plasticizers for poly (lactic acid). *Journal of Applied Polymer Science*, **66**(8), 1507-1513.
4. Jacobsen, S., and Fritz, H. G. (1999) Plasticizing polylactide - The effect of different plasticizers on the mechanical properties. *Polymer Engineering and Science*, **39**(7), 1303-1310.
5. Zhang, L., Goh, S. H., and Lee, S. Y. (1998) Miscibility and crystallization behaviour of poly(l-lactide)/poly(p-vinylphenol) blends. *Polymer*, **39**(20), 4841-4847.
6. Ogata, N., Jimenez, G., Kawai, H., and Ogihara, T. (1997) Structure and thermal/mechanical properties of poly (l-lactide)-clay blend. *Journal of Polymer Science, Part B: Polymer Physics*, **35**, 389-396.
7. Liu, X., Dever, M., Fair, N., and Benson, R. S. (1997) Thermal and mechanical properties of poly(lactic acid) and poly(ethylene/butylene succinate) blends. *Journal of Environmental Polymer Degradation*, **5**, 225-235.
8. Ohkoshi, I., Abe, H., and Doi, Y. (2000) Miscibility and solid-state structures for blends of poly[(S)-lactide] with atactic poly[(R,S)-3-hydroxybutyrate]. *Polymer*, **41**(15), 5985-5992.
9. Juioang, L., Wolcott, M. P., and Zhang, J. (2006) Study of biodegradable polylactide/poly(butylene adipate-co-terephthalate) blends. *Biomacromolecules*, **7**(1), 199-207.
10. Nofar, M., Maani, A., Sojoudi, H., Heuzey, M. C., and Carreau, P. J. (2015) Interfacial and rheological properties of PLA/PBAT and PLA/PBSA blends and their morphological stability under shear flow. *Journal of Rheology*, **59**, 317.
11. Weng, Y.-X., Jin, Y.-J., Meng, Q.-Y., Wang, L., Zhang, M., and Wang, Y.-Z. (2013) Biodegradation behavior of poly(butylene adipate-co- terephthalate) (PBAT), poly(lactic acid) (PLA), and their blend under soil conditions. *Polymer Testing*, **32**, 918-926.

12. Ojijo, V., Ray, S. S., and Sadiku, R. (2013) Toughening of biodegradable polylactide/poly(butylene succinate-co-adipate) blends via in situ reactive compatibilization. *ACS Applied Materials and Interfaces*, **5**, 4266-4276.
13. Pukpanta, P., and Sirisinha, K. (2012) Properties of poly(lactic acid) (PLA)/poly(butylene adipate-co-terephthalate) (PBAT) blends in the presence of antioxidant. *Advanced Materials Research*, **506**, 126-129.
14. Xu, Y. X., and Hanna, M. A. (2005) Preparation and properties of biodegradable foams from starch acetate and poly (tetramethylene adipate-co-terephthalate). *Carbohydrate Polymers*, **59**, 521-529.
15. Quero, E., Muller, A. J., Signori, F., Coltelli, M. B., and Bronco, S. (2012) Isothermal cold-crys- tallization of PLA/PBAT blends with and without the addition of acetyl tributyl citrate. *Macromolecular Chemistry and Physics*, **213**, 36-48.
16. Bhatia, A., Gupta, R. K., Bhattacharya, S. N., and Choi, H. J. (2007) An investigation of melt rheology and thermal stability of poly(lactic acid)/ poly(butylene succinate) nanocompo- sites. *Journal of Applied Polymer Science*, **114**(5), 2837-2847.
17. Mouritz, A., Leong, K., and Herszberg, I. (1997) A review of the effect of stitching on the in-plane mechanical properties of fibre-reinforced polymer composites. *Composites, Part A: Applied Science and Manufacturing*, **28**, 979-991.
18. Saheb, D. N., and Jog, J. (1999) Natural fiber polymer composites: a review. *Advances in Polymer Technology*, **18**, 351-363.
19. Breuer, O., and Sundararaj, U. (2004) Big returns from small fibers: a review of polymer/carbon nanotube composites. *Polymer composites*, **25**, 630-645.
20. Bakis, C., Bank, L. C., Brown, V., Cosenza, E., Davalos, J., Lesko, J., Machida, A., Rizkalla, S., and Triantafillou, T. (2002) Fiber-reinforced polymer composites for construction - state-of-the-art review. *Journal of Composites for Construction*, **6**, 73-87.
21. Fu, S.-Y., Feng, X.-Q., Lauke, B., and Mai, Y.-W. (2008) Effects of particle size, particle/matrix interface adhesion and particle loading on mechanical properties of particulate–polymer composites. *Composites Part B: Engineering*, **39**, 933-961.
22. Sengupta, R., Bhattacharya, M., Bandyopadhyay, S., and Bhowmick, A. K. (2011) A review on the mechanical and electrical properties of graphite and modified graphite reinforced polymer composites. *Progress in Polymer Science*, **36**, 638-670.
23. Ku, H., Wang, H., Pattarachaiyakoop, N., and Trada, M. (2011) A review on the tensile properties of natural fiber reinforced polymer composites. *Composites Part B: Engineering*, **42**, 856-873.

24. Kalia, S., Kaith, B., and Kaur, I. (2009) Pretreatments of natural fibers and their application as reinforcing material in polymer composites - a review. *Polymer Engineering & Science*, **49**, 1253-1272.
25. Xie, Y., Hill, C. A., Xiao, Z., Militz, H., and Mai, C. (2010) Silane coupling agents used for natural fiber/polymer composites: A review. *Composites Part A: Applied Science and Manufacturing*, **41**, 806-819.
26. Thostenson, E. T., Ren, Z., and Chou, T.-W. (2001) Advances in the science and technology of carbon nanotubes and their composites: a review. *Composites Science and Technology*, **61**, 1899-1912.
27. Spitalsky, Z., Tasis, D., Papagelis, K., and Galiotis, C. (2010) Carbon nanotube–polymer composites: chemistry, processing, mechanical and electrical properties. *Progress in Polymer Science*, **35**, 357-401.
28. Mittal, V., Chaudhry, A. U., and Luckachan, G. E. (2014) Biopolymer - thermally reduced graphene nanocomposites: Structural characterization and properties. *Materials Chemistry and Physics*, **147**(1-2), 319-332.
29. Pawar, S. P., Misra, A., Bose, S., Chatterjee, K., and Mittal, V. (2015) Enzymatically degradable and flexible bio-nanocomposites derived from PHBV and PBAT blend: Assessing thermal, morphological, mechanical, and biodegradation properties. *Colloid and Polymer Science*, **293**, 2921-2930.
30. Kumar, M., Mohanty, S., Nayak, S. K., and Parvaiz, M. R. (2010) Effect of glycidyl methacrylate (GMA) on the thermal, mechanical and morphological property of biodegradable PLA/PBAT blend and its nanocomposites. *Bioresource Technology*, **101**, 8406-8415.
31. Jiang, L., Liu, B., and Zhang, J. (2009) Properties of poly(lactic acid)/poly(butylene adipate-co-terephthalate)/ nanoparticle ternary composites. *Industrial and Engineering Chemistry Research*, **48**, 7594-7602.
32. Wu, T.-M., and Wu, C.-Y. (2006) Biodegradable poly(lactic acid)/chitosan-modified montmorillonite nanocomposites: Preparation and characterization. *Polymer Degradation and Stability*, **91**, 2198-2204.
33. Mittal, V., Chaudhry, A. U., and Matsko, N. B. (2014) "True" biocomposites with biopolyesters and date seed powder: Mechanical, thermal, and degradation properties. *Journal of Applied Polymer Science*, **131**(19), doi: 10.1002/app.40816.
34. Patwary, F., Matsko, N., and Mittal, V. (2016) Biodegradation properties of melt processed PBS/chitosan bio-nanocomposites with silica, silicate and thermally reduced graphene. *Polymer Composites*, doi: 10.1002/pc.23947.
35. Wan, Y., Wu, H., Yu, A., and Wen, D. (2006) Biodegradable polylactide/chitosan blend membranes. *Biomacromolecules*, **7**, 1362-1372.
36. Martin, O., and Averous, L. (2001) Poly(lactic acid): Plasticization and properties of biodegradable multiphase systems. *Polymer*, **42**(14), 6209-6219.

37. Mittal, V., and Matsko, N. (2018) Eco-friendly bio-nanocomposites of PHBV/PBAT blend with graphene and vermiculite. In: *Biopolymers Based Advanced Materials*, Mittal, V. (ed.), Central West Publishing, Australia, pp. 57-85.

8

Stable Corrosion-Resistant Chitosan Coatings: Effect of Crosslinking with Silica, Poly(vinyl butyral) Over-coating and Graphene on the Coating Performance

8.1 Introduction

Corrosion of steel is a serious problem affecting the industrial installations, thus, requiring to be minimized in order to reduce maintenance costs. Organic/polymeric coatings in thick layers are widely used for the passive corrosion protection of steel structures [1-2]. For instance, in a recent study, Niratiwongkorn *et al.* [3] reported self-healing polyvinyl butyral (PVB) based organic coating formulations prepared by incorporating polypyrrole-carbon black (PPyCB) composite as an inhibiting pigment. The redox properties and diffusion barrier nature of PPy imparted self-healing to the PVB/PPyCB composite coatings in aggressive environments. Figure 8.1 also shows the Nyquist plots of the composite coatings. The changes in the electrochemical impedance plots indicated that in comparison with PVB, PVB/PPyCB composite coatings imparted much better protection that occurred through the synergistic effect of redox reactions and electrical conductivity of the PPyCB pigments.

In the past few years, the interest of using environmentally friendly paints and coatings has grown tremendously. This trend has occurred not only by the realization that the supply of fossil resources is inherently finite, but also by a growing concern for environmental issues, such as volatile organic solvent emissions and recycling or waste disposal problems at the end of a resin's economic lifetime [4]. This has resulted in the development of coatings formulations from renewable resources such as chitin, chitosan, cellulose, polylactide, etc. Among these polymers, chitosan (Ch) has evolved as a very attractive candidate because this bifunctional natural glucosamine is nontoxic, biodegradable, biocompatible, abundant in the

Gisha Elizabeth Luckachan and Vikas Mittal, The Petroleum Institute (part of Khalifa University of Science and Technology), Abu Dhabi, UAE*
**Current address: Bletchington, Wellington County, Australia*

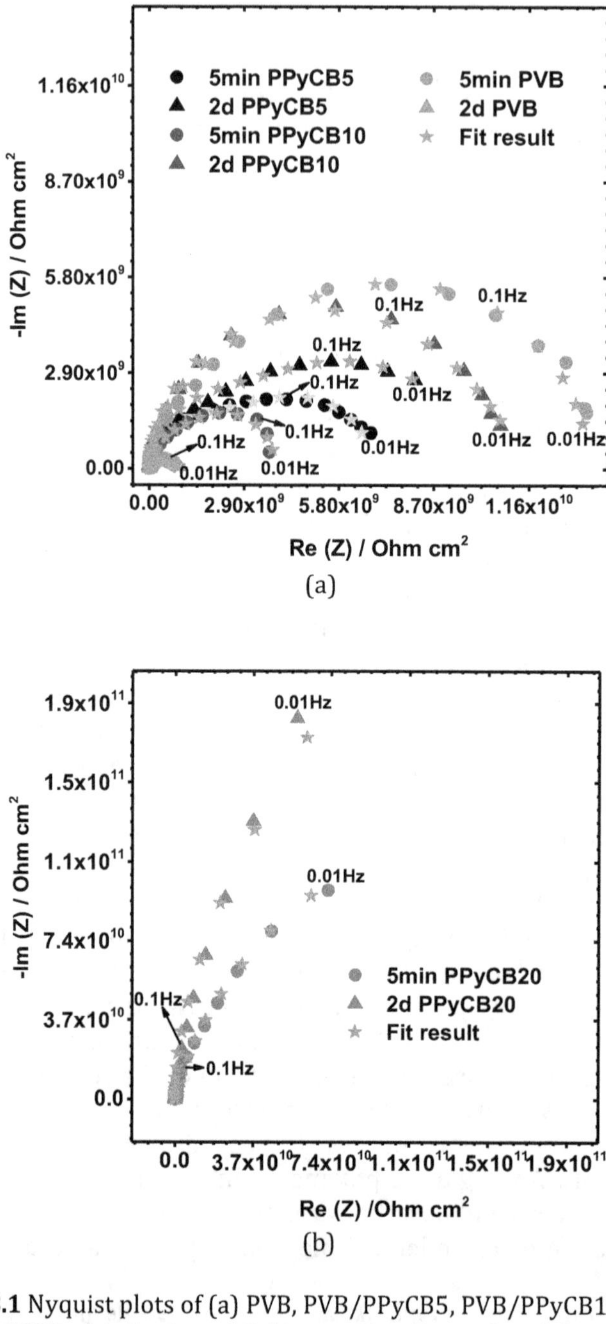

Figure 8.1 Nyquist plots of (a) PVB, PVB/PPyCB5, PVB/PPyCB10 and (b) PVB/PPyCB20 after 5 min and 2 d immersion. Reproduced from Reference 3 with permission from Royal Society of Chemistry.

nature, and produced at low cost [5-10]. Chitosan's affinity towards metals as well as the excellent film forming ability and barrier properties promote its potential in the coating applications [11-14]. In addition, the hydroxyl and amine groups of chitosan can be modified with desired properties needed for effective corrosion protection [15-22]. The lone pairs of electrons on hydroxyl and amine groups of chitosan can effectively inhibit the corrosion processes, thus, resulting in chitosan being a natural corrosion inhibitor [23-27].

Limited number of studies have reported on the development of chitosan based corrosion-resistant coatings for metals [19,28]. However, the critical issue in evolving chitosan as a corrosion barrier is its absorption of large amount of moisture from the atmosphere, thus, forming a hydrogel. This transformation not only leads to biodegradation of the film, but also allows moisture to infiltrate easily into the film, causing its failure as a protective coating [19,28,29]. Sugama *et al.* [19] conducted an extensive study on chitosan coatings to decrease the hydrophilicity and to improve the bond strength of chitosan on the metal surface. The authors reported water-insoluble chitosan biopolymer coatings on aluminum by grafting synthetic poly(itaconic acid) (PIA) onto the linear chitosan chains [19]. The PIA-grafted chitosan polymer coatings displayed a low ionic conductivity, thereby imparting a salt-spray resistance of 694 h for coated Al panels. A dextrin modification of chitosan has also been performed by the same group [28]. However, these measures are not sufficient for long term corrosion protection of chitosan coatings.

The other approach is to improve the physical and chemical properties of chitosan by various crosslinking agents [23,29]. In a recent study, Luckachan and Mittal [23] reported the anti-corrosion coatings of chitosan fabricated by layer-by-layer (lbl) addition of chitosan and poly vinyl butyral (PVB) on mild carbon steel substrate. Effect of chitosan crosslinking on the stabilization process of passive oxide layer as well as the coating stability was studied by incorporating glutaraldehyde in the chitosan layer. The results obtained from electrochemical impedance spectroscopy and Tafel plots clearly showed a superior corrosion protection of lbl coatings generated by 1% glutaraldehyde incorporation in the middle chitosan layer. Figure 8.2 also shows the morphological analysis of the coated substrates. It was confirmed that the infinite array of functional groups on chitosan could act as a chain of nucleation sites from which further growth of iron oxide phase took place. Thus, this

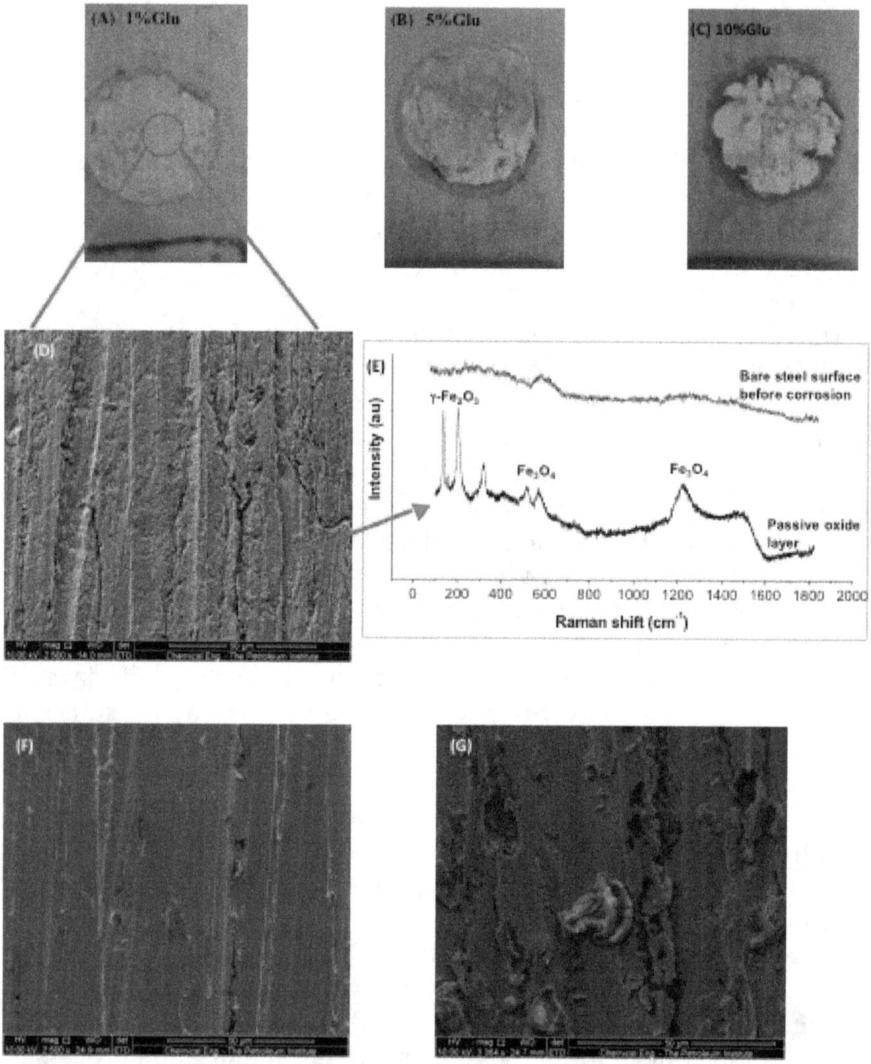

Figure 8.2 Optical photographs of (A) PVB_Ch/1%Glu_PVB, (B) PVB_Ch/5%Glu_PVB and (C) PVB_Ch/10%Glu_PVB coated steel surface after 24 h electrochemical impedance spectroscopy (EIS) measurements in 0.3M salt solution and film removal; (D) SEM image and (E) Raman spectra of the area marked with red circle on optical photograph of steel surface taken after PVB_Ch/1%Glu_PVB coating removal. SEM images of bare steel (F) before and (G) after 24 h EIS measurement in 0.3 M salt solution. Reproduced from Reference 23 with permission from Springer.

process resulted in a uniform layer of iron oxide covered with chitosan backbone on the metal surface. The incorporation of silica as a crosslinking agent has also been specifically reported to enhance the stability of chitosan coatings [29]. Chitosan-silica hybrids are also known to act as immobilization matrices for several molecules and complexes. Sol-gel process is commonly used for the synthesis of such hybrid materials because of its ability to form hybrid organic and inorganic network structures at relatively low temperatures [30,31]. Silica in sol-gel coatings can produce coatings with excellent adhesion between the metallic substrates and subsequent top-coats [32]. It is, therefore, expected that chitosan-silica hybrids would have enhanced interfacial interaction along with improved thermal, mechanical, optical and adsorbing properties. In addition, sol-gel coatings can provide a dense barrier against electrolyte uptake [33]. Lai *et al.* [34] prepared chitosan-silica thick films with *in-situ* silica from 0-100% silica via the sol-gel process to investigate the interfacial interaction of the two phases and its capability to chelate Cu(II) and Fe(III) to a large degree under wide environmental variation. However, efforts for the fabrication of chitosan-silica hybrids and their use for coating applications are required to be further optimized. Moreover, the coating, by itself, is not sufficient to effectively protect the metal substrate against corrosion for long term. The sol-gel coatings contain micro-pores, cracks and areas of low cross-link density which provide pathways for diffusion of corrosive species to the coating/metal interfaces, where the corrosion processes get initiated.

The barrier properties of the coating can also be further enhanced by using suitable reinforcing fillers [35-39]. The addition of filler is generally reported to improve the adhesion strength between the polymer and metal surface as well as to increase the diffusion paths of water and oxygen molecules through the coating, thus, enhancing the corrosion resistance of the coating. Graphene, a two-dimensional sheet of sp^2 carbon atoms, was recently reported as an ideal filler for corrosion inhibition in reinforced polymer coatings because of its barrier nature towards corrosion promoting species [36,37]. It increases the diffusion path of oxygen and ionic species through the coating, and thus enhances the corrosion resistance of the coating. Such a mechanism of corrosion protection has been reported for the graphene incorporated protective coatings of polyvinyl butyral, epoxy, polyurethane and poly(sodium styrene sulfonate) composites [36-39].

In the present work, a detailed study has been conducted for generating thin stable chitosan coatings through sol-gel silica synthesis method, followed by a poly(vinyl butyral) (PVB) over-coating. Electrochemical impedance spectroscopy (EIS) was used to measure the corrosion resistance of the coatings in 0.5 M NaCl solution. In order to evaluate the role of chitosan, corrosion studies of chitosan-silica hybrid coatings without PVB over-coating were also conducted in comparison with pure chitosan, silica and PVB coatings. Effect of nano-fillers like graphene nano-platelets on the coating performance was also evaluated using EIS and polarization measurements.

8.2 Experimental

8.2.1 Materials

Poly(vinyl butyral) with trade name Butvar B-98 (molecular weight 40,000-70,000 g/mol), hydrochloric acid, acetic acid, tetraethoxysilane (TEOS, 98%), isopropanol, methanol and methyltrimethoxysilane (MTMS, 95%) were purchased from Aldrich, Germany and used as received. Chitosan flakes were procured from Bio21, Thailand. Chitosan had a molecular weight of 180,000 g/mol and 90% degree of deacetylation (DD). Single layer graphene nano-platelets were procured from Angstron Materials, USA.

8.2.2 Chitosan-Silica/PVB Hybrid Coatings

Chitosan-silica hybrid coatings were prepared by sol-gel method. Silica sol was prepared first by mixing 0.6 g TEOS with 7.9 g isopropanol and 9.5 g 0.001 M HCl (1:4:16 molar ratio) [40]. The stirring was continued for 90 min at 30 °C. Required amount of chitosan, dissolved in 2 % acetic acid, was added to silica sol under constant stirring. Stirring was continued for 10 min and MTMS (2 times the molar ratio of TEOS) was added to the mixture dropwise to prepare hydrophobic Ch-SiO_2 hybrid. This precursor sol was dip coated twice on the carbon steel substrate and kept at 80 °C for 2 h. PVB over-coating was generated by dissolving 2 g of PVB in 20 ml of methanol. This mixture was dip coated on the surface of Ch-SiO_2 hybrid coating and kept for curing in a vacuum oven at 90 °C over 2 h. Three different types of coatings were prepared by changing chitosan solution corresponding to 10 wt%, 20 wt% and 50 wt% of total silica content and the generated coatings were denoted as 10Ch-

SiO_2/PVB, 20Ch-SiO_2/PVB and 50Ch-SiO_2/PVB respectively. Pure silica coating was prepared by the sol-gel procedure used for hybrid coating synthesis in the absence of chitosan. Pure chitosan coating was also prepared by dissolving 2 g of chitosan in 20 ml of 2% acetic acid and pure PVB coating was prepared by dissolving 2 g PVB in 20 ml of methanol. In order to reinforce Ch-SiO_2 hybrid coatings, 5% graphene was also dispersed in 20 wt% solution of chitosan in acetic acid by bath sonication over 12 h followed by 6 h stirring. The mixture was then probe sonicated (model CPX 750 supplied by Cole Parmer) for 1 h to disperse graphene in chitosan solution. The probe sonicator was operated at 40% amplitude and a pulse of 5 s (5 s on, 5 s off). The mixture was added to silica sol and same procedure was followed to obtain 20Ch-5Gr-SiO_2/PVB hybrid coatings.

8.2.3 Structural Characterization

Structural characterization of Ch-SiO_2/PVB coatings was carried out using Bruker VERTEX 70 FTIR spectrometer attached with a DRIFT accessory. IR acquisitions were achieved between 4000 cm^{-1} - 370 cm^{-1} using OPUS software at 4 cm^{-1} resolution. 128 scans were used for each acquisition. Thickness of the coatings was measured using PosiTector 6000 coating thickness gauge from DeFelsko Corporation, New York. The microstructure of the coatings was analyzed using scanning electron microscope (SEM) (FEI Quanta, FEG250, USA) at an accelerating voltage of 5 kV.

8.2.4 Immersion Test

The edges of the coated coupons were sealed using Nippon epoxy primer so as to avoid seeping of the moisture under the coatings during immersion in sodium chloride solution. After 24 h drying at room temperature, the coupons were immersed in 0.5 M aerated sodium chloride solution for standard corrosion analysis.

8.2.5 Electrochemical Measurements

Electrochemical impedance spectroscopy and polarization measurements were carried out at room temperature in a three-electrode corrosion cell consisting of a saturated calomel reference electrode, a platinum counter electrode and coated steel plates as the working electrode. 0.5 M NaCl was used as the electrolyte solution for the electrochemical analysis. Exposed area of working electrode was 1 cm^2. All measurements were performed on digital electrochemical analyzer (supplied by BioLogic, France). Ultra-low current cables connected to the potentiostat were used for the accurate measurement of the current. This option includes current ranges from 100 nA down to 100 pA with additional gains extending the current ranges to 10 pA and 1 pA. The resolution on the lowest range is 76 aA. Impedance measurements were performed as a function of open circuit potential (E_{OCP}) by applying a sinusoidal potential in frequencies from 10^5 Hz to 10^{-2} Hz, with an amplitude of 15 mV. The impedance plots were fitted using equivalent circuits (shown in Figure 8.6), where pure capacitances were replaced by constant-phase elements (CPE). The software used for the fitting was EC-Lab V10.39. The capacitance values of the different elements in the equivalent circuit were calculated using the following equation:

$$C = Y_0 (\omega_{max})^{n-1}$$

where ω_{max} is the frequency at which the imaginary impedance reaches a maximum for the respective time constant, Y_0 is the magnitude of the CPE and 'n' is the exponential term of the CPE [33]. Polarization measurements were performed by polarizing the working electrode from an initial potential of -250 mV up to a final potential of +250 mV as a function of open circuit potential. A scan rate of 0.1667 mV/s was used for the polarization sweep. To study the reproducibility of the measurements, each set of experiments was repeated three times on freshly coated samples.

8.3 Results and Discussion

In the current study, chitosan based protective coatings of 5-6 μm thickness were prepared by crosslinking with silica and a PVB overcoat. Three coatings, denoted as 10Ch-SiO_2/PVB, 20Ch-SiO_2/PVB and 50Ch-SiO_2/PVB, were prepared in order to study the effect of chitosan concentration, in addition to the role of silica as a crosslinking binder and PVB as an additional barrier to electrolyte.

The IR spectra of pristine chitosan, silica (obtained by hydrolysis and condensation of TEOS) and Ch-SiO_2 hybrid coating with 20 wt% chitosan content are demonstrated in Figure 8.3. It was evident that

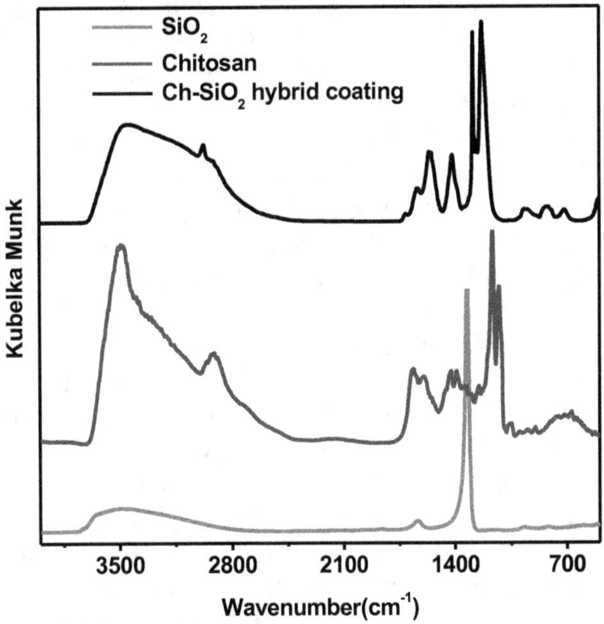

Figure 8.3 IR spectra of silica, chitosan and Ch-SiO_2 hybrid coating with 20 wt% chitosan.

chitosan exhibited characteristic absorbance peaks of amide 1 (C=O stretch) at 1645 cm^{-1} and amide 11 (N-H bending) at 1597 cm^{-1}, along with methyl and methylene group peaks at 1425 cm^{-1} and 1384 cm^{-1} respectively. The band located at 1166 cm^{-1} was related to asymmetric vibrations of CO in the oxygen bridge. The bands at 1150 cm^{-1} to 1050 cm^{-1} were attributed to V_{CO} of the ring COH, COC and CH_2OH of chitosan [22,40-43]. IR spectra of pure SiO_2 exhibited characteristic vibrations of Si-O-Si asymmetric and symmetric stretching at 1318 cm^{-1} and 817 cm^{-1} respectively and Si-O vibrations at 974 cm^{-1} [40]. Amide 1 and amide 11 vibrations of chitosan at 1645 cm^{-1} and 1597 cm^{-1} were changed to intense bands at 1635 cm^{-1} and 1550 cm^{-1} respectively after modification with silica. This is attributed to the interaction of protonated amine group ($-NH_3^+$) of chitosan with negatively charged sites of silica [25]. Chitosan backbone C-O-C bridge and C-O stretch vibrations at 1166 cm^{-1} and 1126 cm^{-1} were shifted to higher wavenumbers of 1229 cm^{-1} and 1175 cm^{-1} respectively. It indicated that silica, existing as crosslinking bridge between chitosan chains, provided extra stability to chitosan coatings by reducing the freedom of chain movement. The hybrid gel structure of Ch-SiO_2 coating was confirmed from the shift of Si-O-Si asymmetric stretching vibration from 1318 cm^{-1} to 1279 cm^{-1}. Si-O bands, formed as a small hump in pure silica, were clearly visible in the hybrid coating spectra at lower wavenumber regions of 960 cm^{-1} and 820 cm^{-1} indicating the existence of a definite interaction between the chitosan and silica phases in the hybrid coatings [40]. The new absorbance at 2977 cm^{-1} and the broadening at 3100 cm^{-1} was also attributed to CH_3 stretching vibrations of MTMS.

The effect of chitosan crosslinking with silica on the corrosion protection of carbon steel was studied by EIS analysis in 0.5 M NaCl solution. Figure 8.4a and b show the Bode and Nyquist plots of Ch-SiO_2 hybrid coatings of 10 % and 20 % chitosan content measured after 5 min and 2 h immersion time. As silica could also provide barrier protection, EIS spectra of pure chitosan and pure silica coatings along with bare steel in 0.5M NaCl solution were collected for comparison (Figure 8.5). The impedance at high frequency region was slightly higher for the silica coatings owing to the barrier effect, however, it dropped towards the low frequency region to a value near to the bare steel which clearly indicated that the protection by silica coating was diminished significantly (Figure 8.5). The amine and hydroxyl groups on chitosan can bind or ionically interact with steel surface via the lone pairs of electrons on the O and N atoms, and are thus able to provide inhibition to steel corrosion [23-25]. It was observed in the Bode plot of chitosan coating with a high low frequency log Z than silica coating (Figure 8.5). However, the characteristic of chitosan to absorb moisture failed the integrity of the chitosan coating on prolonged exposure. Thus, individual coatings of chitosan and silica were not effective as long-term corrosion barrier. Cross linking of chitosan with silica improved the inhibitive nature of chitosan by enhancing its adhesion on the metal surface which was attributed to the high low frequency impedance in the Bode plots of Ch-SiO_2 hybrid coatings (Figure 8.4a). As expected, a high degree of protection was observed at higher content of chitosan. It was obvious due to the increased diameter of the semicircle of 20Ch-SiO_2 hybrid coating in the Nyquist plot measured after 5 min immersion (Figure 8.4b). Bode phase angle plot of both coatings exhibited a single time constant at high frequency region (10^3 Hz - 10^5 Hz), which was reported as an indication of the barrier nature of the coating [18,21]. However, it decreased significantly in the first 2 h and a second time constant attributed to electrochemical reactions on metal surface started to appear at low frequency regions (10^{-1} Hz

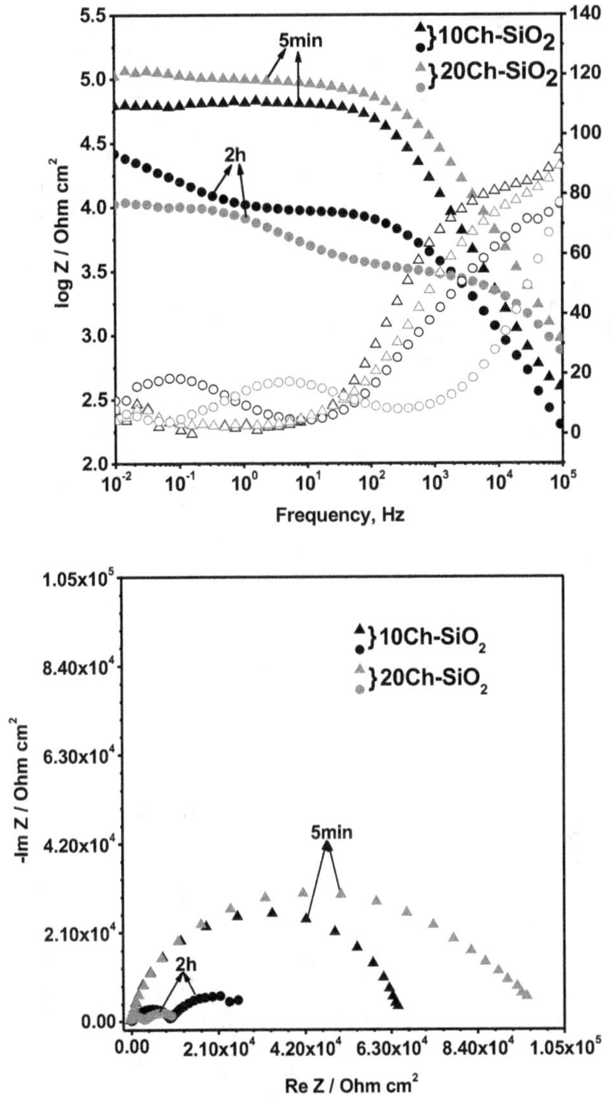

Figure 8.4 (top or a) Bode plots and (bottom or b) Nyquist plots of 10Ch-SiO_2 and 20Ch-SiO_2 hybrid coatings measured at 5 min and 2 h immersion in 0.5M NaCl solution.

to 10^1 Hz) (Figure 8.4a). Two semicircles, contributing respectively to coating response and corrosion process on the metal surface, were obvious in the 2 h Nyquist plots (Figure 8.4b). The observed electrochemical behaviors were studied by fitting the EIS data of the

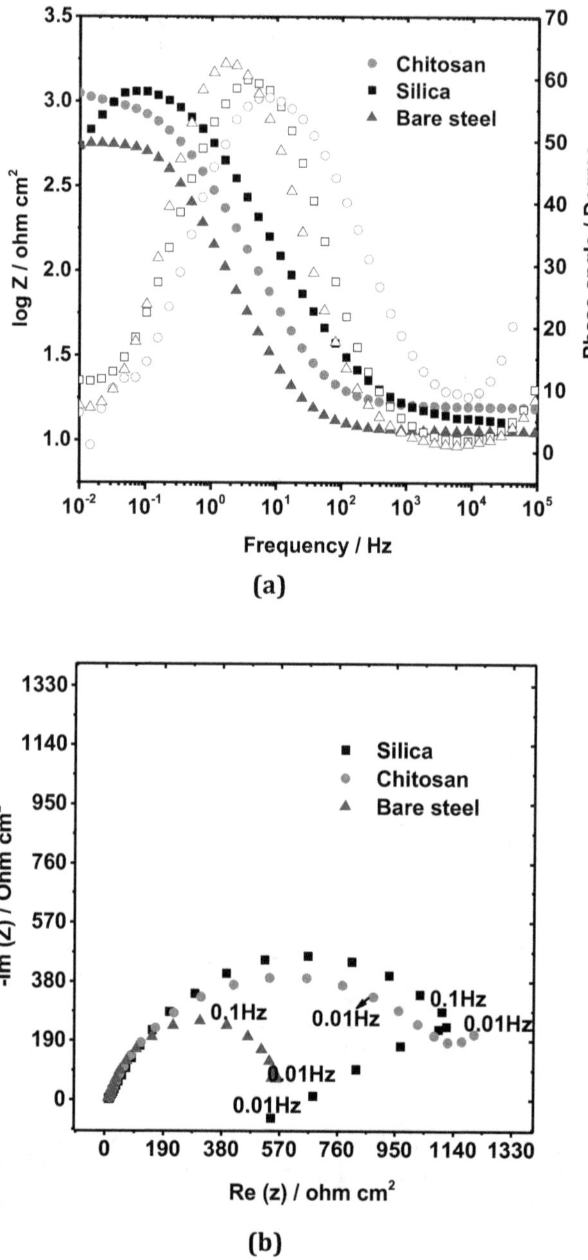

Figure 8.5 EIS plots of pure chitosan coating, pure silica coating and bare steel in 0.5M NaCl solution measured at 2 h immersion. (a) Bode plots and (b) Nyquist plots.

two systems using the equivalent circuits depicted in Figure 8.6. Equivalent circuit in Figure 8.6a contained a single RC circuit representing coating/electrolyte interface, whereas the equivalent circuit in Figure 8.6b contained two RC circuits representing the coating/electrolyte and coating/metal interfaces respectively [44]. The high frequency circuit was ascribed to a coating capacitance (C_c) and pore resistance (R_p) of the sol-gel polymer coating, while the low frequency represented the double layer capacitance (C_{dl}) and charge transfer resistance (R_{ct}) of the underlying metal. Two constant phase elements (Q_c and Q_{dl}) were used instead of pure capacitances (C_c and C_{dl}) for both frequencies. The symbol R_{sol} represented the solution resistance of the bulk electrolyte between the reference and working electrode [44,45].

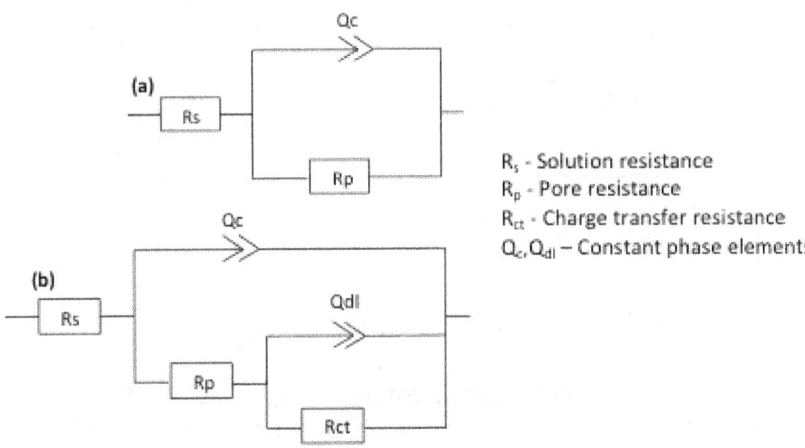

Figure 8.6 Equivalent circuits used for EIS modeling.

Pore resistance (R_p) of a polymeric coating represents resistance of the electrolyte in the pores, cracks and pits in the coating, thus, reflects barrier property of the coating [33]. The high magnitude R_p of 20Ch-SiO_2 hybrid coatings, shown in Table 8.1, therefore, indicated that silica crosslinking of chitosan (evidenced from IR spectra) prevented the ingress of electrolyte by acting as a physical barrier. It was reflected in the coating capacitance which decreased with increasing chitosan content in the hybrid coating. Since coating capacitance accounts for the ability of the coating to hold water [41,42], the reduced coating capacitance of 20Ch-SiO_2 hybrid coating in comparison with 10Ch-SiO_2 hybrid coating, indicated lower permeability of water into the former coating due to the extended cross-linking (Table 8.1). It has been reported earlier that the addition of inorganic fillers reduces barrier effect of the coating if the filler particles are not bound with polymer matrix [41]. In the current study,

Table 8.1 EIS parameters of Ch-SiO_2 hybrid coatings having different percentages of chitosan in 0.5M NaCl solution

Coating	Exposure Time	R_p (Ωcm^2)	C_c (Fcm^2)	Q_{dl}	
				Y_0 (F.s^.n-1)	n
10Ch-SiO_2	5 min	65755	2.238e-3		
	2 h	9585	2.036e-4	0.14e-3	0.68
20Ch-SiO_2	5 min	93175	9.155e-4		
	2 h	3680	5.03e-4	31.54e-6	0.58

Coating	Exposure Time	R_{ct} (Ωcm^2)	C_{dl} (Fcm^2)	Q_{dl}	
				Y_0 (F.s^.n-1)	n
10Ch-SiO_2	5 min				
	2 h	20270	0.023	0.14e-3	0.68
20Ch-SiO_2	5 min				
	2 h	7806	0.073	31.54e-6	0.58

the resistance of Ch-SiO_2 hybrid coatings improved with chitosan content, thus, indicating strong adhesion of silica particles with chitosan. However, continued immersion in 0.5 M NaCl solution decreased coating capacitance of both systems probably due to the deterioration of the coating protectiveness and formation of free passage of electrolyte down to the metal surface. The electrochemical reactions on metal surface were evidenced from the second time constant in the EIS plots (Figure 8.4a & b) and corresponding R_{ct} and C_{dl} values (Table 8.1). Digital image in Figure 8.7a confirmed the presence of corrosion products on the surface of 20Ch-SiO_2 coatings after 7 d immersion in 0.5 M NaCl solution. EIS results along with the digital images suggested that though crosslinking of chitosan with silica enhanced stability and corrosion resistance of chitosan coatings, Ch-SiO_2 hybrid coatings were not suitable as such for the corrosion protection of steel substrates. It might be due to the porous nature of the coatings that was also confirmed through the SEM

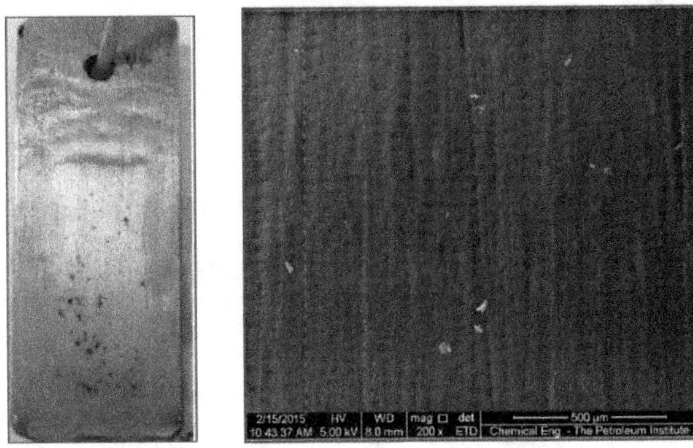

(a) (b)

Figure 8.7 Digital image of 20Ch-SiO_2 coating after 7 d immersion in 0.5M NaCl solution (a) and SEM image of 20Ch-SiO_2 coating on metal substrate before the corrosion analysis (b).

image of the 20Ch-SiO_2 hybrid coating, shown in Figure 8.7b.

In an attempt to improve the stability and corrosion resistance, PVB over-coating layer was applied on Ch-SiO_2 hybrid coatings. As PVB can provide corrosion protection by forming a barrier coating, a PVB coated substrate was also prepared for comparison. EIS plots of this coating, shown in Figure 8.8, exhibited a low frequency log Z of 8.4 Ωcm^2 at the initial time of immersion, which was stable over 3 d. The barrier nature of the coating was observed in bode phase angle plot with a time constant at high frequency region (10^3 Hz to 10^5 Hz). However, small but distinguished second time constant observed at an intermediate frequency of 10^0 Hz to 10^2 Hz exhibited the chances of electrochemical reactions at the metal-coating interface. Continuous immersion in 0.5 M NaCl solution decreased low frequency log Z to a value of 3.4 Ωcm^2 after 11 d exposure which indicated that the coating did not provide any further protection to the underlying metal. The single time constant at the intermediate frequency (10^0 Hz to 10^2 Hz) in the Bode phase angle plot showed the corrosion processes on the metal surface. These results indicated that a thin coating of pure PVB was also not suitable to provide efficient corrosion protection of the underlying metal surface for long periods. It should however be noted that the even though PVB alone could not provide the corrosion protection, its combination with chitosan-silica can be very beneficial for achieving long term corrosion protection.

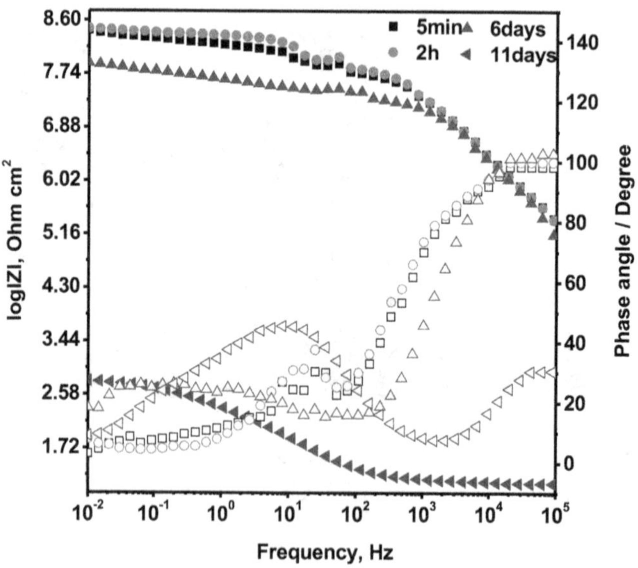

Figure 8.8 Bode plots of pure PVB coating at different time of immersion in 0.5M NaCl solution.

Figure 8.9 shows IR spectra of PVB and PVB over-coated 20Ch-SiO_2 hybrid coating. Presence of PVB was characterized by the butyral units, hydroxyl and acetal groups [42,46]. PVB spectra exhibited C-O-C-O-C stretching vibrations of butyral units at 1167 cm^{-1} and 1012 cm^{-1}. Visible, but less intense peak at 1736 cm^{-1} was related to C=O stretching vibration and peaks at 1006 cm^{-1} and 1242 cm^{-1} resulted from C-O-C stretching vibrations of acetate groups. Intense broad band at 3470 cm^{-1} was assigned to -OH stretching vibrations of hydroxyl group. C-H bending and stretching vibrations of saturated -CH, -CH_2 and -CH_3 were observed at regions of 2850 cm^{-1} to 2720 cm^{-1} and 1500 cm^{-1} to 1300 cm^{-1} respectively [42,46]. Addition of PVB over-coating on Ch-SiO_2 hybrid coating diminished PVB acetate vibrations at 1006 cm^{-1} and chitosan amine vibrations at 1635 cm^{-1}. Also, silica Si-O-Si vibrations at 1279 cm^{-1} shifted to 1268 cm^{-1}. In addition, chitosan backbone vibrations at 1229 cm^{-1} and PVB C-O-C-O-C vibrations at 1167 cm^{-1} were observed to merge to form a single broad band at 1176 cm^{-1}. These shifts indicated the possibility of crosslinking reaction between PVB and Ch-SiO_2 hybrid coating. Further, the broadening of amine and hydroxyl vibrations at 3600 cm^{-1} to 3000 cm^{-1} indicated the hydrogen bond interaction between the polymers. Also, the small hump at 3260 cm^{-1} was attributed to the hydrogen bonded secondary amide N-H stretching vibrations which are expected to form by reaction of PVB acetate and chitosan amine, while curing of coatings at 90 °C [47]. Figure 8.10 shows the digital images of the Ch-SiO_2 /PVB coatings, taken at the first sign of visible corrosion during the immersion in 0.5 M NaCl solution. Black dots covered with red corrosion products appeared on 10Ch-SiO_2/PVB coating after 30 d immersion time (Figure 8.10a), whereas visible corrosion products mainly on the edges of 20Ch-SiO_2/PVB coating appeared after 20 d (Figure 8.10b). This enhanced protection was obtained by the PVB over-coating followed by 90 °C curing that sealed the pores and cracks on Ch-SiO_2 hybrid coatings as observed in the SEM image of 10Ch-SiO_2/PVB coating in Figure 8.10d. Though PVB over-coating enhanced corrosion protection of Ch-SiO_2 hybrid coatings, at high chitosan concentrations, hydrophilic and hydrophobic interfaces formed between chitosan layer and PVB over coating decreased the extent of corrosion protection. It was obvious in the optical image of 50Ch-SiO_2/PVB coating in Figure 8.10c that showed thick corrosion products distributed uniformly on the surface during the initial days of immersion in 0.5 M NaCl solution.

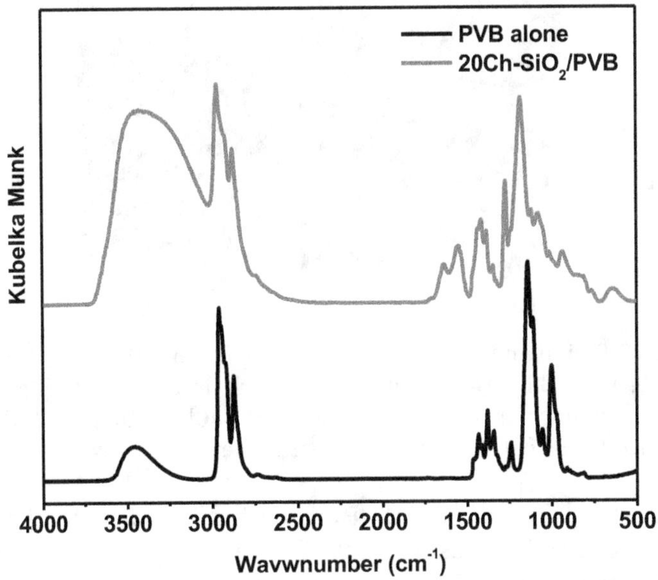

Figure 8.9 IR spectra of PVB and PVB over-coated 20Ch-SiO_2 hybrid coating.

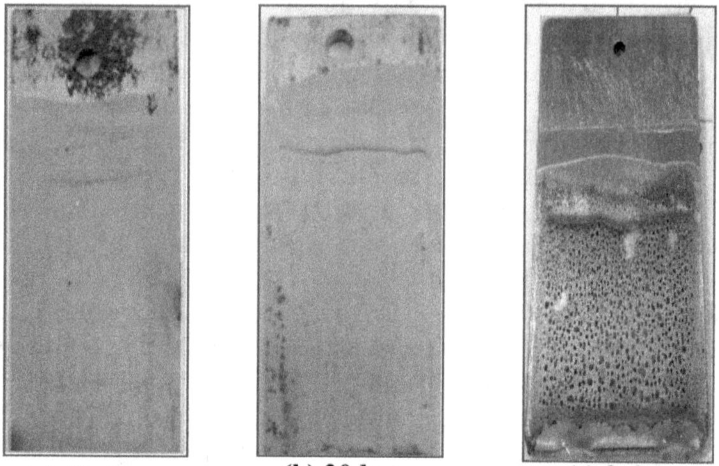

(a) 30 days (b) 20days (c) 2 days

(d)

Figure 8.10 Digital images of (a) 10Ch-SiO_2 /PVB, (b) 20Ch-SiO_2 /PVB and (c) 50Ch-SiO_2 /PVB hybrid coatings taken at the first sign of visible corrosion products during immersion in 0.5M NaCl solution; (d) SEM image of 10Ch-SiO_2/PVB coating on metal substrate before the corrosion tests.

The performance of Ch-SiO_2/PVB coatings in 0.5 M NaCl solution was studied by EIS measurements for periods up to 20 d and the data is represented as Bode plots in Figure 8.11. The coatings were stable on the metal surface over the immersion period, but exhibited different protective abilities. The 10Ch-SiO_2/PVB hybrid coating had a low frequency impedance of 6.8e^7 Ω cm^2 after 20 d immersion (Figure 8.11a). It is generally considered that coatings having a low frequency impedance higher than 10^6 provide good corrosion protection [1]. Therefore, it can be suggested that though the low frequency impedance decreased from 6.1e^9 Ω cm^2 to 6.8e^7 Ω cm^2 for 10Ch-SiO_2/PVB during immersion, this value was still high enough to confirm the superior corrosion resistance of the hybrid coating. The barrier response of the coatings was obvious in the Bode phase angle plots where plateau at high frequency region remained close to 90° for most of the frequency range. In order to study the coating behavior, the EIS plots were modeled using the simple equivalent circuit depicted in Figure 8.6a and the corresponding EIS data are given in Table 8.2. High R_p values (> 10^{8}Ω cm^2) of 10Ch-SiO_2/PVB coating during the initial period of immersion indicated barrier property of the coatings probably due to the crosslinking of Ch-SiO_2 hybrid coating with PVB matrix, as evidenced from IR spectra. R_p values maintained the order of 10^9 over 2 d immersion time. At the same time, coating capacitance (C_c) was also observed to be constant over 2 d, further indicating that coatings were less permeable to water. Continued immersion to a period of 20 d in 0.5 M sodium chloride solution decreased R_p values to about 4.71e^7, however, these were still in the range recommended for good coatings [1]. Furthermore, after an initial decrease at 7 d exposure time, C_c remained almost constant over 20 d, indicating the capacitive behavior of coating. It indicated the diffusion of water molecules throughout the coating instead of localizing as free electrolyte on the metal surface [48].

For 20Ch-SiO_2/PVB coating, the impedance magnitude at low frequency was higher than that of 10Ch-SiO_2/PVB coating during the initial period of immersion (Figure 8.11b). After 2 d, the impedance started to decrease significantly and reached a value of 2.4e^6 Ω cm^2 at 20 d immersion time, which was lower than the value observed for 10Ch-SiO_2/PVB coatings (6.8e^7 Ω cm^2). Modeling EIS plots using appropriate equivalent circuits (Figure 8.6) indicated higher pore resistance (R_p) during the initial time of immersion, thus, confirming better barrier property of 20Ch-SiO_2/PVB coatings over 10Ch-SiO_2/PVB coatings. C_c remained constant for 1 d and the value was also lower than 10Ch-SiO_2/PVB coating, further indicating less permeability of water into 20Ch-SiO_2/PVB coatings. However, continuous immersion in 0.5 M NaCl solution increased C_c to 2.245e^{-4} Fcm^2. It indicated that even though the PVB crosslinking with chitosan and

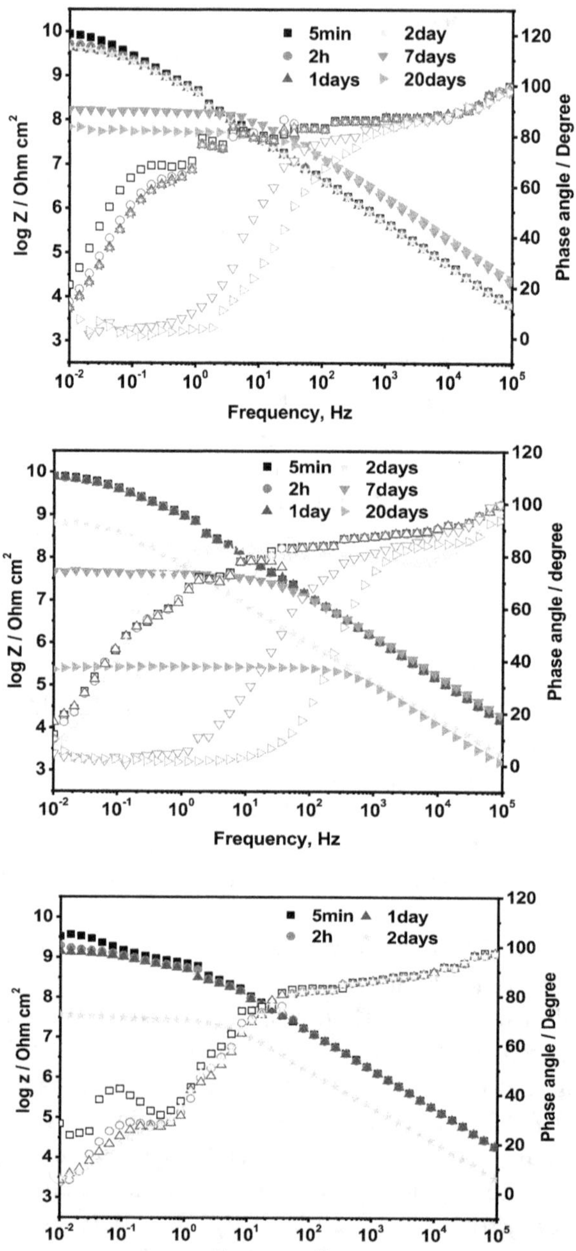

Figure 8.11 Bode plots of 10Ch-SiO_2/PVB hybrid coating (top or a), 20Ch-SiO_2/PVB hybrid coating (middle or b) and 50Ch-SiO_2/PVB hybrid coating (below or c) at different time of immersion in 0.5M NaCl solution.

silica shielded Ch-SiO_2 hybrid coatings from the corrosive agents, the hydrophilic and hydrophobic interphases provided some definite paths for electrolyte ingress through the coating. Evolution in the barrier properties of 20Ch-SiO_2/PVB coating was evidenced further from the Bode phase angle plots where the phase angle plateau at high frequency region decreased to 10^3 Hz at the end of the measurement time (Figure 8.11b).

EIS experiments on 50Ch-SiO_2/PVB coatings could be conducted only for 2 d as black corrosion products were observed to be distributed all over the metal surface (Figure 8.10c). Equivalent circuit in Figure 8.6b was used to model the EIS plots (Figure 8.11c) and the corresponding coating parameters are presented in Table 8.2. As shown in Table 8.2, both low frequency impedance and pore resistance of 50Ch-SiO_2/PVB coating were lower than 10Ch-SiO_2/PVB and 20Ch-SiO_2/PVB coatings. The plateau in the bode phase angle plot remained close to 90° for only up to half of the frequency region used for the EIS measurement (10^2 Hz to10^5 Hz). This behavior combined with decrease in low frequency impedance and R_p suggested that the barrier property of Ch-SiO_2/PVB hybrid coatings decreased with high percentage of chitosan content. Higher value of coating capacitance than the other two systems supported the enhanced water permeability of 50Ch-SiO_2/PVB hybrid coatings. These results led to an assumption that even though silica crosslinking improved the protective nature of chitosan by providing a strong adhesion with metal ions and PVB over-coating, this behavior was effective at low percentages of chitosan, more specifically till 20% of chitosan in the Ch-SiO_2/PVB hybrid coatings. At higher percentages, the hydrophilic characteristics of chitosan reduced its binding with metal and created a phase separation with hydrophobic PVB over-coating, which negatively affected the corrosion resistance behavior of Ch-SiO_2/PVB coatings.

In order to study the effect of graphene platelets on the corrosion inhibitive behavior of Ch-SiO_2/PVB coatings, a composite coating of 20Ch-5Gr-SiO_2/PVB was prepared by incorporating 5% graphene into the chitosan layer of 20Ch-SiO_2/PVB coating. EIS analysis of this sample conducted in 0.5 M NaCl solution for a period of 20 d is shown in Figure 8.12a. The coating exhibited low frequency impedance value in the order of 10^9 Ωcm^2 at beginning of the immersion test, which was maintained for a period of 2 d. Continuous immersion for a period of 7 d decreased this value to an order of 10^7 Ωcm^2, which remained in the same order even after 20 d test period (Table

Table 8.2 EIS parameters of Ch-SiO_2/PVB hybrid coatings having different percentages of chitosan in 0.5M NaCl solution

	Exposure time	Low frequency IZI (Ω cm^2)	R_p (Ω cm^2)	C_c (F cm^2)	Q_c	
					Y_0 $(F.s^{n-1})$	n
	5 min	6.1e9	5.809e9	7.18e-5	0.497e-9	0.95
	2 h	5.1e9	4.768e9	7.498e-5	0.515e-9	0.95
10Ch-SiO_2/PVB	1 day	4.0e9	3.970e9	7.705e-5	0.529e-9	0.95
	2 days	4.2e9	3.876e9	7.805e-5	0.535e-9	0.95
	7 days	1.7e8	1.466e8	1.816e-5	0.167e-9	0.94
	20 days	6.8e7	4.717e7	1.545e-5	0.156e-9	0.93
	5 min	8.03e9	6.9e9	2.898e-5	0.199e-9	0.95
	2 h	8.13e9	6.578e9	2.955e-5	0.207e-9	0.94
20Ch-SiO_2/PVB	1 day	7.74e9	6.449e9	3.176e-5	0.215e-9	0.94
	2 days	6.16e8	6.259e8	4.766e-5	0.306e-9	0.94
	7 days	4.67e7	3.507e7	2.082e-5	0.215e-9	0.92
	20 days	2.39e6	2.007e6	2.245e-4	0.191e-9	0.91
	5 min	3.28e9	7.96e8	6.547e-4	1.42e-8	0.96
50Ch-SiO_2/PVB	2 h	1.47e9	5.94e8	4.073e-4	1.53e-8	0.96
	1 day	9.81e8	3.28e8	3.905e-4	1.45e-8	0.96
	2 days	3.71e7	2.90e7	3.759e-4	1.34e-8	0.97

8.3). For 20Ch-SiO_2/PVB coating, low frequency impedance was observed to decrease to an order of 10^6 after 20 d exposure (Table 8.2). Thus, the high magnitude of low frequency impedance indicated the enhanced corrosion resistance behavior of 20Ch-5Gr-SiO_2/PVB coatings. Equivalent circuit shown in Figure 8.6b was used to model the EIS plots and the parameters corresponding to coating response are shown in Table 8.3. 20Ch-5Gr-SiO_2/PVB coatings exhibited a lower C_c than that of 20Ch-SiO_2/PVB coating, which indicated that the water permeability of the coating decreased further by the incorporation of graphene in chitosan layer. On enhancing the immersion time, coating capacitance increased slightly for 2 d. However, further immersion of the substrate resulted in a significant decrease in the capacitance by an order of three and the magnitude remained constant till 20 d immersion time, indicating the capacitive nature of the coating. It should be noted that this value (15.29e-12 F cm^2) was significantly lower than that of 20Ch-SiO_2/PVB coating (2.245e-4 F cm^2) which is attributed to the higher

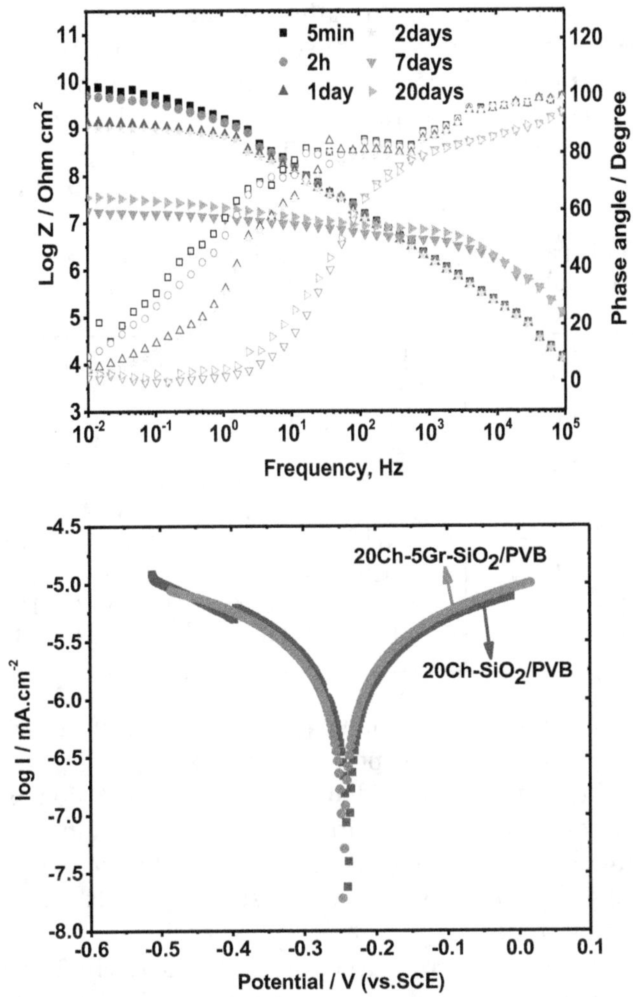

Figure 8.12 Bode plots of 20Ch-5Gr-SiO$_2$/PVB hybrid coating at different time of immersion in 0.5M NaCl solution (top or a). Polarization curves of 20Ch-SiO$_2$/PVB and 20Ch-5Gr-SiO$_2$/PVB hybrid coatings taken after 1 d immersion in 0.5M NaCl solution (bottom or b).

efficiency of graphene in enhancing the water impermeability of the coating. It is reported that graphene incorporation in polymer coatings increases the tortuosity of the diffusion pathways, thus, resulting in effective prevention of the H_2O and O_2 molecules from accessing the substrate surface. Therefore, the decreased C_c values during the 20 d of exposure time would have resulted from enhancement in the tortuosity of the diffusion pathways due to graphene incorporation, which also helped the diffused water molecules to distribute throughout the coating and not accumulating on metal surface as free electrolyte. A schematic representation of the tortuous movement of water through 20Ch-5Gr-SiO_2/PVB coating is given in Figure 8.13. Similar behavior has also been reported by Chang *et al.*

Table 8.3 EIS parameters corresponding to coating response of 20Ch-5Gr-SiO_2/PVB hybrid coatings in 0.5M NaCl solution

	Exposure time	Low frequency IZI (Ω cm^2)	R_p (Ω cm^2)	C_c (F cm^2)	Q_c	
					Y_0 (F.s∧.n-1)	n
20Ch-5Gr-SiO_2/PVB	5 min	6.8e9	5.68e9	0.109e-9	0.109e-9	1
	2 h	5.1e9	3.62e9	0.115e-9	0.115e-9	1
	1 day	1.4e9	1.10e9	0.128e-9	0.128e-9	1
	2 days	1.0e9	8.47e8	0.135e-9	0.135e-9	1
	7 days	3.5e7	7.68e6	16.05e-12	16.05e-12	1
	20 days	1.8e7	4.06e6	15.29e-12	15.29e-12	1

[31] for polyaniline/graphene coating on steel surface, where well dispersed graphene in the polymer matrix enhanced O_2 and H_2O barrier, which in turn provided an effective corrosion protection [31]. Pore resistance of the coating exhibited values in the same order as that of 20Ch-SiO_2/PVB coating with slight decrease in magnitude during the initial period of exposure. Towards the end of the measurement time, magnitude of R_p was higher for 20Ch-5Gr-SiO_2/PVB coatings as compared to 20Ch-SiO_2/PVB. These results indicated that graphene incorporation didn't affect the integrity of the coating, but reinforced the barrier nature of the coating because of its uniform distribution in the Ch-SiO_2 matrix. Corrosion potential (E_{corr}) of 20Ch-SiO_2/PVB and 20Ch-5Gr-SiO_2/PVB coatings obtained from polarization curves measured after 1 d immersion in 0.5 M NaCl solution, shown in Figure 8.12b, exhibited almost the same values of -0.239 V and -0.246 V respectively. However, corrosion current density (I_{corr}) of 20Ch-SiO_2/PVB coatings decreased from 0.983 μA to 0.013 μA by the incorporation of 5% graphene in the chitosan layer, thus, confirming the superior corrosion protection performance in the graphene containing coating. Therefore, it can be suggested from these results that graphene incorporations would be

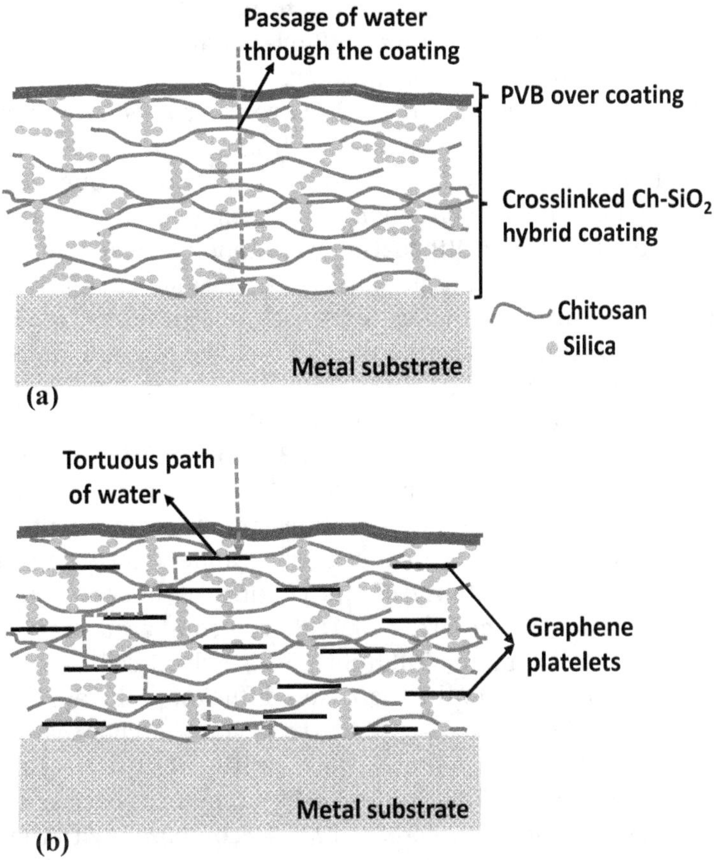

Figure 8.13 Schematic representation of corrosion protection provided by $Ch\text{-}SiO_2$ hybrid coatings in the absence (a) and presence (b) of graphene platelets.

promising as a suitable method to enhance the corrosion protection nature of chitosan-silica hybrid coatings. Further insights into the role of graphene can also be gained by incorporating different amounts of graphene in the coatings as well as by exploring the use of surface functionalized graphene and its derivatives.

8.4 Conclusions

In the current study, $Ch\text{-}SiO_2$ hybrid coatings were prepared by crosslinking with silica using sol-gel process. Though sol-gel coating enhanced the adhesive strength of chitosan, the coating did not provide long-term corrosion protection to underlying metal substrate because of its porous structure. PVB over-coating onto Ch-SiO_2 coatings was effective in sealing the pores and defects on the sol-gel coatings by the interaction of PVB hydroxyl groups with amine and hydroxyl groups on chitosan and silica, which subsequently enhanced the barrier as well as corrosion resistance properties of Ch-SiO_2 sol-gel coatings, even when thin coatings of 5-6 μm thickness were used. 10Ch-SiO_2/PVB coating withstood corrosion over 30 d in 0.5 M NaCl solution. Though high concentrations of chitosan in the coatings decreased corrosion resistance due to its hydrophilic nature, chitosan concentration up to 20% exhibited good inhibition of carbon steel corrosion. At 20% chitosan content, the coatings remained intact for nearly 20 d in 0.5 M NaCl solution. R_p values in the same order of 10^9 obtained for 20Ch-5Gr-SiO_2/PVB as compared to 20Ch-SiO_2/PVB hybrid coatings during the initial period of exposure indicated that graphene incorporation into chitosan didn't deteriorate the coating integrity. Furthermore, graphene incorporation decreased water permeability of the coating, as observed from the lower value of C_c in the order of 10^{-9} Fcm^2 compared with value in the order of 10^{-5} Fcm^2 for 20Ch-SiO_2/PVB hybrid coatings. The behavior was attributed to the increased diffusion pathways for water molecules inside the coating, thus, reducing the chances of water molecules to reach the metal surface as free electrolyte. Enhanced corrosion resistance of 20Ch-5Gr-SiO_2/PVB hybrid coatings was further evidenced from the decreased I_{corr} as compared with the native 20Ch-SiO_2/PVB hybrid coatings. Therefore, it can be concluded that graphene incorporation would be promising to obtain excellent corrosion protection of Ch-SiO_2 hybrid coatings.

Summary of the Results

The present work demonstrates the generation of stable chitosan-silica based protective coatings on carbon steel substrate. Effect of chitosan content, cross-linking with silica, poly(vinyl butyral) (PVB) over-coat as well as graphene reinforcement on the corrosion resistance performance of the coatings in an aggressive environment was studied. Sol-gel process was used to obtain chitosan coatings cross-linked with silica. Though Ch-SiO_2 sol-gel coating enhanced the adhesive strength of chitosan with metal substrate, the coatings were not effective against corrosion for more than 7 d immersion because of the porous structure. An excellent corrosion resistance was obtained by the application of a PVB over-coating on to Ch-SiO_2 hybrid coating, even though the overall coating thickness was still very low (5-6 μm). Superior performance was observed for 10Ch-SiO_2/PVB hybrid coating, as it was stable against corrosion for over 30 d in 0.5 M NaCl solution. Graphene incorporation in the chitosan layer didn't affect coating stability negatively as a pore resistance in the order of 10^9 was observed for 20Ch-5Gr-SiO_2/PVB coatings. Uniformly distributed graphene platelets contributed to the significantly enhanced corrosion resistance by acting as water barrier, which was reflected in the lower coating capacitance and I_{corr} compared with 20Ch-SiO_2/PVB coatings. Such graphene reinforced multilayer coatings exhibit strong potential of commercial application for long-term corrosion protection of different metal substrates.

Acknowledgement

The authors sincerely acknowledge the financial support from Petroleum Institute Gas Processing and Materials Science Research Centre (GRC) for the project GRC-005 Surface Engineered Self-Healing Anti-Corrosion Coatings for Gas Pipelines and Storage Tanks.

References

1. Gonzalez-Garcia, Y., Gonzalez, S., and Souto, R. M. (2007) Electrochemical and structural properties of a polyurethane coating on steel substrates for corrosion protection. *Corrosion Science*, **49**, 3514-3526.
2. Leidheiser, H. (1982) Corrosion of painted metals - a review. *Corrosion*, **38**, 374-383.
3. Niratiwongkorn, T., Luckachan, G. E., and Mittal, V. (2016) Self-healing protective coatings of polyvinyl butyral/polypyrrole-carbon black composite on carbon steel. *RSC Advances*, **6**, 43237-43249.
4. Johannes, T. P., Derksen, F., Cuperus, P., and Kolster, P. (1996) Renewable resources in coatings technology: a review. *Progress in Organic Coatings*, **27**, 45-53.
5. Hang, T. X., Truc, T. A., Olivier, M. G., Vandermiers, C., Guerit, N., and Pebere, N. (2010) Corrosion protection mechanisms of carbon steel by an epoxy resin containing indole-3 butyric acid modified clay. *Progress in Organic Coatings*, **69**, 410-416.

6. Shokry, H. (2009) Corrosion protection of mild steel electrode by electrochemical polymerization of acrylamide. *Chemistry of Metals and Alloys*, **2**, 202-210.
7. Rani, B. E. A., and Basu. B. B. J. (2012) Inhibitors for corrosion protection of metals and alloys: An overview. *International Journal of Corrosion*, Article ID 380217.
8. Lee, H., and Neville, K. (1952) *Handbook of Epoxy Resins*, Mc Graw Hill, USA.
9. Barletta, M., Lusvarghi, L., Mantini, F. P., and Rubino, G. (2007) Epoxy-based thermosetting powder coatings: surface appearance, scratch adhesion and wear resistance. *Surface and Coatings Technology*, **20**, 7479-7504.
10. Srinivasa, P. C., and Tharanathan, R. N. (2007) Chitin/chitosan - safe, ecofriendly packaging materials with multiple potential uses. *Food Reviews International*, **23**, 53-72.
11. Portes, E., Gardrat, C., Castellan, A., and Coma, V. (2009) Environmentally friendly films based on chitosan and tetrahydrocurcuminoid. *Carbohydrate Polymers*, **76**, 578-584.
12. Gupta, B., Saxena, S., Arora, A., and Alam. M. S. (2011) Chitosan/polyethylene glycol coated cotton membranes for wound dressings. *Indian Journal of Fibre and Textile Research*, **36**, 272-280.
13. Liu, B. S., Huang, T. B., Yao, C. H., Fang, S. S., and Chang. C. J. (2009) Novel wound dressing of non-woven fabric coated with genipin-crosslinked chitosan and bletilla striata herbal extract. *Journal of Medical and Biological Engineering*, **29**(2), 60-67.
14. Kumar, G., and Buchheit, R. G. (2006) Development and characterization of corrosion resistant coatings using the natural biopolymer chitosan. *ECS Transactions*, **1**, 101-117.
15. Lundvall, O., Gulppi, M., Paez, M. A., Gonzaleza, E., Zagal, J. H., Pavez, J., and Thompson, G. E. (2007) Copper modified chitosan for protection of AA-2024. *Surface and Coatings Technology*, **201**, 5973-5978.
16. Pang, X., and Zhitomirsky, I. (2007) Electrophoretic deposition of composite hydroxyapatite-chitosan coatings. *Materials Characterization*, **58**, 339-348.
17. El-Sawy, S. M., Abu-Ayana, Y. M., and Abdel-Mohdy, F. A. (2001) Some chitin/chitosan derivatives for corrosion protection and waste water treatments. *Anti-Corrosion Methods and Materials*, **48**, 227-235.
18. Ahmed, R. A., Farghali, R. A., and Fekry, A. M. (2012) Study for the stability and corrosion inhibition of electrophoretic deposited chitosan on mild steel alloy in acidic medium. *International Journal of Electrochemical Science*, **7**, 7270-7282.

19. Sugama, T., and Cook, M. (2000) Poly(itaconic acid)-modified chitosan coatings for mitigating corrosion of aluminum substrates. *Progress in Organic Coatings*, **38**, 79-87.
20. Carneiro, J., Tedim, J., Fernandes, S. C. M., Freire, C. S. R., Silvestre, A. J. D., Gandini, A., Ferreira, M. G. S., and Zheludkevich, M. L. (2012) Chitosan-based self-healing protective coatings doped with cerium nitrate for corrosion protection of aluminum alloy 2024. *Progress in Organic Coatings*, **75**, 8.
21. Carneiro, J., Tedim, J., Fernandes, S. C. M., Freire, C. S. R., Gandini, A., Ferreira, M. G. S., and Zheludkevich, M. L. (2013) Chitosan as a smart coating for controlled release of corrosion inhibitor 2-mercaptobenzothiazole. *ECS Electrochemistry Letters*, **2**(6), C19-C22.
22. Zheludkevich, M. L., Tedim, J., Freire, C. S. R., Fernandes, S. C. M., Kallip, S., Lisenkov, A., Gandini, A., and Ferreira, M. G. S. (2011) Self-healing protective coatings with "green" chitosan based pre-layer reservoir of corrosion inhibitor. *Journal of Materials Chemistry*, **21**(13), 4805-4812.
23. Luckachan, G. E., and Mittal, V. (2015) Anti-corrosion behavior of layer by layer coatings of cross-linked chitosan and poly(vinyl butyral) on carbon steel. *Cellulose*, **22**, 3275-3290.
24. Eduok, U. M., and Khaled, M. M. (2014) Corrosion protection of steel sheets by chitosan from shrimp shells at acid pH. *Cellulose*, **21**, 3139-3143.
25. Umoren, S. A., Banera, M. J., Garcia, T. A., Gervasi, C. A., and Mirifico, M. V. (2013) Inhibition of mild steel corrosion in HCl solution using chitosan. *Cellulose*, **20**, 2529-2545.
26. Yeh, J. M., Chin, C. P., and Chang, S. (2003) Enhanced corrosion protection coatings prepared from soluble electronically conductive polypyrrole-clay nanocomposite materials. *Journal of Applied Polymer Science*, **88**, 3264-3274.
27. Bumgardner, J. D., Wiser, R., and Gerard, P. D. (2003) Chitosan: potential use as a bioactive coating for orthopaedic and craniofacial/dental implants. *Journal of Biomaterials Science, Polymer Edition*, **14**(5), 423-438.
28. Sugama, T., and Jmenez, S. M. (1999) Dextrine-modified chitosan marine polymer coatings. *Journal of Materials Science*, **34**(9), 2003-2014.
29. Al-Sagheer, F., and Muslim, S. (2010) Thermal and mechanical properties of chitosan/hybrid composites. *Journal of Nanomaterials*, Article ID 490679.
30. Yeh, J. T., Chen, C. L., and Huang, K. S. (2007) Synthesis and properties of chitosan/SiO_2 hybrid materials. *Materials Letters*, **61**(6), 1292-1295.

31. Chang, C. H., Huang, T. C., Peng, C. W., Yeh, T. C., Lu, H., Hung, W., Weng, C. J., Yang, T., and Yeh, J. M. (2012) Novel anticorrosion coatings prepared from polyaniline/graphene composites. *Carbon*, **50**, 5044-5051.
32. Metroke, T. L., Parkhill, E. T., and Knobbe, E. T. (2001) Passivation of metal alloys using sol–gel-derived materials - a review. *Progress in Organic Coatings*, **41**(4), 233-238.
33. Zheludkevich, M. L., Serra, R., Montemor, M. F., Yasakau, K. A., Miranda Salvado, I. M., and Ferreira, M. G. S. (2005) Nanostructured sol–gel coatings doped with cerium nitrate as pre-treatments for AA2024-T3: Corrosion protection performance. *Electrochimica Acta*, **51**, 208-217.
34. Lai, S. M., Yang, A. J. M., Chen, W. C., and Hsiao, J. F. (2006) The properties and preparation of chitosan/silica hybrids using sol-gel process. *Polymer - Plastics Technology & Engineering*, **45**(9), 997-1003.
35. Singh, K., Ohlan, A., and Dhawan, S. K. (2012) Polymer-graphene nanocomposites: preparation, characterization, properties, and applications. In: *Nanocomposites - New Trends and Developments*, Ebrahimi, F. (ed.), InTech, Croatia, p. 37.
36. Chaudhry, A. U., Mittal, V., and Mishra, B. (2015) Inhibition and promotion of electrochemical reactions by graphene in organic coatings. *RSC Advances*, **5**, 80365-80368.
37. Chang, K. C., Hsu, M. H., Lu, H. I., Lai, M. C., Liu, P. J., Hsu, C. H., Ji, W. F., Chuang, T. L., Wei, Y., Yeh, J. M., and Liu, W. R. (2014) Room-temperature cured hydrophobic epoxy/graphene composites as corrosion inhibitor for cold-rolled steel. *Carbon*, **66**, 144-153.
38. Li, Y., Yang, Z., Qiu, H., Dai, Y., Zheng, Q., Li, J., and Yang, J. (2014) Self-aligned graphene as anticorrosive barrier in waterborne polyurethane composite coatings. *Journal of Materials Chemistry A*, **2**, 14139-14145.
39. Mayavan, S., Siva, T., and Sathiyanarayanan, S. (2013) Graphene ink as a corrosion inhibiting blanket for iron in an aggressive chloride environment. *RSC Advances*, **3**, 24868-24871.
40. Smitha, S., Shajesh, P., Mukundan, P., and Warrier, K. G. K. (2008) Sol-gel synthesis of biocompatible silica-chitosan hybrids and hydrophobic coatings. *Journal of Materials Research*, **23**, 2053-2060.
41. Lopez, D. Z., Simison, S. N., and De Sanchez, S. R. (2003) The influence of steel microstructure on CO2 corrosion. EIS studies on the inhibition efficiency of benzimidazole. *Electrochimica Acta*, **48**, 845-854.
42. Lian, F., Wen, Y., Ren, Y., and Guan, H. Y. (2014) A novel PVB based polymer membrane and its application in gel polymer electrolytes for lithium-ion batteries. *Journal of Membrane Science*, **456**, 42-48.

43. Silva, S. M. L., Braga, C. R. C., Fook, M. V.L., Raposo, C. M.O., Carvalho, L. H., and Canedo, E. L. (2012) Application of infrared spectroscopy to analysis of chitosan/clay nanocomposites. In: *Infrared Spectroscopy - Materials Science, Engineering and Technology*, Theophile, T. (ed.), InTech, Croatia, pp. 43-62.
44. Akid, R., Gobara, M., and Wang, H. (2011) Corrosion protection performance of novel hybrid polyaniline/sol–gel coatings on an aluminium 2024 alloy in neutral, alkaline and acidic solutions. *Electrochimica Acta*, **56**, 2483-2492.
45. Ruhi, G., Modi, O. P., and Dhawan, S. K. (2015) Chitosan-polypyrrole-SiO2 composite coatings with advanced anticorrosive properties. *Synthetic Metals*, **200**, 24-39.
46. Tripathy, A. R., Chen, W., Kukureka, S. N., and MacKnight, W. J. (2003) Novel poly(butylene terephthalate)/poly(vinyl butyral) blends prepared by in situ polymerization of cyclic poly(butylene terephthalate) oligomers. *Polymer*, **44**, 1835-1842.
47. Socrates, G. (1994) *Infrared and Raman Characteristics Group Frequencies*, 3rd edition, John Wiley & Sons Ltd., USA.
48. Rout, T. K., Bandyopadhyay, N., and Venugopalan, T. (2006) Polyphosphate coated steel sheet for superior corrosion resistance. *Surface and Coatings Technology*, **201**, 1022-1030.

9

PLA - Graphene Nanocomposites: A Brief Review

9.1 Introduction

Polylactic acid (PLA) is a biodegradable, renewable thermoplastic polyester, which has received considerable research attention due to its attractive properties, in comparison with petroleum based polyolefins. High molecular weight PLA is generally prepared by ring-opening polymerization of lactide. Since the basic monomer (lactic acid) is produced via fermentation of agricultural products such as corn and sugar beet, PLA reduces the solid waste disposal problem. PLA presents a set of balanced properties, like mechanical strength and thermal plasticity, along with biodegradability. In addition, its microstructure is also stable and does not undergo structural changes with time. For instance, Figure 9.1 demonstrates the evolution of volumetric expansion/contraction of polycaprolactone (PCL) and PLA as a function of time [1]. Specifically for PLA, the difference between nanofiber diameters in the initial state and aged state was not pronounced, thus, indicating the phase stability with time. PLA also suffers from some disadvantages such as low heat resistance and slow crystallization rate, which limit its applications [2].

In order to overcome the limitations pf PLA and tune the properties of PLA according to end-use applications, different techniques have been developed. One of the commonly used strategy is the blending of the polymer with reinforcing fillers, thus, leading to the generation of reinforced composites. Some representative examples of fillers employed for reinforcement of PLA are bio-fillers like date seed powder (DSP) [3], and other commonly used fillers for generating nanocomposites such as layered silicates [4,5], and carbon nanotubes (CNTs) [6]. For instance, Figure 9.2 shows the storage modulus of the PLA composites as a function of DSP content [3]. The composites with 10% and 20% filler content had similar storage modulus as

Evgenia Georganta, Stavroula Anastasiou and Vikas Mittal, The Petroleum Institute (part of Khalifa University of Science and Technology), Abu Dhabi, UAE*
**Current address: Bletchington, Wellington County, Australia*

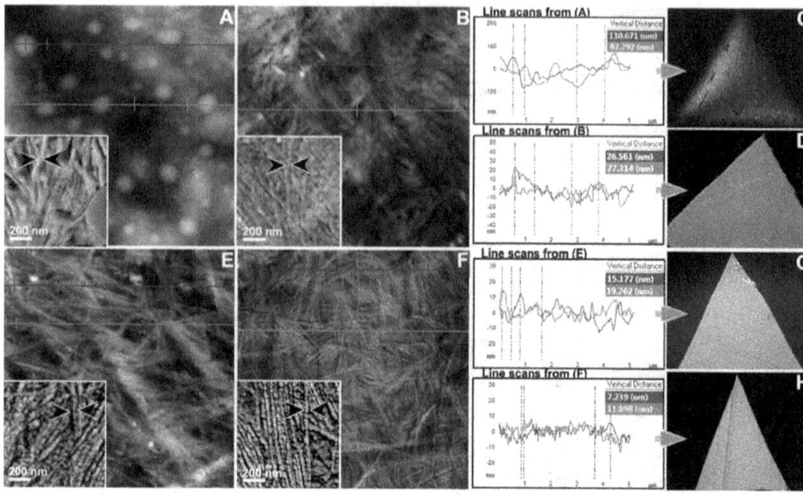

Figure 9.1 Evolution of volumetric expansion/contraction of PCL (A-D) and PLA (E-H) as a function of time. (A, C, and E, G) initial state of the samples (14 days after production), (B, D and F, H) aged stage of the samples (180 270 days after production). Room temperature light microscopy (C, D, G, H) of freshly prepared cross-sections of the corresponding sample blocks with the appropriate age. Reproduced from Reference 1 with permission from Royal Society of Chemistry.

the pure polymer at higher frequencies. At lower angular frequencies, the modulus increased with increasing filler content. The composites with 30% and 40% filler fraction, however, had very different behavior as indicated by their significantly high storage modulus values especially at lower frequencies. The authors observed that the presence of large sized filler particles and non-interaction at the polymer-filler interface resulted in hindrance to the melt shearing process in the composites with higher filler content, thus, underlining the need of effective interface between the polymer and filler particles to achieve optimal composite performance.

Graphene, which consists of one atomic thick sheets of covalently sp^2-bonded carbon atoms in a hexagonal arrangement, has recently received attention of researchers for the generation of polymer nanocomposites [7]. Its choice as a filler is due to its excellent electrical and mechanical properties, which are significantly better than other inorganic filler materials. A single defect-free graphene layer has Young's modulus of ~1.0 TPa, intrinsic strength ~42 N/m, thermal conductivity ~4840–5300 W/(m.K), electron mobility exceeding

25,000 $cm^2/V.s$, excellent gas impermeability and specific surface area of ~2630 m^2/g [7]. All these properties make this material even superior to carbon nanotubes for use in polymer nanocomposites. Thus, a number of studies on polymer nanocomposites based on graphene have been published in a short span of time since its development [8-10]. The parent material for graphene is graphite which is present prolifically in nature. Graphene (and graphite oxide) can be produced from graphite by different methods such as thermal expansion of chemically intercalated graphite, micromechanical exfoliation of graphite, chemical vapor decomposition, and chemical reduction method of graphene oxide. The most commonly used method is the exfoliation of intercalated graphite oxide by introducing graphite oxide suddenly to a higher temperature.

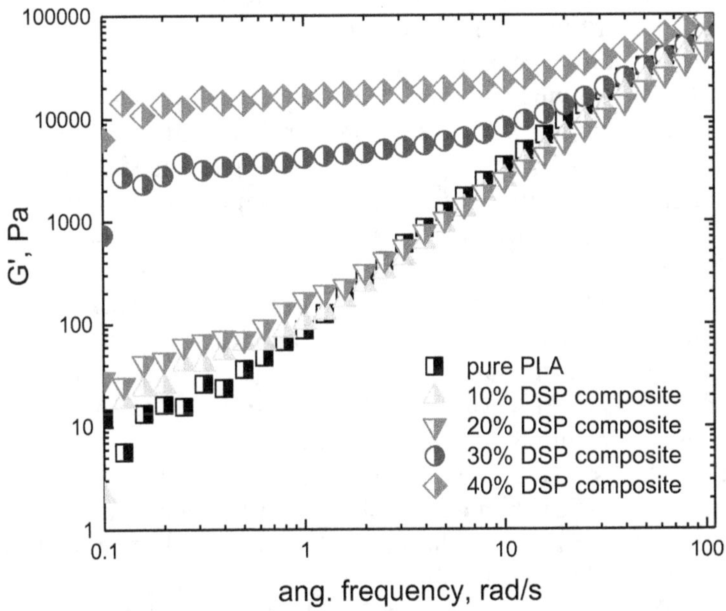

Figure 9.2 Storage modulus of PLA composites as a function of filler (DSP) content. Reproduced from Reference 3 with permission from Wiley.

In this review, composites of PLA with different types of graphene based materials, such as thermally reduced graphene, graphene oxide, graphite, expanded graphite and graphite oxide have been reviewed briefly. Each analogue leads to different extent of polymer-filler interaction and resultant property enhancements of the PLA matrix, thus, it is beneficial to review these types so to make suitable material selection, based on the application. In addition, all of the main techniques for the preparation of PLA/graphene composites, i.e. *in-situ* intercalation polymerization, solution mixing and melt blending methods, have also been reviewed due to the evolution of different degrees of filler exfoliation, and resulting composite properties, while using these techniques.

9.2 Incorporation of Unmodified Filler

In this section, some studies involving the incorporation of unmodified filler in the PLA matrix have been reviewed for mainly enhancement of mechanical and thermal properties of PLA. In a recent study, Mittal *et al.* [10] reported PLA bio-nanocomposites, generated by melt mixing, with varying amounts of graphene. Figure 9.3 shows the TEM images of the PLA nanocomposite with 5% graphene content. The average stack thickness of graphene was observed to reduce significantly and polymer intercalation in the interlayers was observed, the filler distribution was however less optimal with large aggregates present in the polymer matrix, thus, indicating the need of optimizing the polymer-filler interactions to achieve optimal filler dispersion. Figure 9.4 also presents the dynamic mechanical analysis (DMA) of the PLA nanocomposites as a function of temperature. The filler incorporation enhanced the high temperature performance of the nanocomposites. The composites also exhibited a sharp drop near the T_g and reduction of more than two orders of magnitude was observed. The T_g of PLA was also observed to increase by 5-6 °C for nanocomposite with 5% graphene content.

In another study, Kim *et al.* [11] reported polylactide-exfoliated graphite (PLA/EG) nanocomposites, synthesized via melt-compounding at 170-200 °C. The diffractograms of all composites, regardless the EG loading, exhibited a single broad peak, corresponding to PLA, whereas the characteristic peak of the graphite was absent, which indicated that EG was homogeneously dispersed in the polymer matrix, without forming aggregates. TGA thermograms indicated that the thermal stability of the composites was enhanced up to a loading of 3 wt% EG. For instance, the degradation temperature of pure PLA for 5% weight loss was 350 °C, whereas with the addition of 3% EG, the degradation temperature increased to 364 °C. The enhanced thermal stability was attributed to the fact that the filler acted as mass transfer barrier to the high volatile degradation byproducts. The Young's modulus of the composites increased with the increment

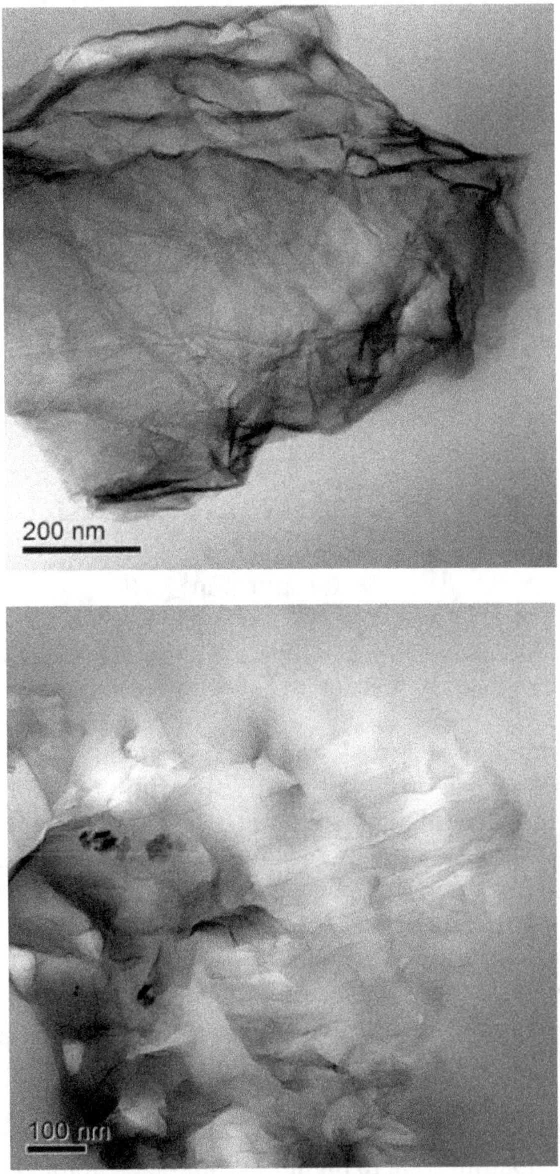

Figure 9.3 TEM micrographs of PLA nanocomposite containing 5 wt% of graphene. The graphene is represented by the dark stacks in the micrographs. Occasionally graphene stacks with multiple sheets were observed. On the other hand, some single layered and few layered sheets were also observed. Reproduced from Reference 10 with permission from Elsevier.

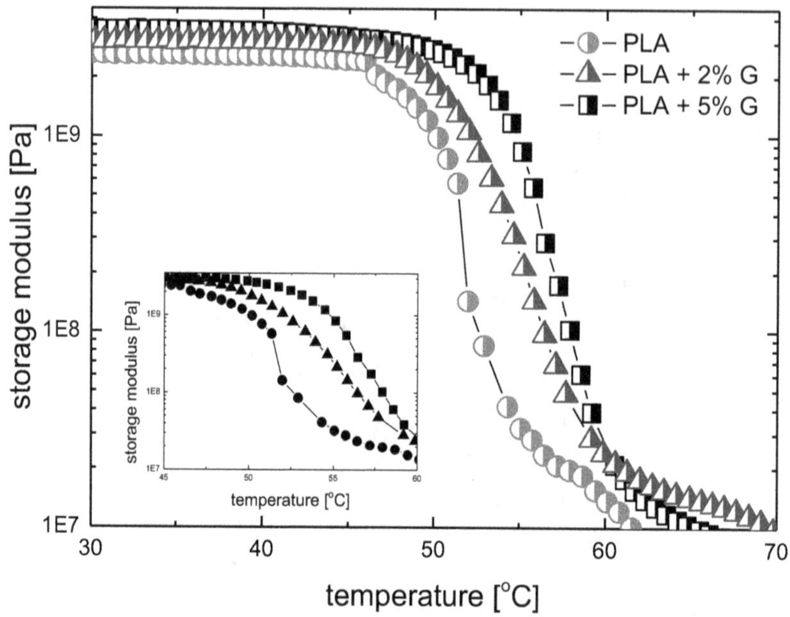

Figure 9.4 Storage modulus of pure PLA and nanocomposites analyzed from DMA. Reproduced from Reference 10 with permission of Elsevier.

of EG up to a loading of 2 wt %, and then remained constant. The tensile strength was also observed to increase from 62 MPa (neat PLA) to ~70 MPa for a loading of 0.5 wt% EG. The elongation at break was observed to decrease with increase in the graphite content, which was attributed to the hardening effect of EG in the PLA matrix. In a similar study, Murariu *et al.* [12] synthesized PLA composites with expandable graphite by melt-blending at 200 °C. The X-ray diffraction analysis confirmed the presence of a broad diffraction peak corresponding to PLA along with the characteristic graphitic peak at 2θ ≈26.5° for all the composite systems, suggesting that the melt-compounding could not completely delaminate the filler nano-layers. The intensity of the graphite peak was also observed to increase with increasing filler loading, due to higher number of graphite layers in the form of aggregates. The addition of up to 12 wt% filler slightly enhanced the thermal stability by increasing the degradation temperature of the composite by 10 °C, compared to the neat PLA. This was attributed to the fact that the dispersion of the filler in PLA created a tortuous path, thus, delaying the diffusion of the volatile decomposition products. It was also observed that the addition of expandable graphite shifted the cold crystallization temperature (T_c) peak to lower values, due to the nucleation effect of the filler, whereas the glass transition temperature (T_g) and peak melting temperature (T_m) were not affected. Also, the crystallinity of the composites initially increased up to a specific filler loading and subsequently decreased possibly due to filler aggregation or poor dispersion. Both tensile strength and elongation at break were observed to decrease with increasing the filler loading. The addition of filler enhanced the rigidity of the composites, by increasing the Young's modulus from 2200MPa (neat PLA) to 3100 MPa, along with increment in the storage modulus as a function of filler loading.

Hassouna *et al.* [13] also reported the melt blending of PLA with EG for the purpose of achieving filler delamination as well as enhanced properties. XRD patterns revealed the presence of diffraction peak of pure graphite at 26.5°. The peak indicated that a complete exfoliation of graphene layers was not achieved and these layers appeared as aggregates. Morphological characterization, however, also confirmed that EG was dispersed homogeneously in the PLA matrix. Also, the dispersion was observed to be better when residence time and screw speed were increased and led to separation and delamination of graphite layers. The thermogravimetric analysis indicated that the addition of EG in the PLA matrix, with a good degree of dispersion, enhanced the thermal stability of the material, as a slight increase of $T_{50\%}$ was observed with the filler incorporation in the PLA matrix. Thermal diffusivity and conductivity of the PLA-EG nanocomposites were also increased compared to neat PLA. In addition, the authors also reported an increment in the storage modulus (E') and glass transition temperature (T_g) of PLA with the addition of EG filler, as compared to the neat PLA.

In a study reporting the use of graphite oxide (GO), Wang *et al.* [14] prepared composites of PLA with GO via solution blending, using DMF as solvent. Different contents of GO were dispersed in DMF under sonication for 2 h and then mixed with PLA solution in DMF for 4 h at 40 °C. After sonication and solvent evaporation, the final PLA/GO composites were obtained by vacuum drying at 80 °C for 4 days. The morphological analysis revealed good dispersion of GO in the PLA matrix, indicating almost full exfoliation of GO. The effect of GO at the non-isothermal melt crystallization temperature was studied by DSC, for two different cooling rates. For a low cooling rate (5 °C/min), the crystallization peak temperature (T_{cc}) was observed to increase, compared to that of PLA, with increasing GO loading up to 1 wt%, whereas it decreased when loading was enhanced further (2 wt%). The degree of crystallinity followed the same trend. At a faster cooling rate (10 °C/min), the crystallization exotherm for the neat PLA was difficult to observe, whereas it was obvious for the case of PLA/GO composites. The enhancement of non-isothermal melt crystallization, especially at the fast cooling rate, indicated that GO acted as a nucleating agent for PLA. Moreover, DSC analysis exhibited that the addition of GO accelerated the isothermal melt crystallization in comparison with the neat PLA. Specifically, the isothermal melt crystallization rate increased with increasing GO loading up to 1 wt%, whereas it decreased for further loading.

In another study, Narimissa *et al.* [15] prepared PLA composites using nano-graphite platelets (NGP) as a filler by melt intercalation process. The melt blending was carried out in an extruder at 180 °C and 40 rpm. The calorimetric analysis revealed that the filler did not affect T_g, which was 59-60 °C. The peak melting temperature also remained constant around 149-151 °C. Both crystallinity and peak crystallization temperature T_c were maximized at 1 wt% of NGP in PLA. The Young's modulus in the composites increased till 3 wt% NGP loading, and subsequently decreased for further loadings. 1 wt% NGP loading resulted in small enhancement of tensile strength, which for further loadings decreased linearly. The elongation at break decreased constantly with any NGP loading, thus, indicating that the composites became more brittle with the addition of filler. The X-ray diffractograms also exhibited a broad peak at $2\theta \approx 16.5°$, corresponding to the semi-crystalline structure of PLA. The characteristic peak for the graphite carbon structure appeared at $2\theta \approx 26°$ in the diffractograms of all composites. This indicated that the melt blending did not completely exfoliate the graphene layers and several platelets remained in the form of aggregates, which has earlier been reported by Murariu *et al.* [12]. It was also illustrated in the diffractograms that the intensity of the characteristic peak of semi-crystalline PLA varied with different NGP loadings.

In another study reporting the use of graphene nano-platelets (GNP), Pinto *et al.* [16] synthesized polylactide-GNP films via solvent-casting. GNP in different concentrations (0.2-0.8 wt%) were dispersed in chloroform through sonication for 2 h and then dispersed in PLA solution with chloroform. Films were produced by spreading the dispersion on plates and drying in a vacuum oven at 40 °C for 10 days. The Young's modulus was observed to increase by 85% for 0.4 wt% GNP loading, whereas for the same loading, the tensile strength increased by 15%. The elongation at break in the composites was not significantly affected. For higher filler loadings, the properties decreased, probably due to the agglomeration effect. The calorimetric analysis also exhibited that the addition of GNP slightly decreased T_g and T_m, in comparison with the neat PLA. However, on comparing with the T_g and T_m of the solvent-cast PLA, both T_g and T_m were enhanced in the composites till 0.4 wt% filler loading.

In a subsequent study, Pinto *et al.* [17] incorporated 0.25 wt% loading of GNP (two types of graphene nano-platelets GNP-C and GNP-M were used in both studies) in PLA by melt-blending at 180 °C. The obtained composites were subjected to hydrolytic degradation for different time lags (2 and 6 months) in phosphate-buffered saline at 37 °C. The morphological analysis revealed good embedding of GNP in the PLA matrix, demonstrating almost complete exfoliation of GNP with the presence of a few agglomerates. The diffraction analysis also identified the characteristic peaks of the filler in the diffraction patterns of the composites, thus, also confirming the absence of complete exfoliation. No difference was observed in the two composites degraded at different time durations, indicating that the presence of the GNP in the polymer was permanent. In addition, no distinguishing peaks were observed in the FTIR spectra of the composites. The thermogravimetric analysis exhibited that the composites samples had a reduction of 10 °C in the onset thermal degradation temperature. The Young's modulus, tensile strength and toughness of both composites were also observed to enhance, in comparison with pure PLA, after six months of degradation, with the GNP-C being the more effective one probably due to its smaller dimensions. The improved performance of the composites was demonstrated by the creep and recovery test as well. The pure PLA exhibited cumulative increase in the permanent creep strain and ruptured after 4 cycles under cyclic creep-relaxation test, while the composites showed stable behavior and did not rupture even after 10 cycles.

In a different study, a comparison between pure PLA and composites of PLA with graphene nano-platelets and functionalized carbon nanotubes (CNT-COOH) as fillers was also reported by Pinto *et al.* [18]. The composites were produced by melt blending followed by compression molding. Morphological analysis showed a good dispersion of GNP in PLA matrix, even though the nano-platelets have the tendency to agglomerate, by forming stacks. The PLA/CNT-COOH demonstrated similar morphology, with some agglomerates. Both composites demonstrated a considerable increase in the Young's modulus compared to pure PLA, i.e. 23% for PLA/GNP and 21% for PLA/CNT-COOH. On incorporation of CNT-COOH in PLA, elongation at break, tensile strength, and toughness were increased by 21.2%, 20.6%, and 22.1% respectively. On the other hand, these properties were reduced for PLA/GNP composites. In another study, Scaffaro *et al.* [19] prepared PLA nanocomposites, comprising graphene nano-platelets as filler and ciproflaxin (CFX) as biocide, through melt-compounding, for antimicrobial applications. Poor adhesion of GnP in the polymeric matrix was observed, with some voids around the filler. However, the filler was observed to be well-dispersed in PLA. The needle-like aggregates of CFX were also noticed. The incorporation of only fuller increased the viscosity, compared to the pure PLA, due to the disruption of the PLA chains movement by the filler. However, for the PLA/filler/CFX nanocomposite, the viscosity was lower than that of pure PLA, possibly due to the PLA-CFX interaction. The incorporation of graphene nano-platelets in the PLA matrix increased the elastic modulus by 40%, compared to the neat PLA, while the tensile strength remained unchanged elongation at break reduced slightly. On the other hand, for PLA/filler/CFX, the stiffness was increased, while the tensile strength and the elongation at break were significantly decreased. These results could be attributed to the addition of CFX which interacted with the polymeric matrix, and caused degradation during compounding. The dynamic mechanical properties of the materials also exhibited differences due to the CFX addition. The incorporation of filler in PLA led to higher storage modulus E', compared to that of neat PLA. Moreover, it was found that the addition of filler increased the working range of the PLA, since the drop of E' took place at slightly higher temperature than that of the neat PLA. However, the storage modulus of PLA/filler/CFX was between the pure PLA and the PLA/filler composite, which is in accordance with the results from the static tensile test, due to the degradation of the PLA matrix caused by CFX. The damping behavior of the PLA/filler composites was reduced, in comparison with neat PLA, due to the fact that the filler platelets imposed restrictions for the motion of the PLA polymeric chains, resulting in a more elastic response of the material. It was noticed that the glass transition temperature T_g was slightly higher than that of the pure PLA. The addition of CFX reduced the damping of the material even more, for the same reason mentioned earlier.

In another study, Cao *et al.* [20] synthesized PLA/graphene nanosheets (GNS) composites using solution mixing method. Lyophilized

GNS powder was dispersed in DMF along with stirring and sonication at 25 °C. Subsequently, PLA was added and the mixture was agitated at 85 °C for 2 h, sonicated at 70 °C for another 2 h and subsequently the mixture was coagulated in methanol. The obtained flocculent was filtered under vacuum and the vacuum dried at 80 °C for 10 h, to generate composite with a loading of 0.2 wt% GNS. The diffractogram of the PLA/GNS composite indicated the absence of the characteristic graphite peak suggesting the absence of aggregation of filler or single-sheet dispersion of GNS in PLA. Addition of GNS improved the thermal stability of the polymer, by more than 10 °C for 5% weight loss due to the tortuous path created by the filler particles, leading to the delay in the permeation of oxygen, and the escape of volatile degradation products and also the char formation. The addition of 0.2 wt% filler loading increased the tensile strength by about 26% and the Young's modulus by 18%. The authors opined that the observed increment in the mechanical performance of the polymer resulted due to the excellent modulus (≈1100 GPa) and intrinsic strength (≈125 GPa) of GNS, along with effective interaction between the polymer and the filler.

In another study, Chen *et al.* [21] prepared PLA/graphene nanocomposites via solution blending. Different graphene contents (0.05-1 wt%) were dispersed in chloroform at room temperature for 2 h. The dispersion was mixed into the PLA solution in chloroform under magnetic stirring to form a homogeneous mixture. The mixture was added to methanol and the precipitated PLA/graphene nanocomposites were collected by centrifugation. The addition of 0.05 wt% graphene content significantly enhanced the thermal stability as compared to neat PLA, whereas further loading exhibited a smaller increase. Similar enhancement in the thermal stability of the nanocomposites was also reported by Murariu *et al.* [12], and attributed the enhancement to the shielding effect of graphene and to the creation of tortuous path by the filler particles, which hindered the diffusion of volatile degradation byproducts. The T_g values of the composites shifted to a higher temperature in comparison to neat PLA, indicating that the filler reduced the mobility of the polymer chains. The cold crystallization temperature of the composites also increased with the increment of graphene loading (except 0.05 wt% filler content). This was attributed to the fact that the incorporation of graphene hindered the segmentation of PLA chains and hence their effective diffusion to the growing crystalline lamella. The addition of filler only slightly affected the values of T_m.

Masterbatch-based melt blending process was reported by Bao *et al.* [22] for synthesizing PLA/graphene nanocomposites. Graphene/PLA masterbatch was synthesized via solution blending. Both graphene and PLA were dispersed/dissolved in tetrahydrofuran (THF) and subsequently mixed and dried. PLA/graphene nanocomposites were synthesized by melt blending at 175 °C using these masterbatches and PLA, followed by hot pressing into sheets. The morphological analysis indicated that the graphene layers were exfoliated. Thus, a successful dispersion was achieved due to the fact that masterbatch prevented graphene layers from re-aggregating. The diffraction studies on the composites exhibited that the crystallinity of the reinforced nanocomposites was improved, indicating that graphene increased the rate of crystallization. The calorimetric analysis also confirmed the finding from X-ray diffraction, due to heterogeneous nucleation effect and molecular segments restriction due to graphene. Up to 0.04 wt% graphene content, the storage modulus of the composites increased (3.38 GPa) due to the reinforcing effect of graphene. In the range of 0.08-0.2 wt% filler content, the value was observed to reduce to 3.22 and 2.99 GPa respectively, due to the elimination effect that damages the interaction between the matrix and filler. Tensile strength of the composites also showed similar trend, as it increased up to 0.08 wt% graphene, followed by its decrease at higher filler loading. Similarly, the hardness of the composites also increased with the filler content. The degradation behavior of the reinforced polymer was similar as the neat PLA but the nanocomposites had lower degradation temperatures. Also, the nanocomposites, except the one with 2 wt% filler, had higher DTG peaks in comparison with neat PLA. This indicated that the degradation process was faster in the nanocomposite.

In a similar study, Li *et al.* [23] synthesized PLA/graphene nanosheets (GS) nanocomposites via solution blending method with chloroform. PLA was dissolved in chloroform at 45 °C. This solution was mixed with GS and continuously stirred for 4 h at 40 °C. The mixed solution was poured into polytetrafluoroethylene mold after sonication. Solvent was evaporated and the nanocomposite film was obtained, which was dried for 4 days at 60 °C under vacuum. The amount of the filler in the samples was raged between 0.01-2.0 wt%. The morphological characterization revealed that GS were homogeneously dispersed on the surface without forming aggregates. Diffraction pattern of PLA with 0.1% filler was observed to be similar as pure PLA. The absence of the diffraction peak at 26° indicated that the GS were exfoliated, which also confirmed that the filler platelets did not aggregate. This observation was also similar to the findings by Cao *et al.* [20]. The addition of GS in PLA increased its thermal decomposition temperature. The weight loss rate of PLA/GSs nanocomposites also decreased as the amount of filler increased in the composites. Generally, the enhancement of thermal properties was attributed to the strong polymer-filler interfacial action. The addition of filler increased the E' of the nanocomposites, especially at high temperatures. For instance, at a particular temperature, pure PLA had storage modulus of 6.87 MPa, which was enhanced to 212 MPa for the composites with 1% filler content. The addition of filler up to 1% weight fraction increased the tensile strength, whereas it was observed to decrease with further enhancement in the filler loading. The amount of filler also correspondingly decreased the elongation at break and toughness of the composites.

Valapa *et al.* [24] synthesized PLA-graphene composites by the solution casting method. PLA was dissolved in chloroform, under continuous stirring for 2 h. Another solution was prepared by dispersing different amounts of graphene (0.1, 0.3 and 0.5 wt% with respect to PLA) in chloroform, under bath sonication for 30 min. Two solutions were mixed and the solution was sonicated (15 min and 30 min), followed by casting on petri-plates and drying for 24 h under ambient conditions. Finally, the synthesized films were peeled off from the petri-plates and dried under vacuum at 40 °C. In the diffractograms of the composites sonicated for 15 min, a characteristic peak for the graphite carbon structure appeared at 2θ 26.5°. This peak disappeared when the composites were sonicated for 30 min, indicating that 15 min of sonication was insufficient for complete exfoliation of graphene sheets. Taking 10% weight loss as a reference point, it was observed that the degradation temperature of the composite with the maximum graphene loading increased by 6 °C compared to the neat PLA. Moreover, the derivative weight loss profile (DTG) indicated that temperature corresponding to the maximum rate of weight loss (T_{max}) was marginally increased from neat PLA, which was attributed to the barrier effect of graphene on PLA. The addition of graphene increased the peak melting temperature (T_m), thus, acting as a nucleating agent for PLA. Glass-transition temperature (T_g) of PLA remained unaffected in the composites. The calculated crystallinity was observed to increase until 0.3 wt% graphene loading. For higher loadings, it was observed to decrease due to aggregation and low degree of dispersion of graphene in PLA. A further indication that the thermal stability of PLA was enhanced by the filler addition is the increase of the activation energy E_a calculated by the Coats-Redfern method. Initially, the tensile strength enhanced from 32 MPa for pure PLA to 40 MPa for the composite with 0.1 wt% filler content. Subsequently, the tensile strength was observed to decrease as the amount of filler was enhanced in the composites. The elongation at break of the composites also exhibited similar trend as tensile strength. Initial addition of graphene enhanced PLA ductility by extending the fracture process, whereas at higher filler loading, the elongation decreased due to defects induced by agglomeration of graphene sheets which made the polymer more brittle.

In a similar study, Papageorgiou *et al.* [25] synthesized nanocomposites of PLLA with graphene oxide via solution casting. Graphene oxide was dispersed in DMF, whereas PLLA was dissolved in chloroform. The two solutions were mixed, stirred and sonicated for 30 minutes. Subsequently, the solvents were evaporated at high temperature. Strength at yield, strength at break, and Young's modulus in the composites were observed to increase with the addition of graphene oxide for up to 1 wt% loading in the PLLA. This could be attributed to the decrease in the heat capacity of the nanocomposites in the glass transition region. A slight reduction in all the properties was observed in the composite reinforced with 2.5 wt% filler, probably due to the formation of aggregates. On the other hand, elongation at break decreased due to the fact that the material became hard and brittle. The X-ray diffraction patterns exhibited the absence of the characteristic filler peak, indicating that the filler was exfoliated in the polymer, which enhanced the mechanical properties further. Both non-isothermal and isothermal melt crystallization rates of the polymer were observed to increase by the addition of the filler. This indicated that the crystallization was faster for the composites in comparison to the neat PLLA. The calorimetric analysis revealed that the values of the T_g were almost the same for all the samples regardless of the filler loading.

In a study reporting the use of in-situ polymerization for the composite generation, Yang *et al.* [26] used thermally reduced graphene oxide (TRG) as the initiator for the melt ring opening polymerization of L-lactide. Molten L-lactide and TRG were mixed at 100 °C with stirring for 5 minutes under inert atmosphere. The mixture was heated to 170 °C and the catalyst was added. The product was dissolved in chloroform, precipitated in methanol, re-dissolved in chloroform and subsequently filtered through PTFE membrane. TRG/PLLA nanocomposites with different filler loadings, (0.01-2 wt%) were, thus, obtained after vacuum drying at 40 °C for 12 h. The X-ray diffractograms of the composites did not show any characteristic peak of TRG, indicating the formation of graphene sheets. A uniform dispersion of the sheets was observed in the polymer with some stacks present throughout the matrix. The morphological analysis also revealed the interfacial interaction between the matrix and filler. A strong interaction between PLLA and filler could be implied due to the absence of both agglomeration and pulling out. The same phenomenon was also observed for the composite with 2 wt% filler content. The thermal degradation temperature of the composites was observed to increase slightly with increasing percentage of filler. Time t_d (the time required for the entire thermal decomposition) was also observed to slightly decrease when the percentage of filler was increased. It could be concluded that the addition of filler may lead to the enhancement of thermal conductivity in this system. The addition of the filler increased the T_g, but the trend was not stable due to possible steric hindrance caused by the filler and the differences in the polymer chain lengths. On the other hand, T_m was clearly increased with increasing percentage of filler. The calorimetric measurements also revealed that the presence of the filler enhanced the crystallization rate, but decreased the crystallinity of PLLA.

Qi *et al.* [27] reported the ternary nanocomposites with polystyrene (PS), PLA and graphene. The authors reported that the conductivity of polystyrene increased from 6.7 × 10-14 to 3.49 S/m, as the amount of graphene was increased from 0.11 to 1.1 vol%. Further, the the conductivity of the PS/graphene composite system could be further increased significantly by the incorporation of polylactic acid (Figure 9.5). It was observed that the volume-exclusion principle led to the localization of graphene in the polystyrene rich phases, thereby, leading to reduction in the percolation threshold from 0.33 to 0.075 vol %.

9.3 Use of Modified Filler/Polymer

Li *et al.* [28] used solution blending method to synthesize PLA nanocomposites with exfoliated graphene (EG). For comparison, the authors prepared the 0.5 wt% filler composites with pristine filler (PLLA/p-EG) and silane treated filler (PLLA/s-EG), and compared their properties with neat PLLA. To prepare the composites, the filler was added in chloroform and sonicated for 2 h. PLLA was added into

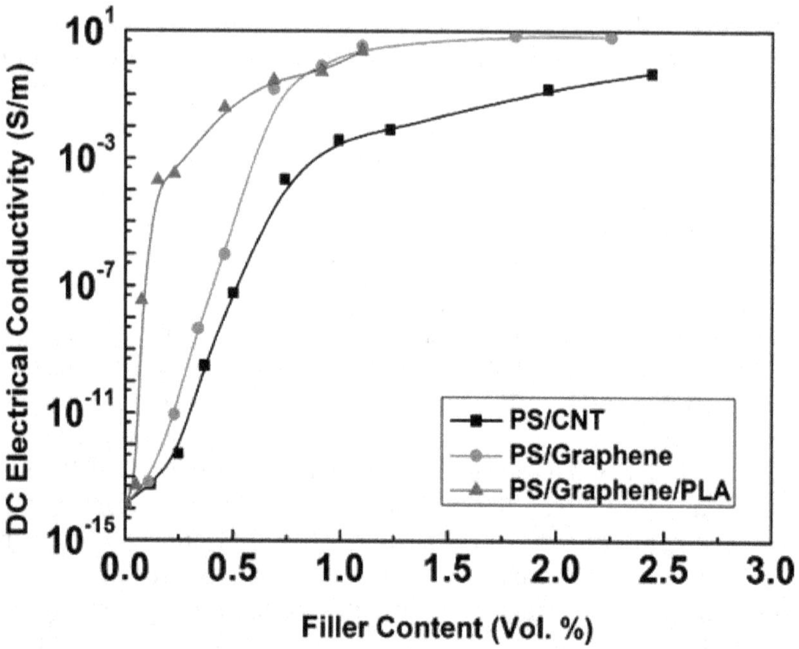

Figure 9.5 Electrical conductivity of PS/graphene/PLA nanocomposites in comparison with PS/graphene and PS/CNT nanocomposites. Reproduced from Reference 27 with permission from American Chemical Society.

the mixture with stirring and sonication. Evaporation of the solvent was carried out for 24 h at room temperature and subsequently the samples were dried in vacuum at 70 °C for 24 h. The reinforced composites were observed to have improved properties as compared to the neat PLLA. This was due to the fact that both the fillers had heterogeneous nucleation effect on PLLA matrix. Tensile strength was better in the case of filler without the silane treatment, whereas the elongation at break and tensile modulus were higher for the filler with silane treatment. For instance, the tensile modulus of PLLA increased from 2.2 GPa to 2.7 and 3.4 GPa for the pristine and silanized filler composites, respectively. On the other hand, the tensile strength of PLLA enhanced from 32.3 MPa to 38.9 and 35.3 MPa for these two composites respectively. Some defects were also observed to be present in the microstructure of the PLLA/p-EG composites. The storage modulus of the composites increased gradually from neat PLLA to PLLA/p-EG and PLLA/s-EG. The fact that EG has high surface area resulted in higher storage modulus in the polymer matrix reinforced with the filler, which was further reinforced by surface modification with silane. The addition of fillers increased the viscosity of the material due to the strong interfacial bonds between the polymeric matrix and filler, which makes the PLLA chains more difficult to move. The glass transition temperature of the composites was also observed to remain unchanged as neat PLLA.

In a recent study, Xu *et al.* [29] synthesized poly(ethylene glycol) (PEG) grafted graphene oxide (GO) (PEGgGO) for the generation of composites with PLLA (Figure 9.6). The main objective of the study was to enhance the crystallization kinetics of PLLA by achieving simultaneous accelerated crystal nucleation and growth. The authors observed that in comparison with the PLLA/GO nanocomposite with 0.5 wt% GO, the PLLA/PEGgGO composite exhibited an increase in the nucleation density and crystal growth rate by 110% and 14.3% at the crystallization temperature. This corresponded to 52.2% relative increase in the final crystallinity, thus, underlining the influence of filler modification on the resultant interaction with the polymer chains as well as composite properties.

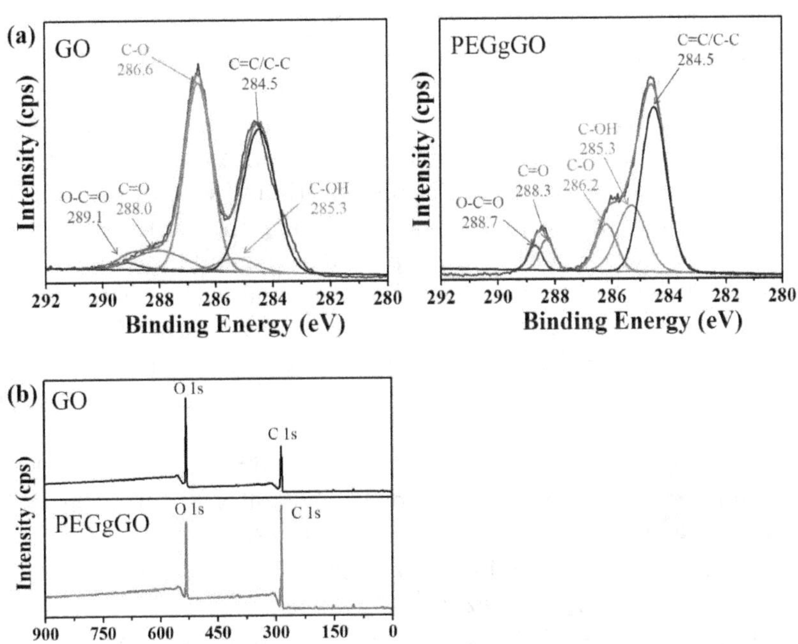

Figure 9.6 C1s XPS spectra (a) and XPS curves (b) of GO and PEG*g*GO. Reproduced from Reference 29 with permission from American Chemical Society.

Tong *et al.* [30] prepared composites of PLA with thermally reduced graphene (TRG) via solution blending. The compound Py-PLA-OH was used as stabilizer for graphene and compatibilizer for the preparation of PLA/TRG composites. Initially, graphene and Py-PLA-OH were dispersed in DMF under stirring and sonication at room temperature, while PLA was dissolved in DMF. The two solutions were mixed together at 60 °C. After coagulation with methanol, PLA/TRG composites were prepared with vacuum drying at 60 °C for 24 h. In order to investigate the effect of Py-PLA-OH, graphene was also dispersed in DMF with and without the compatibilizer. The results showed that the addition of Py-PLA-OH stabilized TRG, prohibiting any precipitation, probably due to the interaction between TRG and Py-PLA-OH. On the other hand, partial precipitation took place when neat TRG was dispersed in the solvent. The good dispersion and homogeneous distribution of TRG in PLA, with the addition of Py-PLA-OH were also confirmed in the morphological analysis. X-ray diffraction analysis, however, exhibited that the diffractograms of both PLA/TRG composites, with and without compatibilizer, had no characteristic peak of the graphite at 26°, indicating absence of TRG aggregates. Both storage modulus and the tensile strength of the PLA/TRG composites were observed to increase, in comparison with the neat PLA. With the addition of the compatibilizer, the increment in the properties was even more significant. For both composites, the elongation at break reduced, but the addition of Py-PLA-OH limited this reduction. For the PLA/TRG composites, the degradation temperature for 5% weight loss decreased due to good thermal conductivity of TRG, whereas the degradation temperature for maximum weight loss increased, due to the mass barrier effect of TRG. The addition of the compatibilizer increased the $T_{5\%}$, approaching the value of the neat PLA and slightly decreased T_{max}, but keeping it at a higher value than that of neat PLA.

In another study, Sun and He [31] reported the synthesis of PLLA nanocomposites using graphene oxide grafted with PDLA as filler (Figure 9.7). The authors synthesized GO-g-PDLA via ring-opening polymerization using modified GO as the initiator. The authors observed that a stereo-complex crystal was formed between the polymer matrix and GO-g-PDLA. The authors also reported lower crystallization activation energy of stereo-complex in the composites and a higher degree of crystallinity in the solution casted samples, as the covalently bonded GO sheets led to the heterogeneous nucleating effect in the composites.

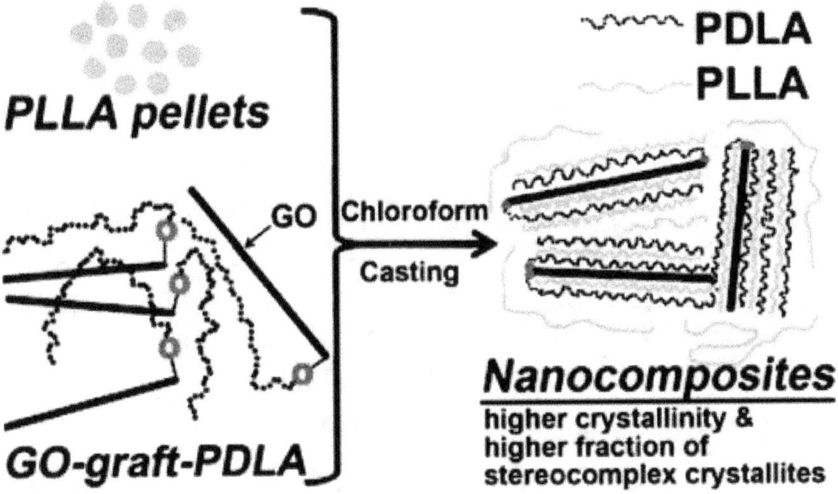

Figure 9.7 Schematic of the synthesis of PLLA nanocomposites with GO-g-PDLA. Reproduced from Reference 31 with permission from American Chemical Society.

Xu *et al.* [32] also reported the graphene oxide-assisted formation of PLA stereo-complex crystals. The authors observed that the graphene oxide nano-sheets, which were decorated by the preferred formation of stereo-complex, led to the simultaneous improvement of gas barrier properties and heat distortion resistance. Figure 9.8 demonstrates the comparison of the performance so as to exhibit the exceptional resistance to heat distortion and gas permeation of composite samples containing precisely decorated nano-sheets.

In another study, Zhang *et al.* [33] used octadecylamine-functionalized graphene (ODAG) as a filler to prepare PLA/ODAG nanocomposites with different ODAG loadings (0.1-1 wt%) via solution blending and casting. ODAG was dispersed in chloroform for 30 minutes under ultrasonication. The filler dispersion was mixed with PLA solution in chloroform under sonication for 30 min at room temperature, poured in plates for solvent evaporation and eventually drying of the prepared PLA/ODAG films was carried out at 75 °C for 48 h. The morphological analysis showed that ODAG was homogeneously dispersed in PLA matrix without any pulling out effects, thus, indicating strong interfacial interaction between the polymer matrix and filler. The addition of ODAG up until a loading of 0.4 wt % decreased the cold crystallization temperature T_{cc} from 123.2 °C (neat PLA) to 110.2 °C. For the same filler loading, the crystallinity increased from

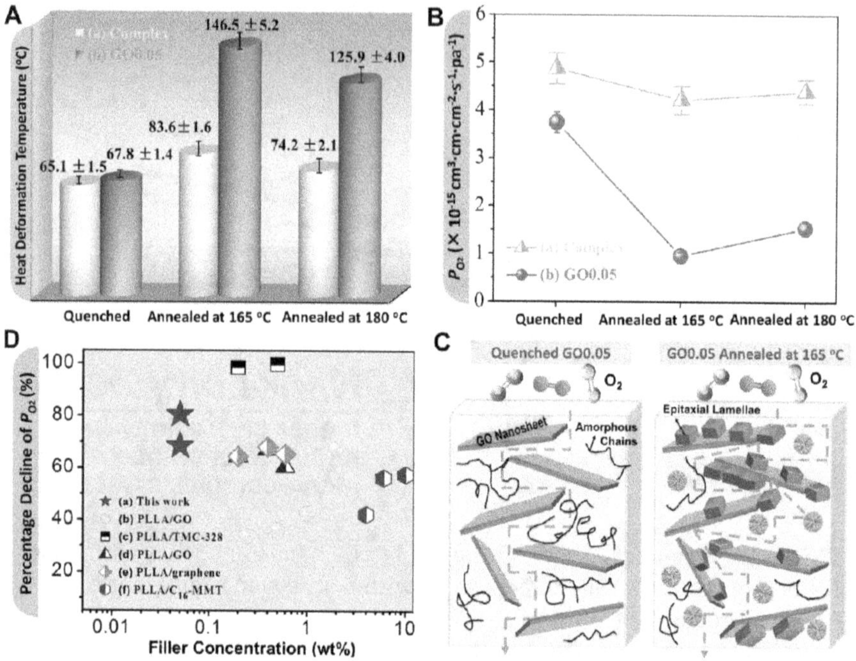

Figure 9.8 (A) HDT and (B) P_{O2} of the quenched and annealed complex and composite films. (C) Schematic illustration of oxygen molecules following a relatively easy path and a tortuous path through quenched and annealed GO0.05. (D) Comparison of percentage decline of P_{O2} between (a) the authors' work and other PLLA composite systems based on (b) GO (c) N,N′,N″-tricyclohexyl-1,3,5-benzenetricarboxylamide (trade name TMC-328, a nucleating agent of PLLA) (d) GO (e) graphene and (f) hexadecylamine-montmorillonite (C16-MMT). Reproduced from Reference 32 with permission from American Chemical Society.

14.2% (neat PLA) to 42.9%, probably due to the nucleation effect of the filler, similar to the earlier reported findings by Wang *et al.* [14]. For higher filler loadings, the crystallinity of PLA was negatively affected, due to the presence of ODAG aggregates. Moreover, the values of T_g and T_m were slightly affected, indicating that the addition of ODAG did not affect the mobility of PLA chains, probably due to flexible long-chain alkyl modification of graphene. Both tensile strength and the Young's modulus increased from 41.5 MPa and 1.9 MPa for the neat PLA at a maximum value of 60.7 MPa and 3 MPa respectively, for 0.2 wt % ODAG loading and subsequently decreased. The elongation at break and toughness also increased at a maximum value of 15% and 687 MJ/m^3 respectively, in comparison to the neat PLA (5%

and 142 MJ/m^3) for 0.6 wt% ODAG loading. Hence, 0.4 wt% could be considered an optimum loading for which all the mechanical properties are enhanced, a finding in correlation with other studies [11]. The addition of 0.4 wt% ODAG increased the temperature for 5% weight loss by about 14 °C, due to ODAG's barrier effect.

In another study, graphene nano-platelets (GnPs) were chemically modified with carboxyl groups and incorporated in polylactic acid [34]. The nanocomposites were generated via solution casting using chloroform as a solvent. 1 wt% of both pristine and functionalized carbon nano-filler were incorporated in PLA. The morphological analysis revealed that no aggregates were formed. Thermogravimetric analysis exhibited that the residue weight at 450 °C was 90.3 wt% and 85.3 wt% for pristine GnPs and functionalized fGnPs, respectively. This difference was attributed to the oxygen containing functional groups attached to the surface of the carbon nano-filler. The diffraction analysis also revealed that the exfoliated layers of fGnPs did not restack, due to the strong interfacial interactions between PLA chains and exfoliated fGnPs. The nanocomposite with 1 wt% of fGnPs were observed to improve tensile strength and modulus by 15.7% and 32.4% respectively. The temperature of 5% weight loss $T_{d,5\%}$ 70% weight loss $T_{d,70\%}$ and the temperature of the maximum rate of the mass loss T_{max} were observed to increase upon filler addition, especially fGnPs, due to the fact that fGnPs platelets act as a mass transfer barrier against thermal decomposition of the nanocomposites.

In a recent study, Zhang *et al.* [35,36] reported the surface grafting of graphene oxide with polyethylene glycol (PEG) by esterification between carboxylic acid functionalized GO (GO-COOH) and end hydroxyl groups of PEG. For generation of the nanocomposites, PLA and GO-g-PEG were mixed with each other using chloroform as solvent, and stirred before electrospinning. The PLA weight fraction in the mixture was 10% weight concentration of PLA in the mixture was fixed at 10%, whereas the fraction of GO-g-PEG by weight of PLA was varied as 1%, 2%, and 5%. The Raman spectra indicated an enhanced dispersion of the filler in the PLA matrix due to the grafting with PEG. In comparison with the pure PLA, the electrospun fibers of the1% GO containing composite exhibited smaller diameters. On the other hand, the diameters of the fibers generated using 2% GO composite were little larger than the fibers obtained from neat PLA. The diameter was observed to decrease as compared to pure PLA, after GO was surface grafted with PEG. Thermal stability of neat PLA was improved on the addition of GO and the extent of improvement was further enhanced on surface grafting of GO with PEG. The authors observed that the incorporation of GO and GO-g-PEG led to improvement in the tensile stress in the composites, but resulted in a slight decrease in ductility. Chieng *et al.* [37] also reported PLA nanocomposites using graphene nano-platelets (xGnPs) as filler and epoxidized palm oil (EPO) as plasticizer, prepared by melt blending method. Tensile strength of PLA/EPO was observed to increases as the xGnP loading increased, reaching to its maximum of 41.1 MPa up to the percentage of 0.3 wt% of xGnP. Further increase in the filler loading decreased the tensile strength probably due to the aggregation of xGnP at higher content. Tensile modulus also increased in 0.1 wt% xGnP composites, as compared to the neat PLA/EPO blend. The cause for the observed enhancement in the mechanical properties could be the fact that xGnPs were dispersed and orientated homogeneously in the polymer matrix and a strong interfacial interaction exists between the two components due to the addition of compatibilizer, which tends to modify the surface of filler to make it more compatible with the polymer. The elongation at break of neat PLA/EPO increased up to 0.3 wt% xGnP content and subsequently decreased. The decomposition onset temperature (T_{onset}), the maximum decomposition temperature (T_{max}) and the decomposition temperature at 50% weight loss (T_{50}) of PLA/EPO/xGnP nanocomposite shifted to higher temperatures. In other words, the nanocomposite is more stable thermally due to the presence of xGnP as a reinforcing filler. The morphological analysis of the composites displayed good uniformity and more homogeneous morphology due to good degree of dispersion of the xGnP.

In another recent study, Keramati *et al.* [38] developed PLA based temperature sensitive shape memory polymer (SMP) system. The dispersion of the filler in the polymer matrix was improved by the addition of a zwitterionic surfactant. PLA nanocomposites with GNPs and surface modified graphene nano-platelets (SMGNPs) were prepared via solution casting using DMF as the solvent. The morphological analysis revealed poor dispersion of GNPs, whereas the SMGNPs were observed to be homogeneously dispersed in the PLA matrix. The enhancement in the crystallization was more noticed in the PLA-SMGNPs composite due to effective graphene dispersion. In addition, the composite with SMGNP had stiff and brittle behavior. Gao *et al.* [39] also reported nanocomposites with 5, 7, 10, 13 and 15 wt% of small graphene nano-platelets (S-GNPs) and large graphene nano-platelets (L-GNPs) in PLA, prepared by melt compounding at 180 °C under nitrogen atmosphere. GNP-S were observed to have tendency to form aggregates, whereas GNP-S revealed more individual flakes. Good dispersion was observed for loadings as high as 10 wt% in the case of GNP-S, while aggregates were observed for the larger GNP-L at 7 wt% content. It was observed that the neat PLA had predominantly amorphous structure. In the diffraction analysis of the composites, the peak for GNP-L was much sharper than GNP-S. The incorporation of both GNP-L and GNP-S increased both the Young's modulus and the tensile strength. The behavior was observed up to 7 wt% filler content and subsequently a decrease in the properties was observed due to aggregation. Overall, 5 wt% GNP content exhibited the best mechanical reinforcement for both small and large systems, with GNP-L nano-platelets providing the highest increase in Young's modulus and tensile strength, and GNP-S exhibiting the highest ductility.

Xu *et al.* [40] studied the isothermal crystallization of PLA induced by graphene nano-sheets, in comparison with carbon nanotubes (Figure 9.9). Both fillers acted as nucleating agents for PLA crystallization and the surface induced crystallization resembled the modified filler surface which interacted effectively with the polymer matrix, thus, creating opportunities for property enhancement. The authors observed that the in the case of pure PLA, $-CH_3$ interchain interactions preceded $-(COC + CH_3)$ interchain interactions during the crystallization. On the other hand, this behavior was reversed, which resulted led to a reduced induction period.

In another study, Chen *et al.* [41] investigated the effect of reduced graphene oxide (RGO) on PLA in the rubbery state. The authors prepared PLA/RGO nanocomposites via a typical solution blending method, using DMF as a solvent. Before the synthesis of the composites, the RGO was modified with N-(aminoethyl)-aminopropyltrimethoxysilane (KH792) in order to achieve homogeneous dispersion and better PLA-RGO interaction. The morphological analysis of the composites revealed that the KH792-RGO was well dispersed in the PLA matrix, suggesting good interfacial interaction and compatibility between the PLA and the functionalized filler, whereas the unmodified RGO formed aggregates. After transition from the glassy to the rubbery state (near 80 °C), the storage modulus of neat PLA dropped drastically to 20 MPa. The addition of only 0.5 wt% KH792-RGO in PLA matrix increased this value to 300 MPa, thus, enhancing the stiffness of PLA. In the temperature range between 80-100 °C, the storage modulus of both the neat PLA and nanocomposites increased, due to the improved crystallinity from cold crystallization of PLA. Further increase of the temperature led to the reduction in the modulus, due to melt crystallization behavior. Moreover, the addition of KH792-RGO in PLA improved significantly the heat distortion resistance.

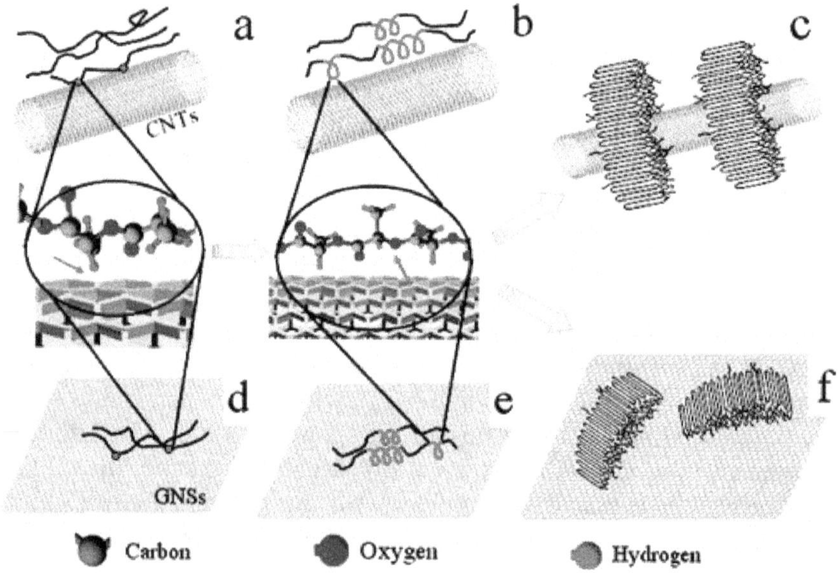

Figure 9.9 Representation of the surface induced crystallization of PLA in the presence of nano-filler. Reproduced from Reference 40 with permission from American Chemical Society.

In a previous work, Chieng *et al.* [42] prepared nanocomposites of PLA with different loadings of graphene nano-platelets (xGnP), using epoxidized palm oil (EPO) as a plasticizer. In the current study, the similar EPO/PLA/xGnP nanocomposites were prepared, but the effect of the xGnP on the thermal properties of PLA was focused [43]. It was observed that the addition of xGnP slightly affected the crystallinity. Also, the cold crystallization temperature T_{cc} and the melting temperature T_m were not affected by the xGnP loading. On the other hand, the glass transition temperature decreased with increasing xGnP loading up to 0.3 wt%. For further filler loadings, T_g had an increasing trend, due to the decreased effect of plasticizer, as it intercalated in the xGnP interlayers. The addition of xGnP was also observed to shift the decomposition temperature to higher values. The observed enhancement of thermal stability was attributed to the fact that xGnP acted as a heat barrier, by accumulating the heat and created a tortuous path, delaying the diffusion of the byproducts.

In another recent study, Chieng *et al.* [44] generated PLA-graphene nanocomposites using EPO and polyethylene glycol (PEG) as plasticizers. The nanocomposites were similar to those synthesized by the authors earlier, however, PEG was added in the current study as plasticizer. PLA and PEG were mixed by melt blending at 170 °C for 10 minutes, followed by the addition of EPO. The fraction of xGnP varied between 0.1 and 1.0 wt%. Both the tensile strength and elongation at break of the reinforced materials decreased as compared to neat PLA/EPO. Generally, both tensile strength and elongation at break did not show much difference when different loadings were incorporated. On the other hand, tensile modulus followed an increasing trend in the beginning followed by a slight decrease. The optimum xGnP loading for tensile strength and modulus enhancement was observed to be 0.1 wt%, whereas for elongation at break, it was noted to be 0.5 wt%. The 50% loss temperature and the degradation temperature of the materials increased with increasing filler loading. This fact was also validated from DTG results, where the temperature that caused main weight losses increased as the loading increased. From these results, it was confirmed that the filler acted as a barrier to reduce the diffusion of the heat in polymer matrices. The morphological analysis for the sample with 0.3 wt% xGnP exhibited good dispersion of the filler inside the matrix, Moreover, a percolated network was formed due to the matrix-filler and filler-filler interaction.

9.4 Conclusions

This review specifically focuses on the synthesis and properties of polylactic acid (PLA)/graphene nanocomposites. PLA is one of the most famous biodegradable linear aliphatic thermoplastic polyesters and its non-toxicity, biocompatibility and biodegradability make it very attractive to be applied in a wide range of products from material science to tissue engineering. On the other hand, graphene has shown excellent potential of application in the polymer composites due to its superior thermal and mechanical properties. As a result, these two materials have attracted a significant research interest to investigate the properties of the nanocomposites that can be derived from their combination. In addition to the review of the recent progresses in the synthesis and properties of PLA/graphene nanomaterials, the role of modification of graphene with different functional groups has also been studied on the resulting microstructure and properties.

It can be concluded that graphene-based nano-fillers enhanced the thermal and mechanical properties of PLA in the nanocomposites. Thermal stability was increased in the majority of the studies, a fact that is obvious by shifting of the degradation temperature to higher values in thermogravimetric analysis. As a result, PLA/graphene nanocomposites exhibited higher heat resistance. Young's modulus and tensile strength of PLA in the composites increased as well, but the elongation at break decreased in most of the studies, thus, indicating the brittleness of the materials. Morphological analysis of the composites confirmed good dispersion of the filler inside the polymeric matrix. Some exceptions were observed where defects appeared on the surface. Calorimetric studies exhibited occasionally a significant difference in the glass transition, melting and crystallization values of the neat PLA and the reinforced PLAs, whereas in the other studies, the values of T_g and T_m remained almost the same. Thus, it indicated that the type of filler and polymer-filler interaction affected the thermal transitions in the PLA nanocomposites. Finally, the surface modification of filler as well as addition of surfactants or compatibilizers in the polymer matrix resulted in enhanced filler dispersion, which in turn led to the enhancement of the composites properties. The bio-nanocomposites based on PLA represent advanced materials with superior properties which can lead to their use in a large number of commercial applications.

References

1. Matsko, N. B., and Mittal, V. (2014) Thermotropic dynamic processes in multiphase polymer systems by (cryo-)AFM. *Soft Matter*, **10**, 5478-5488.
2. Miloaga D. G. (2008) *Nucleating Effect of Exfoliated Graphite Nanoplatelets on Poly(hydroxybutyrate) and Poly(lactic acid) and Their Nanocomposites Properties.* PhD Thesis, Michigan State University, USA.
3. Mittal, V., Chaudhry, A. U., and Matsko, N. B. (2014) "True" biocomposites with biopolyesters and date seed powder: Mechanical, thermal, and degradation properties. *Journal of Applied Polymer Science*, **131**(19), doi:10.1002/app.40816.
4. Bourbigot, S., Fontaine, G., Duquesne, S., and Delobel, R. (2008) PLA nanocomposites: quantification of clay nanodispersion and reaction to fire. *International Journal of Nanotechnology*, **5**, 683-692.
5. Goffin, A.-L., Duquesne, E., Moins, S., Alexandre, M., and Dubois, P.

(2007) New organic–inorganic nanohybrids via ring opening polymerization of (di)lactones initiated by functionalized polyhedral oligomeric silsesquioxane. *European Polymer Journal,* **43**, 4103-4113.

6. Villmow, T., Potschke, P., Pegel, S., Haussler, L., and Kretzschmar, B. (2008) Influence of twin-screw extrusion conditions on the dispersion of multi-walled carbon nanotubes in a poly(lactic acid) matrix. *Polymer,* **49**, 3500-3509.
7. Mukhopadhyay, P., and Gupta, R. K. (2011) Trends and frontiers in graphene-based polymer nanocomposites. *Plastics Engineering,* **32**, 32-42.
8. Cai, D., and Song, M. (2010) Recent advance in functionalized graphene/polymer nanocomposites. *Journal of Materials Chemistry,* **20**, 7906-7915.
9. Rafiee, M. A., Rafiee, J., Wang, Z., Song, H., Yu, Z. Z., and Koratkar, N. (2009) Enhanced mechanical properties of nanocomposites at low graphene content. *ACS Nano,* **3**, 3884-3890.
10. Mittal, V., Chaudhry, A. U., and Luckachan, G. E. (2014) Biopolymer - thermally reduced graphene nanocomposites: Structural characterization and properties. *Materials Chemistry and Physics,* **147**(1-2), 319-332.
11. Kim, I.-H., and Jeong, Y. G. (2010) Polylactide/exfoliated graphite nanocomposites with enhanced thermal stability, mechanical modulus, and electrical conductivity. *Journal of Polymer Science Part B: Polymer Physics,* **48**, 850-858.
12. Murariu, M., Dechief, A. L., Bonnaud, L., Paint, Y., Gallos, A., Fontaine, G., Bourbigot, S., and Dubois, P. (2010) The production and properties of polylactide composites filled with expanded graphite. *Polymer Degradation and Stability,* **95**, 889-900.
13. Hassouna, F., Laachachi, A., Chapron, D., El Mouedden, Y., Toniazzo, V., and Ruch, D. (2011) Development of new approach based on Raman spectroscopy to study the dispersion of expanded graphite in poly(lactide). *Polymer Degradation and Stability,* **96**, 2040-2047.
14. Wang, H., and Qiu, Z. (2012) Crystallization kinetics and morphology of biodegradable poly(l-lactic acid)/graphene oxide nanocomposites: Influences of graphene oxide loading and crystallization temperature. *Thermochimica Acta,* **527**, 40-46.
15. Narimissa, E., Gupta, R., Bhaskaran, M., and Sriram, S. (2012) Influence of nano-graphite platelet concentration on onset of crystalline degradation in polylactide composites. *Polymer Degradation and Stability,* **97**, 829-832.
16. Pinto, A. M., Cabral, J., Tanaka, D. A. P., Mendes, A. M., and Magalhaes, F. D. (2013) Effect of incorporation of graphene oxide and graphene nanoplatelets on mechanical and gas permeability properties of poly(lactic acid) films. *Polymer International,* **62**, 33-40.

17. Pinto, A. M., Goncalves, C., Goncalves, I. C., and Magalhaes, F. D. (2016) Effect of biodegradation on thermo-mechanical properties and biocompatibility of poly (lactic acid)/graphene nanoplatelets composites. *European Polymer Journal,* **85**, 431-444.
18. Pinto, V. C., Ramos, T., Alves, A. S. F., Xavier, J., Tavares, P. J., Moreira, P. M. G. P., and Guedes, R. M. (2017) Dispersion and failure analysis of PLA, PLA/GNP and PLA/CNT-COOH biodegradable nanocomposites by SEM and DIC inspection. *Engineering Failure Analysis,* **71**, 63-71.
19. Scaffaro, R., Botta, L., Maio, A., and Gallo, G. (2017) PLA graphene nanoplatelets nanocomposites: Physical properties and release kinetics of an antimicrobial agent. *Composites Part B: Engineering,* **109**, 138-146.
20. Cao, Y., Feng, J., and Wu, P. (2010) Preparation of organically dispersible graphene nanosheet powders through a lyophilization method and their poly (lactic acid) composites. *Carbon,* **48**, 3834-3839.
21. Chen, Y., Yao, X., Zhou, X., Pan, Z., and Gu, Q. (2011) Poly(lactic acid)/graphene nanocomposites prepared via solution blending using chloroform as a mutual solvent. *Journal of Nanoscience and Nanotechnology,* **11**, 7813-7819.
22. Bao, C., Song, L., Xing, W., Yuan, B., Wilkie, C. A., Huang, J., Guo, Y., and Hu, Y. (2012) Preparation of graphene by pressurized oxidation and multiplex reduction and its polymer nanocomposites by masterbatch-based melt blending. *Journal of Materials Chemistry,* **22**, 6088-6096.
23. Li, X., Xiao, Y., Bergeret, A., Longerey, M., and Che, J. (2014) Preparation of polylactide/graphene composites from liquid-phase exfoliated graphite sheets. *Polymer Composites,* **35**, 396-403.
24. Valapa, R. B., Pugazhenthi, G., and Katiyar, V. (2015) Effect of graphene content on the properties of poly (lactic acid) nanocomposites. *RSC Advances,* **5**, 28410-28423.
25. Papageorgiou, G. Z., Terzopoulou, Z., Bikiaris, D., Triantafyllidis, K. S., Diamanti, E., Gournis, D., Klonos, P., Giannoulidis, E., and Pissis, P. (2014) Evaluation of the formed interface in biodegradable poly(l-lactic acid)/graphene oxide nanocomposites and the effect of nanofillers on mechanical and thermal properties. *Thermochimica Acta,* **597**, 48-57.
26. Yang, J.-H., Lin, S.-H., and Lee, Y.-D. (2012) Preparation and characterization of poly (L-lactide)–graphene composites using the in situ ring-opening polymerization of PLLA with graphene as the initiator. *Journal of Materials Chemistry,* **22**, 10805-10815.
27. Qi, X.-Y., Yan, D., Jiang, Z., Cao, Y.-K., Yu, Z.-Z., Yavari, F., and Koratkar N. (2011) Enhanced electrical conductivity in polystyrene nanocomposites at ultra-low graphene content. *ACS Applied Materials &*

Interfaces, **3**(8), 3130-3133.

28. Li, W., Shi, C., Shan, M., Guo, Q., Xu, Z., Wang, Z., Yang, C., Mai, W., and Niu, J. (2013) Influence of silanized low-dimensional carbon nanofillers on mechanical, thermomechanical, and crystallization behaviors of poly (L-lactic acid) composites - A comparative study. *Journal of Applied Polymer Science*, **130**, 1194-1202.
29. Xu, J.-Z., Zhang, Z.-J., Xu, H., Chen, J.-B., Ran, R., and Li, Z.-M. (2015) Highly enhanced crystallization kinetics of poly(L-lactic acid) by poly(ethylene glycol) grafted graphene oxide simultaneously as heterogeneous nucleation agent and chain mobility promoter. *Macromolecules*, **48**(14), 4891-4900.
30. Tong, X.-Z., Song, F., Li, M.-Q., Wang, X.-L., Chin, I.-J., and Wang, Y.-Z. (2013) Fabrication of graphene/polylactide nanocomposites with improved properties. *Composites Science and Technology*, **88**, 33-38.
31. Sun, Y., and He, C. (2012) Synthesis and stereocomplex crystallization of poly(lactide)-graphene oxide nanocomposites. *ACS Macro Letters*, **1**(6), 709-713.
32. Xu, H., Wu, D., Yang, X., Xie, L., and Hakkarainen, M. (2015) Thermostable and impermeable "nano-barrier walls" constructed by poly(lactic acid) stereocomplex crystal decorated graphene oxide nanosheets. *Macromolecules*, **48**, 2127-2137.
33. Zhang, L., Li, Y., Wang, H., Qiao, Y., Chen, J., and Cao, S. (2015) Strong and ductile poly(lactic acid) nanocomposite films reinforced with alkylated graphene nanosheets. *Chemical Engineering Journal*, **264**, 538-546.
34. Rostami, A., Nazockdast, H., and Karimi, M. (2016) Graphene induced microstructural changes of PLA/MWCNT biodegradable nanocomposites: rheological, morphological, thermal and electrical properties. *RSC Advances*, **6**, 49747-49759.
35. Zhang, C., Wang, L., Zhai, T., Wang, X., Dan, Y., and Turng, L.-S. (2016) The surface grafting of graphene oxide with poly(ethylene glycol) as a reinforcement for poly(lactic acid) nanocomposite scaffolds for potential tissue engineering applications. *Journal of the Mechanical Behavior of Biomedical Materials*, **53**, 403-413.
36. Zhang, C., Salick, M. R., Cordie, T. M., Ellingham, T., Dan, Y., and Turng, L.-S. (2015) Incorporation of poly(ethylene glycol) grafted cellulose nanocrystals in poly(lactic acid) electrospun nanocomposite fibers as potential scaffolds for bone tissue engineering. *Materials Science and Engineering: C*, **49**, 463-471.
37. Chieng, B. W., Azowa, I. N., Then, Y. Y., and Loo, Y. Y. (2016) Plasticized and nanofilled poly (lactic Acid) nanocomposites: Mechanical, thermal and morphology properties. *Materials Science Forum*, 429-433.
38. Keramati, M., Ghasemi, I., Karrabi, M., Azizi, H., and Sabzi, M. (2016)

Incorporation of surface modified graphene nanoplatelets for development of shape memory PLA nanocomposite. *Fibers and Polymers,* **17**, 1062-1068.

39. Gao, Y., Picot, O. T., Bilotti, E., and Peijs, T. (2017) Influence of filler size on the properties of poly(lactic acid) (PLA)/graphene nanoplatelet (GNP) nanocomposites. *European Polymer Journal,* **86**, 117-131.
40. Xu, J.-Z., Chen, T., Yang, C.-L., Li, Z.-M., Mao, Y.-M., Zeng, B.-Q., and Hsiao, B. S. (2010) Isothermal crystallization of poly(L-lactide) induced by graphene nanosheets and carbon nanotubes: A comparative study. *Macromolecules,* **43**(11), 5000-5008.
41. Chen, P., Wang, Y., Wei, T., Meng, Z., Jia, X., and Xi, K. (2013) Greatly enhanced mechanical properties and heat distortion resistance of poly (L-lactic acid) upon compositing with functionalized reduced graphene oxide. *Journal of Materials Chemistry A,* **1**, 9028-9032.
42. Chieng, B. W., Ibrahim, N. A., Wan Yunus, W. M. Z., Hussein, M. Z., and Silverajah, V. (2012) Graphene nanoplatelets as novel reinforcement filler in poly (lactic acid)/epoxidized palm oil green nanocomposites: Mechanical properties. *International Journal of Molecular Sciences,* **13**, 10920-10934.
43. Chieng, B., Ibrahim, N., Wan Yunus, W., Hussein, M., and Loo, Y. (2014) Effect of graphene nanoplatelets as nanofiller in plasticized poly(lactic acid) nanocomposites. *Journal of Thermal Analysis and Calorimetry,* **118**, 1551-1559.
44. Chieng, B. W., Ibrahim, N. A., Yunus, W. M. Z. W., Hussein, M. Z., Then, Y. Y., and Loo, Y. Y. (2015) Reinforcement of graphene nanoplatelets on plasticized poly (lactic acid) nanocomposites: Mechanical, thermal, morphology, and antibacterial properties. *Journal of Applied Polymer Science,* **132**(11), doi:10.1002/app.41652.

10

Mechanical and Thermal Properties of Poly(butylene succinate)/Graphene Nanocomposites

10.1 Introduction

The disposal problem of the petroleum wastes and their lethal environmental impacts have prompted the development, production and application of the biodegradable polymers. Poly(butylene succinate) (PBS), a typical synthetic biodegradable aliphatic polyester, is one such biopolymer which has attracted significant research interest [1]. Two main routes are generally employed for the synthesis of PBS. The first route is the trans-etherification process that is performed using the succinate diesters, whereas the second route employs the direct esterification process using diacids. The most common method of preparation of PBS is the esterification of succinic acid with 1,4-butanediol [2-3]. It has potential applications as a packaging material that can alleviate environmental pollution caused by commodity polymers since it is not only biodegradable but also has a good elongation at break along with other physical properties similar to polyethylene. The thermoplastic PBS has a good melt processability that can also be compared to that of polyethylene [3]. PBS also possesses good thermal resistance and resistance from the decomposition by bacteria and fungi. PBS is a semi-crystalline polymer with a crystallinity of 35 to 40%. In addition, PBS has a T_g of -32 °C and a melting point of 115 °C. Among the industrially available biodegradable polymers, PBS is well known for its reasonable production cost and good performance characteristics [4-6]. However, it still lacks the mechanical and thermal properties required for industrial applications. Therefore, incorporating other suitable materials into the PBS matrix may be an effective way to enhance the utility of PBS. PBS can also generate spherulites because of the slower crystallization rate, which can be considered as

Abdallah Yassin, Latifa Al Ghailani and Vikas Mittal, The Petroleum Institute (part of Khalifa University of Science and Technology), Abu Dhabi, UAE*

**Current address: Bletchington, Wellington County, Australia*

an obstacle to its end-use applications. The addition of nucleating agents and other additives like fillers may be necessary to control this behavior.

One of the routes to circumvent the limitations of PBS is to blend the polymer with other biopolymers so as to synergistically combine the properties of individual polymers. In one such study, Ravati and Favis [7] reported ternary blends of PBS with PLA, PBAT and PCL in order to improve the limitation of low elongation of PBS. PCL was chosen due to its flexibility, whereas PLA was incorporated to enhance bio-based content in the blends. PBAT was chosen due its capability to enhanced toughness and elongation at break. The authors observed partial wetting for the ternary blend based on PBS/PLA/PCL, as PLA droplets were observed to self-assemble at the interface of PBS and PCL. On the other hand, the blend based on PBS/PLA/PBAT exhibited complete wetting with tri-continuous morphology. In comparison, the ternary blend PBS/PBAT/PCL had combined morphology, where both partial and complete wetting were observed. Mittal *et al.* [8] also reported ternary blends of PBS, PCL and PBAT. The tensile modulus of the blends was observed to increase as the PBS content was enhanced. Specifically, the $PCL_{50}PBAT_{10}PBS_{40}$ blend exhibited highest modulus of 408 MPa, however, its elongation was only 6% as compared to 3% for pure PBS. On the other hand, the blend $PCL_{50}PBAT_{25}PBS_{25}$ exhibited more optimal performance with a strain at break of 194% and tensile modulus of 345 MPa. Figure 10.1 also demonstrates the morphology of the $PCL_{33}PBAT_{33}PBS_{33}$ ternary blend. As evident, the PBS phase was surrounded by PBAT. Among all the polymers, PCL appeared to form a continuous phase and was well mixed with PBAT. The IR and X-ray analysis confirmed no change in the crystalline peak positions of PCL, though a reduction in its degree of crystallinity was observed with increasing PBS content in the blends. On the other hand, structural changes were observed in PBS in the blends. Also, increasing the content of PBS in the blends exhibited higher intensity crystalline signals for PBS, indicating increment in its crystallinity, though the crystal structure was modified. It was also indicated by the small increment in the peak melting temperature of PBS as a function of PBS content in the blends. Though the thermal stability of the blends was negatively impacted by the addition of PBS, the blends were still stable for high temperature processing conditions. Overall, it was concluded that the obtained morphology leads to optimal enhancement in the thermal and mechanical properties of the polymer.

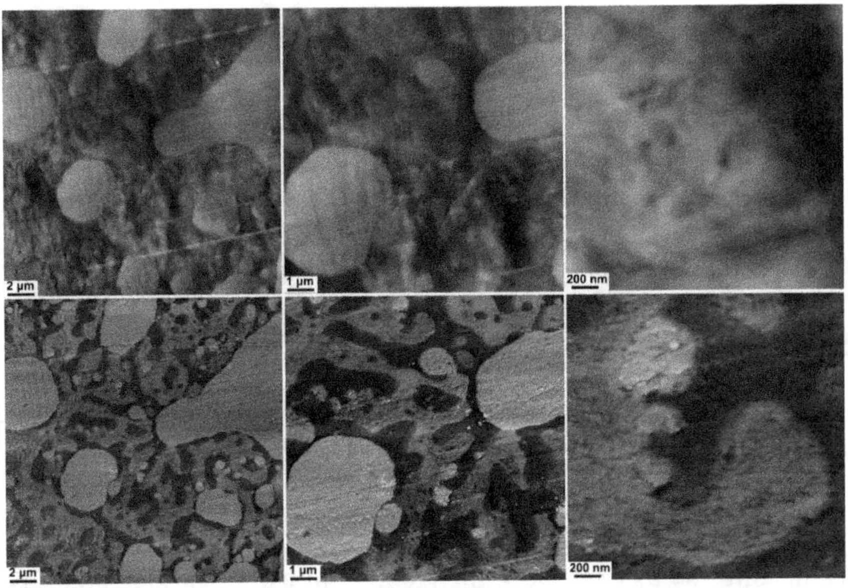

Figure 10.1 AFM height (top row) and phase (bottom row) images of $PCL_{33}PBAT_{33}PBS_{33}$. In the images, the dark phase corresponds to PBAT (largely amorphous), whereas the lighter phase represents PCL (semi-crystalline with good degree of crystallinity). PBS is observed as the bright/white phase (also contains high degree of crystallinity). Reproduced from Reference 8.

Other commonly used strategy for the enhancement of the polymer performance is the incorporation of functional fillers, which not only enhances the mechanical and thermal performance, but may also occasionally positively impact the biodegradation performance. There are a number of studies on PBS nanocomposites with fillers like TiO_2, silica, clay, carbon black, CNT but very limited studies with graphene or graphene oxide. Among biodegradable polymers, PBS has received an increasing amount of attention for generating composites because of its biodegradability, abundant renewable source, and excellent processability, as also mentioned earlier [9-17]. There are many modification attempts that are aimed at improving the mechanical and thermal properties of PBS. The other studies have also reported the incorporation of cellular materials such as sisal fiber, starch, wheat straw, flour, rice, and husk among others to reinforce PBS to produce the low-density, low cost and biodegradable composites that are non-toxic [2,3,18]. Findings from previous work [15-17,19] are persuasive that graphene is able to make polymeric materials much stronger and more resistant to degradation. Graphene, which consists of one atomic thick sheets of covalently sp^2-bonded carbon atoms in a hexagonal arrangement, has recently received attention of researchers for the generation of polymer nanocomposites [20]. Its choice as a filler is due to its excellent electrical and mechanical properties, which are significantly better than other inorganic filler materials. A single defect-free graphene layer has Young's modulus of ~1.0 TPa, intrinsic strength ~42 N/m, thermal conductivity ~4840-5300 W/(m.K), electron mobility exceeding 25,000 cm^2/V.s, excellent gas impermeability and specific surface area of ~2630 m^2/g [20]. All these properties make this material even superior to carbon nanotubes for use in polymer nanocomposites. Thus, a number of studies on polymer nanocomposites based on graphene have been published in a short span of time since its development [21-23]. In this context, the current review deals with the preparation and properties of PBS/graphene nano-sheets nanocomposites so as to explore their potential use in various applications as well as to replace the existing non-degradable polymers.

10.2 PBS/Graphene Nanocomposites

Various methods have been reported in the literature for the generation of PBS/graphene nanocomposites. The filler surface modification and/or compatibilizer have been reported in the literature to enhance the interaction between the matrix and filler particles, as these interactions are of immense significance for optimal enhancement of the composite properties. However, the review of the literature specifically reporting the generation of PBS/graphene nanocomposites reveals that these strategies for component compatibilization have not been commonly used for generating these composites. Thus, there is a strong potential to generate PBS-graphene nanocomposites by optimizing the filler-matrix interactions through optimal use of compatibilizers and filler surface treatments.

Wang *et al.* [24] and Pallathadka *et al.* [25] reported the use of melt intercalation method for the generation of PBS/graphene nanocomposites. The procedure followed by Pallathadka *et al.* [25] was as follows: prior to the melt mixing process, PBS resin was dried in a vacuum oven at 80 °C overnight in order to remove any trace of moisture, which might cause its potential hydrolytic degradation during its melt processing. The PBS/graphene nanocomposites were prepared using a co-rotating twin screw-extruder. The mixing temperature was kept around 170 °C at a screw speed of 270 rpm. Wan and Chen [26] reported the use of solution intercalation method for the generation of PBS-graphene nanocomposites. Graphite oxide powder was dispersed in DMF under continuous stirring and ultrasonication to obtain a uniform 0.6 wt/vol% of filler/DMF dispersion. The authors subsequently mixed different amount of filler/DMF dispersion with the pre-prepared PBS/chloroform solution in order to obtain different filler concentrations. The mixture was stirred further, followed by casting onto petri-dishes and oven-drying at 60 °C. Similarly, Wang *et al.* [27] also reported the PBS nanocomposites using the solution mixing method. Hydrazine was used to chemically reduce the aqueous graphene oxide (GO) nano-sheet dispersion to obtain graphene nano-sheets (GNS). Fabrication of the PBS/GNS composite was achieved by dispersing 50 mg of GNS powder in 50 mL of chloroform with the assistance of sonication at room temperature and addition of 9.95 g of PBS resin to the suspension. Mechanical stirring at 50 °C and sonication were followed with precipitation by methanol. The precipitate was filtered and vacuum dried at 80 °C to obtain the nanocomposite. Using same procedure, composites containing 1 wt% and 2 wt% GNS were also prepared. In-situ intercalation method was also used in many studies to generate the PBS nanocomposites with graphene [28-30]. For instance, Wang *et al.* [29] dispersed graphene oxide in distilled water, followed by the addition of 1,4-butanediol and succinic acid. Subsequently, the mixture was kept at 200 °C for 2 h and at 230 °C for 4 h under vacuum for in-situ formation of PBS. The obtained nanocomposites were subsequently washed and dried.

Figure 10.2 demonstrates the TEM images of the PBS/graphene nanocomposites generated by Pallathadka *et al.* [25] using melt mixing approach. The micrographs confirmed that the graphene nano-sheets were homogeneously dispersed in the PBS matrix. Thus, the use of melt mixing technique even without the use of any externally added compatibilizer resulted in uniform mixing of the polymer with the filler. As a result, the composites also exhibited improved properties. For instance, the modulus of the composite (measured using nano-indentation) with 2% filler was 1.32 GPa, as compared to 1.15 GPa for pure PBS. Similarly, the hardness of the composites as also observed to improve to a value of 0.09 GPa, compared to 0.07 for pure PBS. Figure 10.3 also shows the storage and loss modulus of nanocomposites as a function of filler's fraction as well as angular

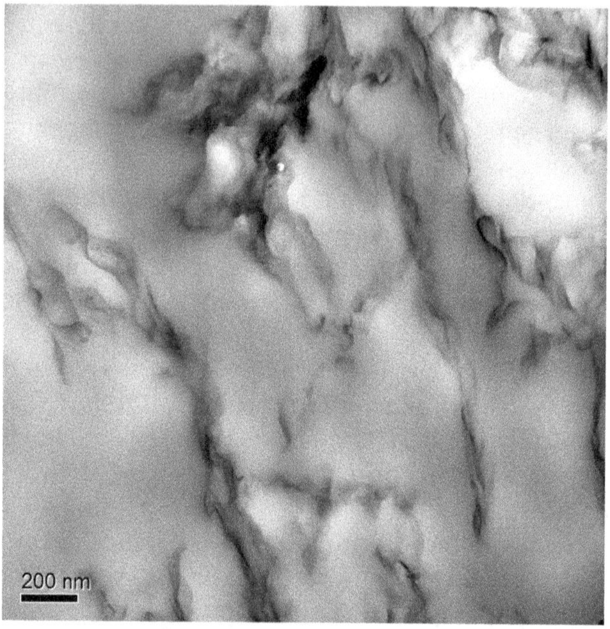

(a)

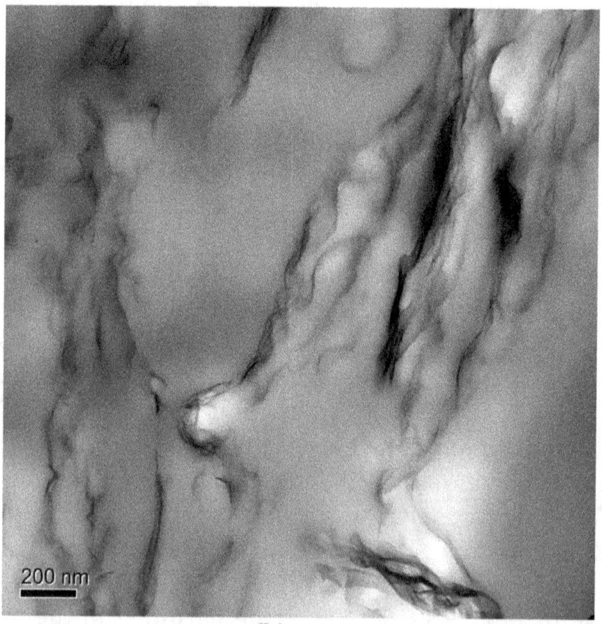

(b)

Figure 10.2 TEM micrographs of PBS/graphene nanocomposites (a) 2 and (b) 5 wt%. Reproduced from Reference 25 with permission from Wiley.

frequency. The composites were observed to have significant enhancement in modulus till 3 wt% filler content. In fact, the composite with 3 wt% filler exhibited frequency independent behavior. However, on further increasing the filler loading to 5%, reduction in the modulus was observed probably due to the presence of large filler aggregates. In addition, the polymer exhibited lower storage modulus than loss modulus, due to the dominance of viscous nature. However, the composites exhibited higher storage modulus than loss modulus, the extent of which also increased with filler content, thus, indicating strong elastic nature due to the hindrance to the chain movement caused by the presence of exfoliated graphene nano-sheets.

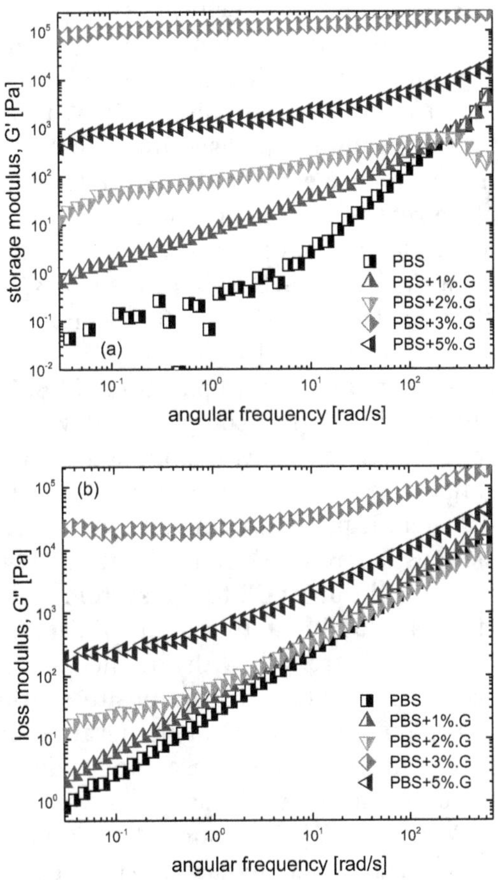

Figure 10.3 (a) Storage and (b) loss modulus of PBS/graphene nanocomposites as a function of filler content and angular frequency.

Wang *et al.* [27] reported that the addition of 0.5 wt% of graphene nano-sheets (GNS) in PBS using the solution blending process increased the tensile strength to 34.3 MPa, while the addition of 1 wt% of GNS yielded a tensile strength of 36.4 MPa. On the addition of 2 wt% GNS, the tensile strength increased to 37.2 Mpa (which was 21% higher). The storage modulus was also increased by 24% on the addition of the 2 wt% of graphene nano-sheets. The significant increase in the tensile strength of the nanocomposites was attributed to the uniform dispersion of GNS in the polymer matrix and the interfacial interactions between the two components. The degree of crystallinity (X_c) also decreased from 46.4% for pure PBS to 45.8, 43.8 and 42.9% on the addition of 0.5, 1 and 2 wt% respectively [27]. However, the change in the crystallinity of PBS was little and did not affect the mechanical properties of the composites. This indicated that the interfacial interactions and the homogeneous distribution of the graphene nano-sheets in the polymer matrix led to increased tensile strength and modulus [27]. Wan and Chen [26] reported that tensile strength increased by almost double the value in the composites, which were also generated using the solution based process. The reason for the good mechanical properties was attributed to the availability of oxygen-containing functional groups on the graphene surface and carboxyl groups of PBS macromolecular chains which, thus, led to good interfacial interactions between the graphene and PBS phases [27].

Yan *et al.* [30] reported in-situ preparation of PBS/graphene and addition of cellulose (CE) to the composite. It was observed that high tensile strength of 122 MPa was obtained with the use of 1% graphene, in comparison to the 55 MPa for neat CE-PBS. The reason for the observed significant enhancement in the tensile strength comparing to the tensile strength obtained with PBS/graphene only [26,27] could be the addition of CE to the system and/or the use of in-situ polymerization [30]. In another study, Jin *et al.* [28] reported a decrease in the tensile strength with the increase of the wt% of graphene, though in-situ method of composite synthesis was employed. The authors opined that the agglomeration of graphene at higher filler content might be the reason for reduced mechanical properties. A reduction on the tensile strength was also observed by Wang *et al.* [24]. Wang *et al.* [27] reported the Young's modulus of pure GPa as 0.5 GPa, which was increased to 0.6 GPa on the addition of the 0.5 wt% GNS. Also, the increase of the graphene oxide content to 1 and 2% caused the modulus of PBS to enhance by 40% and 70%

respectively. Roughly similar increase in the Young's modulus was reported by Wan and Chen [26] by using the same synthesis method as well as the graphene content. Yan *et al.* [30] reported an increase of 366.9% in the Young's modulus with only 1% of graphene. Such a significant increase can again be related to the mixture of CE with the PBS-graphene. The elongation at break also increased from 17.9% (pure PBS) to 18.5 and 19.0% on the addition of 1.0 and 2.0 wt% content respectively [27]. The increment in the elongation at break reported by Wan and Chen [26] was double the magnitude than observed in [27]. In comparison, Jin *et al.* [28] reported that after adding 0.5% of graphene to the PBS, the elongation at break increased from 280% to 560%. A negative impact on the elongation at break with different percentage of graphene was also reported [24,29,30]. Figure 10.4 also shows the X-ray diffractograms of the in-situ generated PBS-graphene composites by Wang *et al.* [29]. The authors observed that incorporation of 3% GO led to an increase of 43% in tensile strength and 45% in storage modulus. The authors attributed these increments to the strong interfacial interactions between the polymer and uniformly dispersed GO nano-sheets. Thus, different degree of filler dispersion and property enhancement was observed in the composites generated with different synthetic methodologies, indicating that all the techniques have advantages and can be chosen for the composite generation based on the requirement.

In another study by Sisti *et al.* [31], PBS and poly(L-lactide) (PLL) were mixed with GO using thermo-pressed molding technique. The

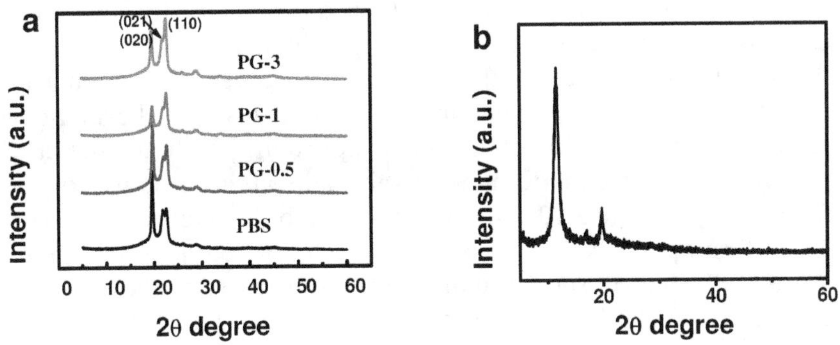

Figure 10.4 (a) XRD pattern of polymer nanocomposites, (b) XRD pattern of GO. Reproduced from Reference 29 with permission from American Chemical Society.

incorporation of both 90 and 120 ppm of GO to PBS-PLL system led to an increment in tensile strength by 17 and 27% respectively. Similarly, the elongation of break was also observed to increase by 30% and 47% for these composites respectively. In another study by Abidin *et al.* [32], PBS was grafted with octadecylamine-functionalized graphene oxide (GO-ODA) using a solution blending method to generate novel PBS/GO-ODA nanocomposites. The GO-ODA content in the composites varied between 0.5-1 wt% (Table 10.1). The tensile strength of the composite reached the maximum at 0.5 wt% filler loading with significant 30% improvement compared to neat PBS. Furthermore, the addition of GO-ODA considerably improved the Young's modulus by up to 165% from 662 MPa to 1755 MPa at 0.5 wt% filler content. On increasing the filler content to 1%, the enhancement of tensile modulus was reduced to 52%. Elongation at break of PBS/GO-ODA nanocomposite gradually reduced with increasing the loading of GO-ODA as the effect of improved rigidity of the composite [32].

Halpin-Tsai model is widely used to predict the young modulus (E) of the nanocomposites [33]. Wan and Chen [26], Wang *et al.* [27] and Yan *et al.* [30] used Halpin-Tsai model to simulate the Young's modulus (E) of PBS/graphene nanocomposites, and exhibited a good agreement between the theoretical and experimental results for the composites.

Table 10.1 shows the comparison of the percentage enhancement in the mechanical properties (Young's modulus, tensile strength and elongation at break) of the studies reporting PBS/graphene composites using different synthesis methods. As also observed earlier, the method for the incorporation of GO in the polymer to form nanocomposites impacted the resulting properties.

The similar effect of filler was also observed on the thermal properties of the PBS/graphene nanocomposites. The thermal degradation analysis of the composites reported by Pallathadka *et al.* [25] revealed the enhancement in thermal stability as a function of filler content. Figure 10.5 shows the thermogravimetric analysis of the PBS composites. Both onset as well as peak degradation temperature of PBS was observed to increase with increasing the filler content, upto 5 wt%. As mentioned earlier, the filler was uniformly distributed in the polymer even at 5 wt% content. Thus, the uniform filler dispersion led to corresponding enhancement in the thermal stability due to efficient heat transfer from the polymer to the filler nano-sheets, thus, providing thermal shielding effect to the polymer

Table 10.1 Comparison of the mechanical properties of the PBS/graphene nanocomposites reported in literature

Synthesis (preparation method of the PBS/G)	graphene wt%	polymer	% increase in Young's modulus	% increase in (ultimate) tensile strength	% increase in elongation at break	Ref.
Solution based process (mechanical and sonication)	(1) (2)	poly(butylene succinate)	(40) (70)	(18.2) (21)	(3.4) (6.1)	27
Solution based process (solution blending and compression molding method) DMF	(1) (2)	poly(butylene succinate)	(38.1) (70.5)	(37) (53.1)	(5.1) (12.1)	26
In-situ polymerization	(1) (3)	poly(butylene succinate)	--	(31.1) (43.2)	(-81.3) (-93.3)	29
In-situ polymerization	(0.3) (0.5)	poly(butylene succinate)	--	(-33.3) (-40.7)	(125) (100)	28
In-situ polymerization (ionic liquid)	(0.5) (1)	cellulose-poly(butylene succinate	(223.4) (366.9)	(85.5) (121.8)	(-32.1) (-55.5)	30
Melt-blended in a twin-roller	(1) (2)	poly(butylene succinate)/Melamine polyphosphate (flame retardant)	--	(-6.2) (-1.9)	(-31.8) (-46.9)	24
Thermo-pressed molding technique	(0.01) (0.012)	poly(butylene succinate)/poly(l-lactide)	-	(17) (27)	(30) (46.6)	31
Solution based process with sonication	(0.5) (1)	poly(butylene succinate	(165) (52)	(30) (-25)	(-35.5) (-54)	32

chains. The polymer composites also retained enhanced thermal stability as a function of time at a particular temperature, thus, also confirming the durability of the polymer composites under extreme conditions. The enhancement in thermal stability of the composites was also revealed from the viscosity of the composites as a function of angular frequency at 170 °C as well as strain sweep analysis at a temperature of 170 °C, as shown in Figure 10.6. The viscosity of the composites increased significantly with increasing filler fraction, up to 3 wt%. It indicated enhanced ability of polymer to withstand shear and temperature. Similarly, the strain sweep analysis also indicated enhanced resistance as well as modulus of the polymer as the filler content was increased till 3 wt%. Further increasing the filler content, however, led to a reduction in these properties.

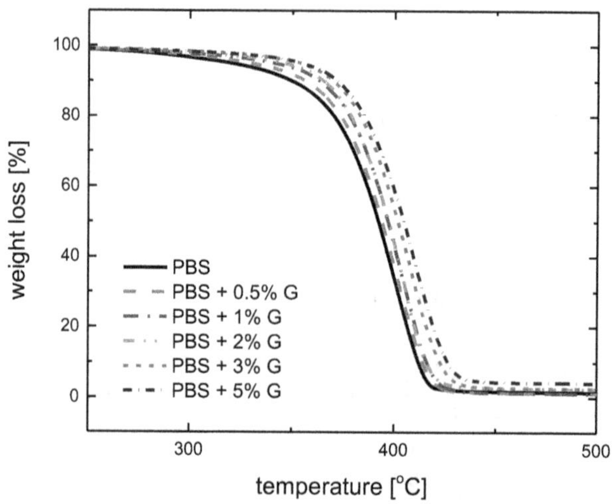

Figure 10.5 TGA thermograms of PBS/graphene nanocomposites as a function of filler content.

Wang *et al.* [27] observed that the crystallinity of the nanocomposites was not significantly affected by the addition of the GNS filler to PBS. The peak melting temperature (T_m) was observed to be slightly affected by the addition of the graphene nano-sheets in PBS. For instance, the peak melting temperature of pure PBS was 108.1°C, whereas it was increased to 109.2 °C on the addition of 2 wt% of GNS. The melting enthalpy exhibited significant changes on the addition of GNS. As the extent of GNS was enhanced in the composites, the melting enthalpy was correspondingly decreased. Pure

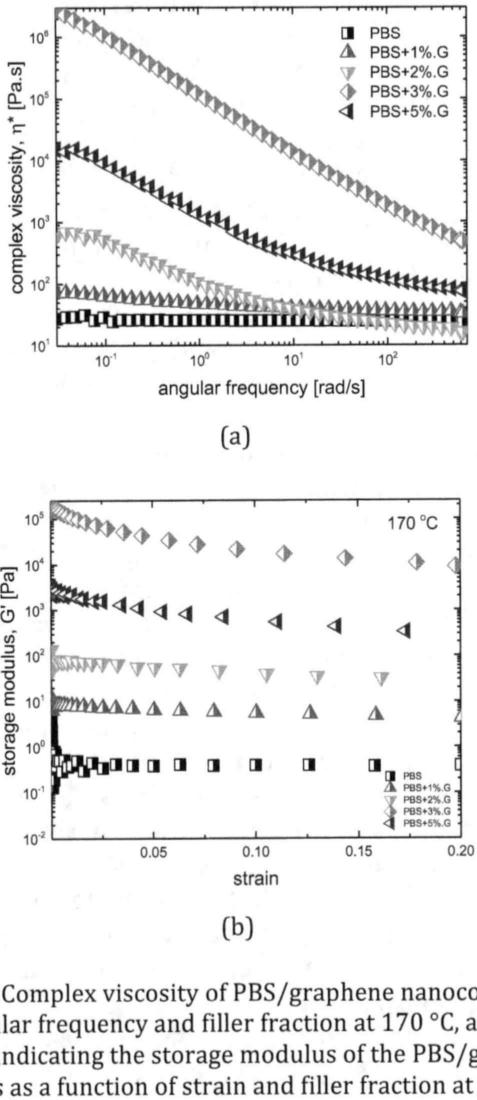

(a)

(b)

Figure 10.6 (a) Complex viscosity of PBS/graphene nanocomposites as a function of angular frequency and filler fraction at 170 °C, and (b) strain sweep analysis indicating the storage modulus of the PBS/graphene nanocomposites as a function of strain and filler fraction at 170 °C.

PBS had a melting enthalpy of 92.7 J/g, which was decreased to 91.5 J/g on the addition of 0.5 wt% filler. Further, the addition of 1 wt% and 2 wt% filler decreased the melting enthalpy to 87.5 and 85.8 J/g respectively. The "tortuous path" effect of graphene was suggested to be the reason of the improvement in the thermal stability, where the graphene sheets played a role in delaying the escape of char formation and degradation products. The incorporation of graphene in any amount had no significant effect on the glass transition temperature (T_g) [27]. Wan and Chen [26] observed that the addition of GO in PBS caused increment in the melting enthalpy as a function of filler fraction. The addition of 0.5 wt% GO resulted in the increment of the melting enthalpy to 70 J/g from 66.5 J/g for pure PBS. The addition of 1 wt% and 2 wt% GO resulted in the corresponding increment in the melting enthalpy to 73.9 and 76.6 J/g respectively. A slight increase in peak crystallization temperature (T_c) was reported, from 70.3 °C for PBS to 75.4 °C in the composites incorporated with 2 wt% graphene. The crystallinity of PBS was also observed to increase steadily from 60.3% to 63.5, 67.0 and 69.2% when the weight fraction of GO was 0.5, 1.0 and 2.0 wt%. The peak melting temperatures in the nanocomposites were within ±2 °C of the peak melting temperature of pure PBS.

As mentioned earlier, Pallathadka *et al.* [25] concluded that thermally reduced graphene nano-sheets had heterogeneous nucleation effect on the crystallization of PBS, therefore, the addition of graphene nano-sheets induced the crystallization of PBS. Nevertheless, the T_m of pure PBS (~111°C) was not affected by the addition of graphene nano-sheets. This was attributed to the small size and low content of graphene nano-sheets in the nanocomposites. Furthermore, the melting and crystallization peaks in PBS/graphene nanocomposites were observed to be narrower than in neat PBS signifying a narrower crystallite size distribution in the nanocomposites.

Bosq *et al.* [34] also analyzed the crystallization kinetics of PBS/graphene nanocomposites. The authors observed that the crystallization from the melt and from the glass was enhanced in the presence of graphene (Figure 10.7). By studying the activation energies of the melt and glass crystallization (Figure 10.7), the authors concluded that the growth rate of PBS crystals was higher in the presence of graphene nano-sheets.

Table 10.2 shows the comparison of the extent of enhancement in the thermal properties (melting temperature T_m, and degree of crystallinity X_c) of the literature studies reporting the synthesis of PBS/graphene composites with different synthesis methods.

Patwary *et al.* [35] also reported graphene based composites using a blend of PBS with chitosan (CS) flakes. The filler and polymer materials were physically premixed in a small bowel before feeding into the extruder. To generate nanocomposites, compounding temperature of 170 °C, screw speed of 100 rpm and mixing time of 15 min were used. PBS:CS (60:40) nanocomposites with graphene contents ranging from 2 wt% to 10 wt% were generated. The resulting

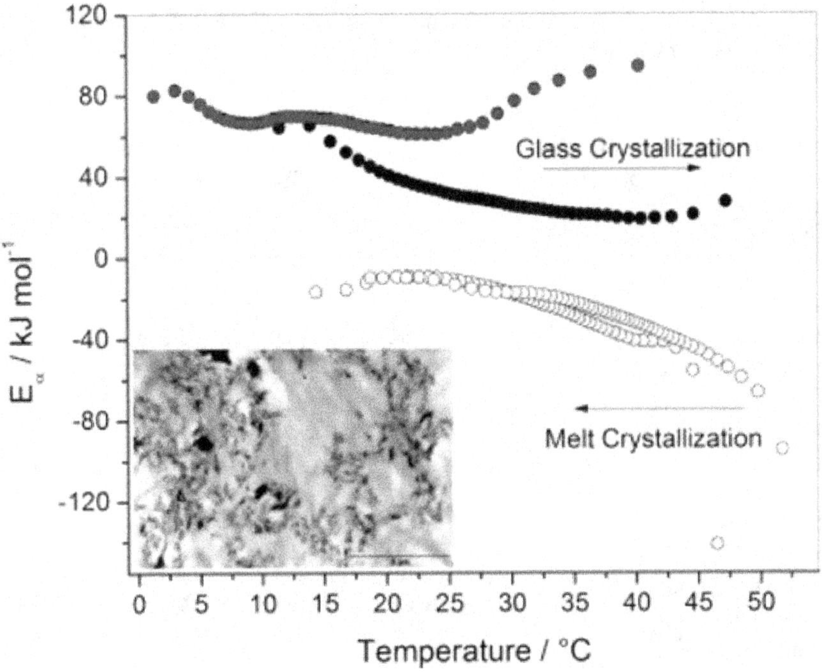

Figure 10.7 Glass crystallization and melt crystallization of PBS/graphene nanocomposites. Reproduced from Reference 34 with permission from American Chemical Society.

nanocomposites were embedded in soil for different durations and the thermal properties of the nanocomposites were studied as a function of embedding time, as demonstrated in Figure 10.8. As evident from the thermograms in Figure 10.8a corresponding to the PBS:CS nanocomposites with 2% filler content, the peak melting point (T_m) shifted to higher temperatures for samples embedded for 15 and 30 d. Subsequently with increased embedding time of 45 and 60 d, the peak melting point exhibited no change or minor decrease in its value. Similar phenomenon was observed for all filler concentrations. This indicated that the composites went through some structural change during increasing period of soil burial. The increase in T_m in the initial degradation phase indicated that there might be an increase in overall crystallinity which is also expected due to the fact that microbes degrade preferentially the random amorphous component of the polymer, resulting in an increase in the fraction of the crystalline phase. Presumably, by 30 to 45 d, easily accessible amorphous portion of the composites was degraded by

Table 10.2 Comparison of the thermal properties of the PBS/graphene nanocomposites

Synthesis (preparation method of the PBS/G)	**graphene %**	**polymer**	**% increase in T_m**	**% increase in degree of crystallinity**	**Reference**
Solution based process (mechanical and sonication)	(1) (2)	poly(butylene succinate)	(0.2) (1)	(-5.6) (-7.5)	27
Solution based process (solution blending and compression molding method)	(1) (2)	poly(butylene succinate)	(-0.9) (-0.9)	(11.1) (14.8)	26
In-situ polymerization	(0.3) (0.5)	poly(butylene succinate)	(15.8) (19.1)	(-2.3) (-6.6)	28
Melt-blended in a twin-roller	(1) (2)	poly(butylene succinate)/Melamine polyphosphate	(-1.6) (-1.7)	(-8.7) (-21.1)	24
Thermo-pressed molding technique	(0.01) (0.012)	poly(butylene succinate)/poly(l-lactide)	(-0.16) (-1.2)	(-1.5) (-3.3)	31
Melt blending (twin screw extruder)	(1) (2)	poly(butylene succinate)	(0.18) (0.18)	(-20.5) (-21.5)	25

the microbes followed by attack on the ordered or crystalline regions which caused the decrease in T_m and crystallinity, as also observed in Figure 10.8b. The TGA thermograms of blend and composites samples in nitrogen atmosphere exhibited two mass loss regions, one with a peak degradation temperature of 310 °C attributed

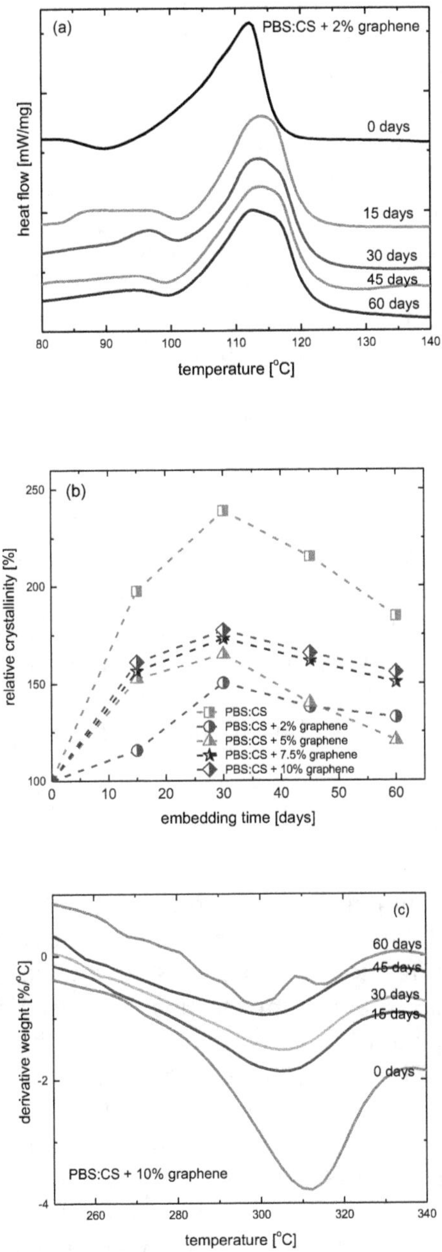

Figure 10.8 (a) DSC thermograms, (b) change in crystallinity and (c) TGA thermograms of the PBS:CS/graphene composites as a function of embedding time. Reproduced from Reference 35 with permission from Wiley.

to chitosan degradation and other with a peak degradation peak around 390 °C due to PBS. In the thermograms for the samples after embedding, no change in the degradation mechanism was observed. However, the degradation temperature of the lower temperature peak was observed to be shifted to lower temperatures as the biodegradation time was enhanced, as shown in Figure 10.8c for nanocomposites with 10% filler content. The observed reduction in degradation temperature indicated the changes happening during the biodegradation process specifically reduced the thermal resistance of chitosan. This also corresponded with the earlier findings of increased crystallinity of PBS as well as preferential degradation of chitosan in the nanocomposites. In the TGA-MS analysis in the air atmosphere for the PBS:CS+5% graphene nanocomposite in Figure 10.9, CO_2 (m/z = 44) was observed to evolve around 310 °C, 360-370 °C and 490-510 °C. The evolution of various components was observed to occur at lower temperature as a function of embedding time during degradation of both chitosan and PBS. This is observed because the aerobic oxidation produces functional groups, like COOH, at the early stage of biodegradation that usually have lower thermal stability and, thus, result in an overall lower degradation temperature. After 30 d of embedding time, the evolution of components due to degradation was observed to occur at same temperatures as the samples before burial, thus, confirming the earlier findings.

10.3 Conclusions

PBS is a versatile biopolymer with many advantageous properties such as biodegradability, processability, toughness, ease of synthesis, etc., which confirm its potential of use in a wide variety of application areas. It has superior properties compared to different biopolymers, and, in addition, can also compete with many traditional non-biopolymers. There is a need to further enhance its property profiles in order to expand the spectrum of its applications. Especially, enhancing the mechanical performance and thermal resistance would make PBS compete with many engineering polymers. For other applications, enhancement of toughness is also needed. Thus, different strategies have been evolved targeting the required properties. Many binary and ternary blends have been reported with PBS by mixing it with other biopolymers to synergistically combine the positive features of different individual polymers

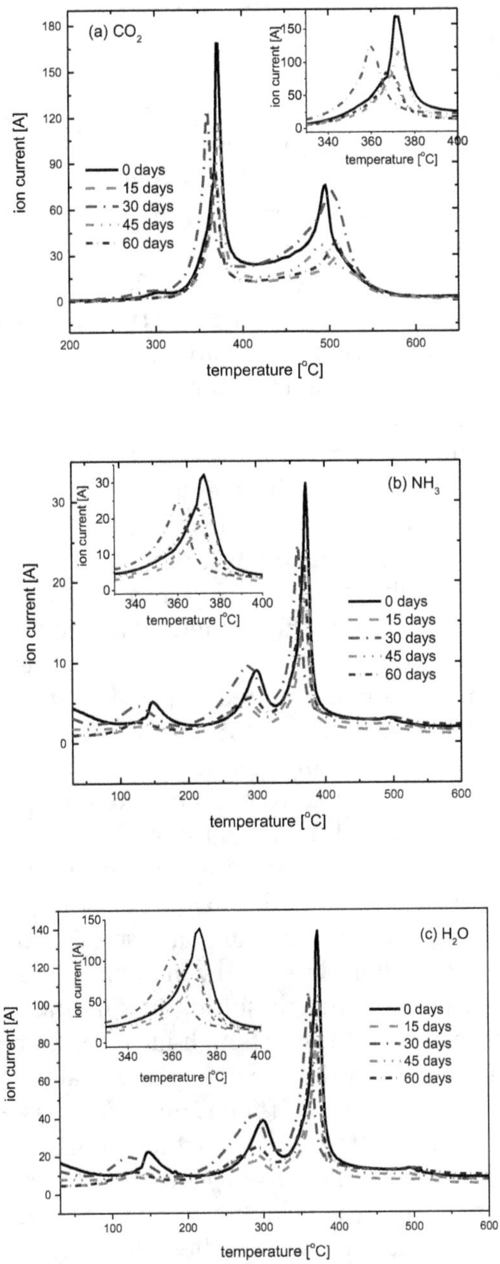

Figure 10.9 TGA-MS spectra of (a) CO_2 (b) NH_3 and (c) H_2O evolution from PBS:CS+5% graphene nanocomposites as a function of soil burial time. Reproduced from Reference 35 with permission from Wiley.

This leads to an optimal enhancement in properties, while also keeping the biodegradability of PBS. However, blending suffers from many challenges like phase separation as well as time dependent changes in the component properties. Incorporation of a variety of organic and inorganic fillers has also been developed to enhance the mechanical properties of PBS, especially tensile strength, modulus and thermal properties. This review has specifically focused on these aspects of PBS nanocomposites with graphene and its derivatives, owing to the potential of graphene to significantly enhance the composite properties at lower filler fractions. In the reviewed studies, analysis of mechanical and thermal properties characterized using tensile test, dynamic mechanical analysis (DMA), shear rheology, differential scanning calorimetry (DSC), and thermo-gravimetric analysis (TGA) was carried out.

The review of the various studies revealed that different methods of composite synthesis, i.e. melt intercalation, solution mixing as well as in-situ polymerization, could be successfully employed for the generation of PBS/graphene nanocomposites. All the synthesis methods exhibited their specific advantages and depending on the processing conditions used, these methods resulted in different composite morphologies and resultant physical properties. Even the use of melt mixing was demonstrated to provide uniform delamination of the graphene platelets in PBS matrix, thus, underlining the need to optimize the processing conditions. Overall, the addition of graphene to PBS was observed to enhance the tensile modulus and strength of the polymer and the extent of their increase was a function of filler fraction in the composites, though some property reduction was also observed at higher loadings of filler. With effective strategies of polymer-filler intermixing and stress transfer from the polymer chains to the filler platelets, even enhancement in the elongation was also achieved, thus, resulting in simultaneous enhancement of toughness and stiffness. Though most of the reported studies were generated without the use of any externally added compatibilizer or surface modification of filler, however, it is expected that employment of these strategies can further enhance the extent of property improvement. Similar to mechanical properties, the thermal performance of PBS was also enhanced on the addition of graphene or its derivatives. In many studies, the filler acted as nucleating agent and impacted the crystallization and degree of crystallinity of polymer in the composites, thus, further impacting the mechanical properties. Some changes were also observed in the peak melting and crystallization temperature of PBS in the presence of filler, though these depended on the polymer-filler interactions and the state of filler dispersion in the polymer. The thermal degradation behaver was also observed to become superior in the presence of filler as the onset and peak degradation temperatures of PBS shifted to higher values in the presence of filler, as compared to pure PBS. In addition to other properties, graphene also impacted the biodegradation behavior of PBS.

Overall, incorporation of graphene in PBS represents a functional strategy which can lead to overall enhancement of PBS properties, thus, enabling it to become competitive with many traditional and engineering polymers. Further strategies of polymer grafting on the filler surface or polymerization initiated from the filler surface can also be explored in order to achieve covalently bound polymer-filler hybrid materials which would have more durability while in service.

References

1. Yang, B., Ni, H., Huang, J., and Luo, Y. (2014) Effects of poly(vinyl butyral) as a macromolecular nucleating agent on the nonisothermal crystallization and mechanical properties of biodegradable poly(butylene succinate). *Macromolecules*, **47**(1), 284-296.
2. Fujimaki, T. (1998) Processability and properties of aliphatic polyesters, 'BIONOLLE', synthesized by polycondensation reaction. *Polymer Degradation and Stability*, **59**, 209-214.
3. *Polymers, Polymer Blends, Polymer Composites and Filled Polymers: Synthesis, Properties and Applications*, Mikitaev, A., Ligidov, M., and Zaikov, G. (eds.), Nova Science Publishers, USA (2006).
4. *Polymer Nanocomposites: Synthesis, Characterization and Modeling*, Krishnamoorti, R., and Vaia, R. A. (eds.), ACS Symposium Series, USA (2001).
5. *Nanocomposite Science and Technology*, Ajayan, P. M., Schadler, L. S., and Braun, P. V. (eds.), Wiley VCH, Germany (2004).
6. Bikiaris, D. N. (2013) Nanocomposites of aliphatic polyesters: An overview of the effect of different nanofillers on enzymatic hydrolysis and biodegradation of polyesters. *Polymer Degradation and Stability*, **98**, 1908-1928.
7. Ravati, S., and Favis, B. D. (2013) Tunable morphologies for ternary blends with poly(butylene succinate): Partial and complete wetting phenomena. *Polymer*, **54**(13), 3271-3281.
8. Mittal, V., Manna, Y., and Matsko, N. (2017) Ternary blends of PCL,

PBAT and PBS: Morphology and properties. In: *Biopolymers Based Advanced Materials*, Mittal, V. (ed.), Central West Publishing, Australia, pp. 87-114.

9. Ray, S. S., Okamoto, K., and Okamoto, M. (2003) Structure-property relationship in biodegradable poly (butylene succinate)/layered silicate nanocomposites. *Macromolecules*, **36**, 2355-2367.
10. Shih, Y. F. (2009) Thermal degradation and kinetic analysis of biodegradable PBS/multi-walled carbon nanotube nanocomposites. *Journal of Polymer Science, Part B: Polymer Physics*, **47**, 1231-1239.
11. Papageorgiou, G. Z., Achilias, D. S., and Bikiaris, D. (2007) Crystallization kinetics of biodegradable poly (butylene succinate) under isothermal and non-isothermal conditions. *Macromolecular Chemistry and Physics*, **208**, 1250-1264.
12. Huang, J., Lu, X., Zhang, N., Yang, L., Yan, M., Liu, H., Zhang, G., and Qu, J. (2014) Study on the properties of nano-TiO_2/polybutylene succinate composites prepared by vane extruder. *Polymer Composites*, **35**, 53-59.
13. Snowdon, M., Mohanty, A. K., and Misra, M. (2015) Melt processing and characterization of bionanocomposites made from poly(butylene succinate) bioplastic and carbon black. *Macromolecular Materials and Engineering*, **300**, 118-126.
14. Mittal, V., Patwary, F., and Matsko, N. (2016) Biodegradation properties of poly(butylene succinate) and chitosan nanocomposites. *SPE Plastics Research Online*, doi:10.2417/spepro.006441
15. Han, S. I., Lim, J. S., Kim, D. K., Kim, M. N., and Im, S. S. (2008) In situ polymerized poly(butylene succinate)/silica nanocomposites: Physical properties and biodegradation. *Polymer Degradation and Stability*, **93**(5), 889-895.
16. Bian, J., Han, L., Wang, X., Wen, X., Han, C., Wang, S., and Dong, L. (2010) Nonisothermal crystallization behavior and mechanical properties of poly(butylene succinate)/silica nanocomposites. *Journal of Applied Polymer Science*, **116**, 902-912.
17. Vassiliou, A. A., Bikiaris, D., El Mabrouk, K., and Kontopoulou, M. (2011) Effect of evolved interactions in poly (butylene succinate)/fumed silica biodegradable in situ prepared nanocomposites on molecular weight, material properties, and biodegradability. *Journal of Applied Polymer Science*, **119**(4), 2010-2024.
18. Papageorgious, G., and Bikiaris, D. (2005) Crystallization and melting behavior of three biodegradable poly(alkylene succinates). A comparative study. *Polymer*, **46**, 12081-12092.
19. Mittal, V. (2014) Functional polymer nanocomposites with graphene: A Review. *Macromolecular Materials and Engineering*, **299**, 906-931.
20. Mukhopadhyay, P., and Gupta, R. K. (2011) Trends and frontiers in graphene-based polymer nanocomposites. *Plastics Engineering*,

32, 32-42.

21. Cai, D., and Song, M. (2010) Recent advance in functionalized graphene/polymer nanocomposites. *Journal of Materials Chemistry,* **20**, 7906-7915.
22. Rafiee, M. A., Rafiee, J., Wang, Z., Song, H., Yu, Z. Z., and Koratkar, N. (2009) Enhanced mechanical properties of nanocomposites at low graphene content. *ACS Nano,* **3**, 3884-3890.
23. Mittal, V., Chaudhry, A. U., and Luckachan, G. E. (2014) Biopolymer - thermally reduced graphene nanocomposites: Structural characterization and properties. *Materials Chemistry and Physics,* **147**(1-2), 319-332.
24. Wang, X., Hu, Y., Song, L., Yang, H., Yu, B., Kandola, B., and Deli, D. (2012) Comparative study on the synergistic effect of POSS and graphene with melamine phosphate on the flame retardance of poly(butylene succinate). *Thermochimica Acta,* **543**, 156-164.
25. Pallathdka, P. K., Koh, X. Q., Khatta, A., Luckachan, G. E., and Mittal, V. (2015) Characteristics of biodegradable poly(butylene succinate) nanocomposites with thermally reduced graphene nanosheets. *Polymer Composites,* doi:10.1002/pc.23824
26. Wan, C., and Chen, B. (2012) Reinforcement of biodegradable poly(butylene succinate) with low Loading of graphene oxide. *Journal of Applied Polymer Science,* **127**, 5094-5099.
27. Wang, X., Yang, H., Song, L., Hu, Y., Xing, W., and Lu, H. (2011) Morphology, mechanical and thermal properties of graphene-reinforced poly(butylene succinate) nanocomposite. *Composite Scince and Technology,* **72**, 1-6.
28. Jin, T.-X., Liu, C., Zhou, M., Chai, S.-g., Chen, F., and Fu, Q. (2014) Cystallization, mechanical performance and hydrolytic degradation of poly(butylene succinate)/graphene oxide nanocomposites obtained via in situ polymerization. *Composites: Part A,* **68**, 193-201.
29. Wang, X. W., Zhang, C. A., Wang, P. L., Zhao, J., Zhang, W., Ji, J. H., Hua, K., Zhou, J., Yang, X. B., and Li, X. P. (2012) Enhanced performance of biodegradable poly(butylene succinate)/graphene oxide nanocomposites via in situ polymerization. *Langmuir,* **28**(18), 7091-7095.
30. Yan, C. Y., Ren, P. G., Zhang, Z. P., Wang, H., and Li, Z. M. (2015) In-situ preparation and characterization of highly oriented graphene oxide/cellulose-poly(butylene succinate) teranary composite films. *Cellulose,* **22**, 1243-1251.
31. Sisti, L., Belcari, J., Mazzocchetti, L., Totaro, G., Vannini, M., Giorgini, L., Zucchelli, A., and Celli, A. (2016) Multicomponent reinforcing system for Poly(butylene succinate): Composites containing poly(L-lactide) electrospun mats loaded with graphene. *Polymer Testing,* **50**, 283-291.

32. Abidin, A. S. Z., Yusoh, K., Jamari, S. S., Abdullah, A. H., and Ismail, Z. (2016) Enhanced performance of alkylated graphene reinforced polybutylene succinate nanocomposite. *AIP Conference Proceedings*, **1756**, doi:http://dx.doi.org/10.1063/1.4958766.
33. Halpin, J. C., and Kardos, J. L. (1976) The Halpin-Tsai equations: A review. *Polymer Engineering and Science*, **16**(5), 344-352.
34. Bosq, N., Guigo, N., Aht-Ong, D., and Sbirrazzuoli, N. (2017) Crystallization of poly(butylene succinate) on rapid cooling and heating: Toward enhanced nucleation by graphene nanosheets. *Journal of Physical Chemistry C*, doi:10.1021/acs.jpcc.7b02887
35. Patwary, F., Matsko, N., and Mittal, V. (2016) Biodegradation properties of melt processed PBS/chitosan bio-nanocomposites with silica, silicate, and thermally reduced graphene. *Polymer Composites*, doi:10.1002/pc.23947.

Index

A

B

C

D

E

F

H

I

K

L

M

N

O

P

R

S

T

V

W

Y